The Economics of Agricultural Prices

The Economics of Agricultural Prices

Peter G. Helmberger

Department of Agricultural Economics
University of Wisconsin—Madison

Jean-Paul Chavas

Department of Agricultural Economics
University of Wisconsin—Madison

PRENTICE HALL, Upper Saddle River, New Jersey 07458

Library of Congress Cataloging-in-Publication Data

Helmberger, Peter G.
The economics of agricultural prices / Peter G. Helmberger, Jean
-Paul Chavas.
p. cm.
Includes index.
ISBN 0-13-372640-1
1. Agricultural prices. 2. Agricultural—Economic aspects.
I. Chavas, Jean-Paul. II. Title.
HD1447.H45 1996
338.1′3—dc20 95-18566
CIP

Acquisition eidtor: *Charles Stewart*
Editorial/production supervision: *Maes Associates*
Copy editor: *William O. Thomas*
Cover designer: *Jayne Conte*
Manufacturing buyer: *Ilene Sanford*

Simon & Schuster/A Viacom Company
Upper Saddle River, New Jersey 07458

Printed in the United States of America

10 9 8 7 6 5 4 3 2 1

ISBN 0-13-372640-1

PRENTICE-HALL INTERNATIONAL (UK) LIMITED, *London*
PRENTICE-HALL OF AUSTRALIA PTY. LIMITED, *Sidney*
PRENTICE-HALL CANADA INC., *Toronto*
PRENTICE-HALL HISPANOAMERICANA, S.A., *Mexico*
PRENTICE-HALL OF INDIA PRIVATE LIMITED, *New Delhi*
PRENTICE-HALL OF JAPAN, INC., *Tokyo*
SIMON & SCHUSTER ASIA PTE. LTD., *Singapore*
EDITORA PRENTICE-HALL DO BRASIL, LTDA., *Rio de Janeiro*

TO

Catherine R. Geier
Mary S. Jones
Peter J. Helmberger
Timothy J. Helmberger

AND

Eloisa D. Chavas

Brief Contents

Contents

Preface

This text incorporates recent advances in economics that are indispensable to understanding how agricultural markets operate and how prices are determined. Areas where such advances have occurred include the theory of the firm under price and production uncertainty, the rational formation of expectations, the pricing of heterogeneous outputs, and the critical role of commodity storage. The more traditional areas, such as international trade, industrial organization, and farm programs, remain as relevant as ever. Synthesizing the new discoveries with traditional approaches allows a more rigorous analysis of the special characteristics of agriculture, shedding new light on the economic functioning of the agricultural sector and the determination of agricultural prices. We believe that agricultural economics instructors who are eager to include the results of recent economic research in their instructional programs will find the book of considerable value. Indeed, a major motivation for writing it has been to help reduce the gap between what is taught in the classroom and what is discussed in the economic literature.

The breadth and depth of coverage in this text surpass, we believe, those of competing texts, encouraging students to develop a holistic view of agriculture that cuts across production, marketing, and policy. Instead of splitting agricultural economics into separate fields, possibly subjecting students to overspecialization, we urge that instructional programs emphasize interdependency, synthesis, and coherence. The study of agricultural prices, although important in its own right, is an excellent vehicle for this purpose.

Studying agricultural prices is also an excellent way for students to learn that real-world observations lead inexorably to the need to frame and analyze sets of assumptions sometimes absent in standard microeconomics textbooks. Students are given the chance to adapt basic tools learned in their courses in economic theory and mathematics to the analysis of real-world markets and to think analytically and rigorously about economic phenomena. Put differently, this book will help students to learn to think like economists, which is, after all, the fundamental role of textbooks and courses in applied economics.

Intended for students who have had a solid course in differential calculus and in intermediate microeconomic theory, this book is appropriate mainly for upper-division undergraduates and graduate students. The material is presented with an eye on flexibility from the user's standpoint. To assure accessibility to upper-division undergraduates, the bulk of the text makes liberal use of graphic exposition, simple functional forms, and numerous arithmetic examples. We have refrained from including considerable quantities of data, statistical analyses, and real-world case studies on the argument that most agricultural economists, with their customary interests in applied research and/or extension, will be in a good position to organize supplementary material that is both current and of the greatest relevance to their students. At the graduate level, instructors and students can make good use of appendices that provide more rigorous and general analyses of subjects found in the chapters. Because this book covers topics that are essential for a proper understanding of the functioning of agricultural markets, economists charged with analyzing these markets will be particularly interested in reading the chapters dealing with risk, storage, and farm policy.

We thank Drs. Carlos-Federico Perali, Maria Luisa Ferreria, and Jung-Sup Choi and Professors Bruce Gardner and Marilyn Whitney for many helpful comments and suggestions on earlier drafts of this book. We also thank Ms. Karen Denk for typing innumerable drafts of the manuscript. The authors acknowledge financial support from the College of Agricultural and Life Sciences and the University of Wisconsin—Madison.

Peter G. Helmberger
Jean-Paul Chavas

The Economics of Agricultural Prices

PART I

Farm Production Decisions, Output Supply, and Input Demand Under Uncertainty

1

Introduction

According to competitive price theory, both the price and level of output are determined by the intersection of demand and supply. That being the case, we may inquire why the real price of crude oil rose by more than 100 percent between 1978 and 1980. Was it because of a large, unanticipated increase in the demand for oil? Was it because a great Mideast earthquake destroyed many of the world's oil wells, swallowing alive thousands of small oil producers in the process? The professor of economics, eager to show the usefulness of economic theory in understanding how the real world works, might quickly point out that competitive price theory does not apply to crude oil, at least not in the short run; instead, we should invoke the theory of cartels. The increase in the price of crude oil referred to was, after all, engineered by the Organization of Petroleum Exporting Countries (OPEC); it was not the result of many small sellers interacting in a market with many small buyers in a period characterized by a sharp increase in demand and/or a sharp decrease in supply. The economist might well go on to point out that competitive price theory cannot be applied indiscriminantly to all product markets. It can only be applied to markets with structures or characteristics that are reasonably in line with the markets envisaged in the theory. Well aware that the qualifying phrase "reasonably in line" is vague and is likely to raise questions on the part of students, some of whom may be inclined to think that the study of economic theory is a waste of time in any case, the economist might suggest that those interested in applying competitive price theory would do well to consider agriculture instead of crude oil. Fine. Let's do that.

In the spring of 1982, 81.9 million acres were planted to corn in the United States. Planted corn acreage fell to 60.2 million acres in 1983. This 26 percent decline cannot be explained on the basis of competitive demand and supply shifts. For example, the price of corn rose from $2.60 in the spring (May) of 1982 to $3.03 in 1983. The price of farm inputs rose only slightly, and the price of soybeans, corn's main competing crop, actually fell from $6.27 per bushel in the spring of 1982 to $6.06 in 1983. Planting conditions were normal in both years. How can this remarkable decline in the nation's leading crop be explained? An economist not familiar with agriculture might be given some time to read up on the subject and to do a little homework. It soon will be discovered that the U.S. government has had a voluntary program for corn producers for many years. The program offers inducements to farmers to idle corn acreage, thus elevating the price of corn. Land idled under the 1983 corn program equaled 32.2 million acres. Our hypothetical professor of economics would need to confess that, because of government intervention, competitive price theory as it stands does not apply to corn; and for the same reason, the theory does not apply to sorghum, barley, oats, wheat, cotton, rice, tobacco, peanuts, sugar, milk, wool, and several other farm products we could mention.

Now it might be argued that competitive price theory would apply to agriculture if only the government stopped intervening in farm markets. It is true that, in farm industries, sellers number in the thousands. Barriers to entry are negligible relative to the barriers in other markets, such as automobiles and banking. Farm products are also graded and lumped together in more or less homogeneous lots. If competitive price theory does not apply here, it may be argued, it does not apply anywhere. Even so, the applicability of price theory to farm product markets is problematic. For example, according to the typical theory textbook, the supply curve shows how much producers would be willing to produce at alternative prices. In most cases, however, when farmers commit inputs to a production process, they do not know what the price will be when the process is completed, when a crop is harvested, for example. We might suppose at this juncture that a bit of sweat would appear on the forehead of our hypothetical theorist, who might lamely admit, "Well, farmers must be supposed to respond to expected price, not actual price." Expected price? The typical textbook treatment of competitive price theory says nothing about expected prices. If farmers really do make production plans on the basis of expected prices, then shouldn't theory explain how expected prices are determined?

This brief foray into the economics of oil and agriculture should not be construed as an attack on economic theory. The very essence of theory, after all, is abstraction and simplification. Certainly, the models set forth in this book can be faulted for not explicitly taking into account factors (the cost of acquiring knowledge, for example) that may be very important in explaining real-world behavior. Theories rarely tell us everything we need or would like to know. Our discussion of oil and corn is intended not so much as criticism of theory as a warning that the study of real-world markets often involves the need to modify theory, sometimes in substantial ways, to increase its relevance and usefulness.

The student who proposes to study the economics of agriculture will assuredly discover phenomena that do not appear to be explained very well by any of the models found in the typical economic theory textbook. Technological relationships often involve significant lags between the time when inputs are applied and outputs materialize. Random elements, such as the weather, frequently confound production plans. Biological processes coupled with random elements make control of product quality difficult. Farm output is heterogeneous, not homogeneous as assumed in competitive theory. Production lags also give rise to uncertainty and the need to explain how expected values for prices and other variables get determined. The spatial and seasonal aspects of farm production call attention to the importance of transportation and commodity storage. The nature and consequences of farm production activities go a long way toward explaining marketing institutions, such as futures markets and cooperatives, and the nature of marketing functions, such as grading, processing, transportation, storage, and the like. Although the farmer may be our oyster from the viewpoint of science, the sea in which the farmer thrives or turns belly up is a turbulent one, buffeted by waves of prosperity and depression that can only be understood if we examine the interdependency between farm markets and the rest of the economy. As already suggested, economic analysis of agriculture must recognize a variety of farm programs that affect in important ways the performance of farm markets, often giving rise to effects that would not be expected on the basis of standard competitive price theory.

Agricultural economics and, indeed, any applied field of economics have two broad tasks. The first is to frame and analyze the implications of alternative sets of assumptions designed to explain the performance of the specific real-world markets of interest. The second involves testing hypotheses for empirical validity and quantifying behavioral relationships, such as demand and supply functions.

The first broad task entails making observations on actual markets and market activities and making assumptions underlying some analytical framework deemed appropriate for the analysis of a real-world situation. Description is an integral part of the work. Importantly, the assumptions made may be very different from those encountered in courses in economic theory. Becoming an economist involves recognizing the need to modify theory and developing the skills to do so. This first task, which tends to be theoretical in nature and provides the student with a marvelous opportunity to apply the tools of analysis forged in courses in economic theory, is mainly what this book is about.

The second broad task of agricultural economics, centering on hypothesis testing and quantification, requires a knowledge of research methods. Researchers sometimes confess that economics research is an art, not in the sense of that which is beautiful, but in the sense of that which requires the skills and techniques acquired through practice. Admission to the associated guild of artisans requires completion of graduate courses, dissertations, and advanced degrees. We must learn to speak a peculiar language and learn what sins of omission and commission are acceptable to guild members, particularly to the Poo-Bahs who control publication of papers in prestigious journals. Research methods and their application are the appropriate subject matter of undergraduate courses to a limited degree only. Undergraduate degrees are not ordinarily viewed as research degrees. Accordingly, this book does not center on research. It may be argued, however, that modification of theory to better fit specific economic phenomena of interest is a necessary preliminary to the collection and analysis of data. We might hope that the present book not only helps the student think like an economist, but also that it lays useful groundwork for those aspiring to do economic research.

Before turning to other matters, it should be noted that the two tasks of agricultural economics set forth above should not be thought to be independent one from the other. The student should recognize the evolutionary nature of economics, and indeed of all sciences, and the endless interplay between theorizing, on the one hand, and the testing for empirical relevance, on the other.

In the remainder of this chapter, we first provide an overview of the main subject matter of the book. Attention is then centered on basic concepts and analytical techniques that the student may have seen before, but that constitute necessary background for the material that follows. References are provided at the end of the chapter for additional reading, as for other chapters.

Chapter 2 introduces the concepts of production lags, uncertainty, and risk. On the basis of the hypothesis of rational expectations, price expectations are treated as phenomena to be explained by models of farm output markets. The interdependence between production decisions and the formation of expectations is given a central role. Chapter 3 analyzes decoupling consumption expenditures and profit, product diversification, and hedging as three alternative strategies farmers might use to decrease risk. A simple model of a futures market is developed, and the effects of futures prices on the performance of farm markets are analyzed. Although Chapter 4 centers on the derivation of input demand functions, the more important topic is the interdependence among farm output and input markets, a topic that is of basic importance in analyzing the benefit–cost implications of farm programs and of changes in exogenous factors. Resource pricing under uncertainty is also modeled, with the pricing of farmland treated as an illustrative example. Chapter 5 takes up the derived demand for farm output, competitive pricing in the presence of product heterogeneity and grading, and export demand. Chapter 6 introduces the concepts of supply and demand for commodity storage (stocks) and shows the important role of storage in the stabilization of commodity prices and consumption. Chapter 7 centers on the basics of benefit–cost analysis. The relevance of consumer and producer surplus is established on the basis of the criterion of willingness to pay. This chapter develops concepts that are used in the remainder of the book. Chapter 8 analyzes noncompetitive markets and emphasizes the importance of market structure and conduct in the determination of the performance of markets in the food distribution system. Chapters 9 and 10 describe and analyze the effects of important farm programs. Modeling farm programs is shown to be indispensable in the analysis of agricultural pricing.

Appendixes are included in this book for various purposes. Those given at the end of this chapter review the concept of elasticity and the technique of adding (subtracting) curves horizontally, as is often done when analysis moves from individual decision makers to the aggregate for a market. These appendixes are intended as handy reviews of material students likely will have seen before. In contrast, the appendixes given at the end of the book cover much of the material given in the chapters, but at a higher level of generality and rigor. They are intended mainly for graduate students with a good background in mathematics, including some linear algebra. While the end-of-book appendixes complement the chapters, the main body of the book can be read independently and should be accessible to a large audience.

1.1 CAUSALITY AND INTERDEPENDENCE IN ECONOMICS

Several basic concepts that will be used throughout the book are now considered. A convenient place to start is the elementary theory of supply and demand. In Fig. 1.1, the demand curve D shows how much consumers would be willing to buy at alternative prices. The supply curve S shows how much firms would be willing to supply at alternative prices. The price P_0, given by the intersection of D and S, is said to be the *equilibrium price* because, at this price, consumers are able to buy as much as they desire, given their utility functions and budget constraints; firms are able to sell as much as they want, given their desire to maximize profits and the constraints imposed by production functions and input prices. Importantly, at price P_0 all the market participants are able to achieve their constrained objectives—utility in the case of the consumer and profit in the case of the firm. No other price has this property. For any price less than P_0, consumers would want to buy more than firms would be willing to sell. The standard argument is that firms would soon discover that consumers are willing to pay a higher price. The price would be bid up. Any price in excess of P_0 would find firms unable to sell as much as they like; price cutting would cause the market price to fall. Importantly, the student should recognize that the theory consists of more than a graph showing the intersection of supply and demand. In addition to the curves, there is a story to be told, a story about dynamics, disequilibria, bidding processes, rising inventories, and queues of consumers. The theory of supply and demand describes a process whereby a market achieves equilibrium; it offers a tentative explanation of how a market works.

Were this all, supply and demand theory would be of slight interest. What makes the theory of basic importance is the notion of *demand and supply shifters*. From first principles, demand shifters include population, money incomes, prices of related goods, tastes, and preferences. Supply shifters include input prices, technol-

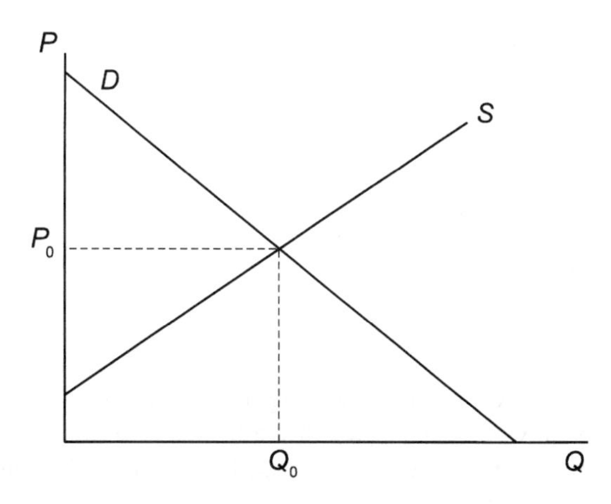

Figure 1.1 Demand and supply curves.

ogy, and the number of firms. Shifters, by definition, are variables that move demand or supply or both. They are basic to hypotheses that purport to explain changes in market price and quantity. A rise in population, for example, increases demand and elevates both price and output. Notice that the increase in price does not cause population to grow. Causality flows in one direction only. Similarly, a technological discovery that lowers production costs expands supply, causing price to fall and output to expand. Again, the causality flows in one direction only.

Demand and supply shifters are examples of *exogenous variables*. A variable that appears in a theory is said to be exogenous if its value is determined by processes that are *not* described by the theory. Figure 1.1 and the story that goes with it do not tell how population is determined. The values of the exogenous variables (shifters) and the changes in these values are taken as given.

In the elementary theory of supply and demand, as sketched above, price and quantity are examples of *endogenous variables*. A variable that appears in a theory is said to be endogenous if the theory explains or describes processes that determine its value for given values of exogenous variables. Importantly, it is very often the case that processes that determine the value of one endogenous variable determine simultaneously the value of at least one other endogenous variable. In such areas, the endogenous variables are said to be *jointly dependent* or *jointly determined*. In the above theory, price P and output Q are jointly dependent endogenous variables. It is nonsense to say that the value of one endogenous variable determines the value of another. The flow of causality is one way, from the exogenous variables, that is, the supply and demand shifters, to the endogenous variables. Ask yourself the following question: Does a high price cause output to expand? Well, it depends. An increase in demand does indeed elevate price, with output rising as well, but a contraction of supply causes the price to rise and the output to fall.

Although we make considerable use of graphic analysis in this book, mathematical analysis is often convenient and sometimes indispensable. An algebraic treatment of supply and demand will start the student down the road of thinking about economic models in terms of equation systems and will allow the further definition of basic terms. Consider, therefore, the following supply–demand model:

$$Q_d = \beta_0 - \beta_1 P_d \qquad \text{demand} \tag{1-1}$$

$$Q_s = \gamma_0 + \gamma_1 P_s \qquad \text{supply} \tag{1-2}$$

$$Q_s = Q_d = Q \qquad \text{equilibrium condition} \tag{1-3}$$

$$P_d = P_s = P \qquad \text{equilibrium condition} \tag{1-4}$$

The betas and gammas are to be interpreted as nonnegative constants called parameters. Parameters will always be viewed as nonnegative in this book unless specifically noted otherwise. The s and d subscripts indicate supply function values and demand function values, respectively. The variables P and Q are to be interpreted as the equilibrium price and quantity. The six equation system (count the equality signs) contains six endogenous variables: Q_d, P_d, Q_s, P_s, P, and Q. We may reduce the size of the model through substitution, that is, by using equations to eliminate variables. If we use equations to eliminate Q_d, P_d, Q_s, and P_s, we have

$$Q = \beta_0 - \beta_1 P \tag{1-5}$$

$$Q = \gamma_0 + \gamma_1 P \tag{1-6}$$

This supply–demand model is stated in terms of the equilibrium price and quantity. The resulting two equation system may be solved as follows:

$$P = \frac{\beta_0 - \gamma_0}{\beta_1 + \gamma_1} \tag{1-7}$$

$$Q = \frac{\beta_0\gamma_1 + \beta_1\gamma_0}{\beta_1 + \gamma_1} \tag{1-8}$$

The expression (solution) for P given by Eq. (1-7) corresponds to P_0 in Fig. 1.1. Similarly, the expression for Q given by Eq. (1-8) corresponds to Q_0.

We next introduce two exogenous variables, Z_1 and Z_2.

$$Q = \beta_0 - \beta_1 P + \beta_2 Z_1 \tag{1-9}$$

$$Q = \gamma_0 + \gamma_1 P - \gamma_2 Z_2 \tag{1-10}$$

Let Z_1 equal per capita income. Because β_2 is positive, we are implicitly assuming that the good is a normal good like skis, instead of an inferior good like cheap clothing. (By definition, the demand for a normal good rises with increases in income.) Also, let Z_2 equal the price of an input. The model given by Eqs. (1-9) and (1-10) is an example of a *structural model*. A structural model describes the processes that determine the values of endogenous variables. It is through structural models that economic theories are specified. The signs of the parameters are often predicted. For example, the demand for normal goods *must* be downward sloping. A structural model always contains at least one equation with more than one endogenous variable.

We next solve this system for P and Q as follows:

$$P = \frac{1}{\beta_1 + \gamma_1}\left(\beta_0 - \gamma_0 + \beta_2 Z_1 + \gamma_2 Z_2\right) \tag{1-11}$$

$$Q = \frac{1}{\beta_1 + \gamma_1}\left(\beta_1\gamma_0 + \gamma_1\beta_0 + \gamma_1\beta_2 Z_1 - \beta_1\gamma_2 Z_2\right) \tag{1-12}$$

The model given by Eqs. (1-11) and (1-12) is an example of a *reduced form* model. A reduced form model is the solution of a structural model. Each equation contains no more than one endogenous variable. Reduced form models are important because they often allow the formulation of hypotheses that are merely implicit in structural models. Typically, a structural model is specified on the basis of theoretical arguments. Then, the analyst finds, if possible, the reduced form model. By finding and evaluating the signs of the partial derivatives of the endogenous variables with respect to the exogenous variables in the reduced form model, the analyst is often able to frame hypotheses that specify how endogenous variables are affected by changes in exogenous variables. For example, because $\partial P/\partial Z_1 > 0$ and $\partial Q/\partial Z_1 > 0$, we predict that an increase in per capita income would increase both price and output. Because $\partial P/\partial Z_2 > 0$ and $\partial Q/\partial Z_2 < 0$, we predict that an increase in the price of an input would cause price to rise and quantity to fall.

As a final topic in this discussion of causality in economics, we note that, in the analysis of real-world phenomena, economists are often concerned about the values of parameters that appear in economic theory. For example, agricultural economists often emphasize the importance of the elasticity of demand. In structural and reduced form models, parameters are important because they condition or determine the impacts of exogenous variables on endogenous variables. As often happens, the value of a certain parameter, whether it is positive or negative or whether it is larger than or equal to one, can play a strategic role in determining the effects of exogenous shocks on the performance of a market (prices, outputs, and inputs).

1.2 RANDOM VARIABLES AND PROBABILITY DISTRIBUTIONS

Courses in microeconomic theory, even at the graduate level, often focus entirely on comparative-static analysis, analysis that examines and compares markets in alternative positions of equilibrium as in the previous section. Such analysis is subject to several limitations that become serious in the study of agricultural economics. What is required is analysis that allows for random or stochastic elements and deals more explicitly with market dynamics. The nature of such analysis will become clear as we progress. As a primer, however, it is necessary to introduce some tools of analysis common to the field of statistical inference.

A variable is said to be *random* if it takes on different numerical values with relative frequencies or probabilities that we assume are known or estimable. Random variables are often said to be stochastic; they are often referred to as stochastic variables. Such variables may be continuous or discrete. The annual quantity of rainfall is an example of a continuous variable that might be treated as random. A random variable that assumes distinct values only is called discrete. Examples will be given shortly. In this chapter, we assume that all variables are discrete, but the results apply to the continuous case as well.

Consider tossing a fair coin. We adopt the convention that a random variable is assigned the value zero when the coin turns up heads and one when it turns up tails. We let X represent this discrete random variable with $X_1 = 0$ and $X_2 = 1$. Subscripts indicate specific values. If the coin is tossed many times, we would expect that X equals zero half the time and one otherwise. We say the probability of getting heads, that $X_1 = 0$, is one-half, and similarly for tails. The probabilities sum to one if we ignore those instances when the coin, upon landing, balances on its edge.

As a second example, consider the throw of a die on the convention that the value of a variable X equals the number of dots on the upward face after the die comes to rest. In this case, the discrete random variable X takes on six different values with equal probabilities.

As a third example, consider the amount of rainfall at a specific geographic location. Since water is perfectly divisible, annual rainfall may be thought of as a continuous variable. Even so, we can measure rainfall using a discrete variable as follows: If the rain is less than one-tenth of an inch, we say no rain fell. We say 0.1 inch fell, if the rain is between 0.1 and 0.2 inch. Rainfall is thus measured in intervals of one-tenth of an inch of rain, starting from zero and ranging upward. In year t we may be unable to predict accurately how much rain will fall in the subsequent year, year $t + 1$, but, on the basis of an historic record, we could make statements such as this: The probability of getting 10.9 inches of rain in year $t + 1$ is one-sixteenth. Clearly, the amount of rain that falls each year may be measured as a discrete random variable. The probabilities associated with various levels may be calculated from historical records.

Expectation and Variance

Let X be a random variable that takes on n distinct values. We say that $X = X_i$, letting i range over the values $1, 2, 3, \ldots, n$. The probability that $X = X_i$ is given by p_i for all i. The *expected value or mean* of the random variable, $E(X)$, is defined as follows:[1]

$$E(X) = p_1X_1 + p_2X_2 + \cdots + p_nX_n \tag{1-13}$$

[1]Let X be a continuous random variable. The probability density function $f(X)$ for X is defined as

$$1 = \int_{-\infty}^{+\infty} f(X)\, dX$$

Then,

$$E(X) = \int_{-\infty}^{+\infty} Xf(X)\, dX$$

As a simple example, consider the triangular density function such that $f(X) = 0$ for $X < 0$ and $f(X) = 0$ for $X > +2$. For $0 \le X \le 2$, $f(X) = (1/2)X$. Therefore, $E(X) = 4/3$. The interested student should draw a graph and experiment with creating other density functions. In this textbook, the student will not go too far awry if she or he imagines that the random variables of interest are discrete, although they can take on an extremely large number of values.

The *variance* of X is given by

$$V(X) = p_1[X_1 - E(X)]^2 + p_2[X_2 - E(X)]^2 + \cdots + p_n[X_n - E(X)]^2 \quad (1\text{-}14)$$

The expression $[X_i - E(X)]$ is defined as the deviation of the random variable from its expected value. It can be shown that

$$V(X) = E[X - E(X)]^2$$

$$= E(X^2) - [E(X)]^2$$

The concepts of expected value and variance are of basic importance in this book. Trivially, the expected value and variance of a constant equal the constant itself and zero, respectively. We pause again, however, to consider examples.

In the case of tossing a coin, the expected value or mean $E(X)$ equals (½)0 plus (½)1, or 0.5. The variance $V(X)$ equals $(½)(0 - 0.5)^2$ plus $(½)(1 - 0.5)^2$, which equals 0.25. As a useful exercise, the student should calculate $E(X)$ and $V(X)$ for the case of throwing a die.

To gain further understanding of expectation and variance, we next consider a gambling house that offers the following game: A fair coin is tossed once. The customer, who may play the game once only, wins $200 if a head appears. The customer pays $100 if a tail appears. Let X equal the expected gain or payoff. The expected value of X equals $50, equaling (½)$200 – (½)$100. The variance equals $22,500. (Customers are said to be averse to risk if they would pay less than $50 to play the game; those who like to incur risk would pay more than $50.) We expect many people would be willing to play this game for a fee of $50 or less.

Now consider a second game in which the customer gets $50,000 if a head appears. Otherwise, if a tail appears, the customer pays the house $49,900. The expected value of X in this game is $50, the same as before. Still, we expect few people would play this game even if there were no charge. Many people could afford to lose $100 or so, as in the previous game, but a $49,900 loss would be a calamity. The trouble with the latter game in comparison with the first is that the variance of the payoff X is enormous, equaling almost $2.5 billion. This example suggests that in the study of random or stochastic processes it is often useful to know more about the probability distribution of a random variable than simply its expected value or mean. In Chapter 2 it will be shown further that the variance of economic returns is of great importance in many applications.

Covariance and Correlation

Additional concepts of considerable importance in the chapters that follow apply to the case where two or more random variables are of interest. Given two random variables Y and X, we may ask whether they vary together or independently. What is often useful is a measure of the degree of association between them, and in this connection the concepts of covariance and correlation will prove useful. *Covariance* is defined as follows:

$$\begin{aligned}\text{Cov}(Y, X) &= E[(Y - E(Y))(X - E(X))] \\ &= E(YX) - E(Y)E(X)\end{aligned} \tag{1-15}$$

In the simple case when Y takes on only two values, Y_1 and Y_2, and X takes on two values, X_1 and X_2, we have

$$\begin{aligned}\text{Cov}(Y, X) = p_{11}&[(Y_1 - E(Y))(X_1 - E(X))] \\ + p_{12}&[(Y_1 - E(Y))(X_2 - E(X))] \\ + p_{21}&[(Y_2 - E(Y))(X_1 - E(X))] \\ + p_{22}&[(Y_2 - E(Y))(X_2 - E(X))]\end{aligned} \tag{1-16}$$

where p_{ij} equals the probability that Y equals Y_i *and* X equals X_j, where $i = 1, 2$ and $j = 1, 2$. The result given by Eq. (1-16) can be generalized to the case when both Y and X take on a large number of values. If large Y deviations often occur at the same time that large X deviations occur, the association between Y and X will be positive, and the covariance between the two variables will tend to be relatively large. If large Y deviations tend to occur with large negative X deviations, the association between the two variables will be negative. No association is, of course, possible. As a numerical example, consider the tossing of two coins such that, for one coin, $Y = 1$ if heads appear or zero otherwise, and similarly for the other coin, where $X = 1$ or $X = 0$. In this important special case it can easily be shown that $\text{Cov}(Y, X)$ equals zero and that $E(YX) = E(Y)E(X) = +\,0.25$. More generally, if two random variables vary independently one from the other, then $\text{Cov}(Y, X) = 0$, which implies [using Eq. (1-15)] that $E(YX) = E(Y)E(X)$.

Continuing the example of tossing two coins, suppose a poltergeist is present with the power to make the two coins show the same side 80 percent of the time, such that the four combinations $(Y_1 = 1, X_1 = 1)$, $(Y_1 = 1, X_2 = 0)$, $(Y_2 = 0, X_1 = 1)$, and $(Y_2 = 0, X_2 = 0)$ occur with probabilities 0.4, 0.1, 0.1, and 0.4, respectively. Then

$$\begin{aligned}\text{Cov}(Y, X) = 0.4&(1 - 0.5)(1 - 0.5) \\ + 0.1&(1 - 0.5)(0 - 0.5) \\ + 0.1&(0 - 0.5)(1 - 0.5) \\ + 0.4&(0 - 0.5)(0 - 0.5)\end{aligned}$$

Thus $\text{Cov}(Y, X) = 0.15$. Alternatively, using the definition of covariance given by Eq. (1-15), we see that $E(YX) = 0.4$ and the product $E(Y)E(X) = 0.25$. The difference, of course, equals 0.15, the covariance.

The concept of *correlation*, closely linked with that of covariance, is defined as

$$\rho(Y, X) = \frac{\text{Cov}(Y, X)}{\sigma_Y \sigma_X} \tag{1-17}$$

where $\sigma_Y = \sqrt{V(Y)}$ and $\sigma_X = \sqrt{V(X)}$ are *standard deviations*, defined to be the square root of the variance. Variances are thus often expressed as σ_Y^2 or σ_X^2. The idea of correlation is used frequently in everyday conversations largely because of the existence of causal relationships among variables. The greater the pressure is on the accelerator, the greater the speed of the car. The better the weather is, the higher the crop yield. Equation (1-17) simply provides a formal definition of a widely used notion. It can be shown (see DeGroot, 1975) that $-1 \le \rho(Y, X) \le 1$. Importantly, the correlation coefficient is unit free, that is, a pure number. This is why the correlation coefficient is often preferred to covariance as a measure of the association between two variables.

The student should show in the example of tossing two coins in the presence of a poltergeist that the correlation between Y and X, $\rho(Y, X)$, equals $+ 0.6$. The student should also verify that, if the poltergeist could assure that the same sides of the two coins always appeared, the correlation coefficient would equal 1. This is called perfect positive correlation. If the opposite sides of the two coins always appear, the correlation coefficient would equal –1. This is a case of perfect negative correlation.

We next consider linear functions of random variables. Suppose that $Y = f(X, Z)$, where X and Z are random. Then Y is random as well, and it is extremely useful for analytical purposes to know how the mean and variance of Y can be found given the means and variances of X and Z and the covariance between X and Z [i.e., $\text{Cov}(X, Z)$]. At this juncture we simply state a number of results together with a few examples, leaving proofs to DeGroot (1975).

If $Y = a + bX + cZ$, then

$$E(Y) = a + bE(X) + cE(Z) \tag{1-18}$$

where a, b, and c are constants. Recall that the expected value of a constant is the constant itself. Also, we have

$$V(Y) = b^2V(X) + c^2V(Z) + 2bc\,\text{Cov}(X, Z) \tag{1-19}$$

It is often useful to compare the variability of one variable with that of another. For this purpose a useful measure of variability, expressed in percentage terms, is the *coefficient of variation CX*, defined as

$$CX = \frac{\sigma_x}{E(X)} 100 \tag{1-20}$$

where $\sigma_x = \sqrt{V(X)}$ is the standard deviation of X, as before. The expected values and variances of different variables may be expressed in different units, such as bushels of corn and hundredweights of milk. The coefficient of variation has the advantage of being a unit-free measure. As such, it allows comparing the variability of corn production with that of milk production.

A Stochastic Economic Model

Consider the following model:

$$P = 10 - Q$$

$$Q = H$$

where the first equation is the demand for wild mushrooms and where H equals harvested mushrooms. Half the time, the weather is good and $H = 8$. Half the time, the weather is bad and $H = 4$. We cannot solve the model for an equilibrium price because the model is stochastic; there is no equilibrium price. We can, however, calculate both the expected price and the variance of price. Since $P = 2$ with $p = 0.5$ and $P = 6$ with $p = 0.5$, we have $E(P) = 4$. Alternatively, we find that $E(H) = 0.5(8) + 0.5(4) = 6$ and, using Eq. (1-8), we have

$$E(P) = 10 - E(H)$$

$$= 4$$

Turning to the variance of price, we use the definition of variance given by Eq. (1-14); thus

$$V(P) = 0.5(2 - 4)^2 + 0.5(6 - 4)^2$$

$$= 4$$

Alternatively, we calculate the variance of H, which equals 4.0, and use Eq. (1-19) to again show that $V(P)$ equals 4.

Letting total revenue be given by TR, where $TR = PH$, we can readily calculate the expectation and the variance of TR using basic definitions. We note that $TR = 24$ with $p = 0.5$ and $TR = 16$ with $p = 0.5$. Hence, $E(TR) = 20$ and $V(TR) = 16$. Notice that, since $TR = 10H - H^2$, the expectation and variance of TR cannot be estimated straightforwardly using Eqs. (1-18) and (1-19) because these two formulas hold only for linear functions.

For an application of the coefficient of variation, we find that $CP = 50$ percent and $CQ = 33.3$ percent. Hence price is more variable than consumption.

We next consider the covariance of Q and P. The probability that $Q = 4$ *and* $P = 6$ equals one-half, and similarly, for $Q = 8$ *and* $P = 2$. The probability that our two random variables will take on a pair of values other than the two cited here is zero. The covariance is therefore given by

$$Cov\,(Q,\,P) = \frac{1}{2}(4-6)(6-4) + \frac{1}{2}(8-6)(2-4)$$

$$= -4.0$$

Analyzing the equation for $\rho(Q, P)$, we find that $\rho = -1.0$. The correlation is perfect because P is determined by the value of $H = Q$ and by nothing else. Notice how different this case is from the simple coin tossing example, where both the covariance and the correlation between Y and X equaled zero.

Before leaving this example, we should note what happens to $\rho(Q, P)$ if demand is unstable. Let $P = 10 - Q + e$, where e is a random demand shifter that is sometimes positive and sometimes negative, but where price is always finitely large. Perhaps consumer preferences are unstable because consumers are subject to caprice. Sometimes they like mushrooms a lot and sometimes they don't, and no one knows why. The important point is that e is a random variable that shifts demand. Then the correlation between Q and P need not be perfect. It need not even be negative or nonzero. It could happen, for example, that when $H = 4$ demand tends to be weak, and when $H = 8$ demand tends to be strong. Then the correlation between yield and price could be positive.

Continuing this example, we suppose that $e = +1$ with probability p equal to 0.2 and $e = -1$ with probability p equal to 0.8. Let e and Q vary independently. We then have four possible market outcomes as follows:

Q	e	P	p
4	+1	7	0.1
4	−1	5	0.4
8	+1	3	0.1
8	−1	1	0.4

The student should verify that $E(Q) = 4$, $E(e) = -0.6$, and $E(P) = 3.4$ and that $V(Q) = 4$, $V(e) = 0.64$, and $V(P) = 4.64$. Notice that the variance results are consistent with Eq. (1-19) since $\text{Cov}(Q, e) = 0$. The covariance between output and price may be computed as follows:

$$\text{Cov}(Q, P) = (0.1)(7-3.4)(4-6) + (0.4)(5-3.4)(4-6) + (0.1)(3-3.4)(8-6) + (0.4)(1-3.4) \qquad (8-6)$$

Thus the covariance between Q and P equals -4.0. The correlation $\rho(Q, P)$ equals -0.9285. The introduction of the random shifter of demand destroyed the perfect negative correlation between quantity and price.

The student unfamiliar with the concepts of expectation, variance, covariance, and correlation should work through the exercises given at the end of this chapter with great care. Farm commodity markets are subject to important random shocks, such as the weather. A good grasp of the concepts introduced in this section is imperative to an understanding of how these markets perform in the real world.

1.3 A FAMILY FARM MODEL

The economic agents visualized in economic theory have limited affinity with their real-world counterparts. This is particularly true of the entrepreneur, a decision maker who seems more like a robot than a human being, solving complicated optimization problems in the blink of a robotic eye. Interestingly, in long-run models of perfect competition, all the total revenue is paid to factors of production, with normal returns paid to the suppliers of equity capital. For all his or her prodigious feats of calculation, the entrepreneur receives nothing. Notice, also, that in the typical large corporation decisions are the product of complicated organizations, not the work of a single person who might be thought of as an entrepreneur.

To make theory conform more realistically to agriculture, we define a *family farm* as a farm in which the family or household supplies labor, equity capital, and management. We make no distinction between managerial labor and the labor used to plant crops, feed animals, milk cows, and so on. Nor do we insist that the head of the household make the decisions. Final authority for decision making may be diffused throughout the entire family. Also, to simplify analysis, we assume that once a household organizes a farm, that is, enters farming, all the family labor is fixed in farming. More general formulations that allow for off-farm work are left to the advanced literature. Finally, although we suppose that the family farm is free to hire labor, hired labor and family labor are not perfect substitutes.

Because the household's capital and labor are both fixed in the farm, the usual concept of total fixed cost must be modified. It is convenient to begin by considering the definition of profit in a short-run model:

$$\pi = TR - TVC - TFC \tag{1-21}$$

where π equals profit, TR and TVC equal, respectively, total revenue and total variable cost, and TFC equals total fixed cost. Although TVC varies with the level of output, TFC, as its name suggests, does not. Throughout this book, unless otherwise noted, quantities are flow variables. They are defined on a per unit of time basis, output per year, for example. Total revenue is calculated by multiplying price times output. Total variable cost equals the sum of the products of variable input quantities times their respective prices. Total revenue and minimized total variable cost can be shown to be functions of output.

Equation (1-21) can also be written as

$$\pi = QR - TFC$$

where $QR = TR - TVC$ is defined as *quasi-rent*. Since quasi-rent is the net return over variable cost, it can be interpreted as the amount of money available to remunerate the fixed factors, that is, as the potential return to fixed equity capital and fixed family labor.

We will in what follows break up total fixed cost into two components, one attributable to fixed capital and the other to fixed family labor. The fixed cost of capital is denoted by CC (cost of capital). The fixed cost of family labor is denoted by TE (transfer earnings). Some brief discussion of these components is in order, but it should be stressed that, for the moment, we ignore uncertainty and all the complications that uncertainty entails.

To define cost of capital (CC), we suppose that a household has \$100 at the beginning of the current year (year 1), but owns no other assets. (We choose \$100 to simplify calculations.) The household plans to retire at the end of some future year, year T. The \$100 could be invested in a money market fund at an annual rate of interest given by i. How much money would be available at the end of year T? The formula that can be used to answer this question is given by

$$M_T = M_0(1 + i)^T$$

where M_0 is the initial amount of money invested at the beginning of year 1 and M_T is the amount to be withdrawn at the end of year T. In our example, M_0 = \$100. If we let i equal 0.05 and T equal 20, M_T equals \$265.33.

We next suppose that the household has the alternative of purchasing a farm complete with land and machinery. We assume for the moment that the farm would be operated *entirely* by hired labor. At the end of each year, the household receives quasi-rent (QR), which is deposited in the money market fund.[2] How much money will have accumulated in the money market fund under this investment option at the end of year T? To see the answer, let QR_t equal the quasi-rent to be invested in the money market fund at the end of each year t, $t = 1, 2, \ldots, T$. Also, let QR_t be constant for all t. At the end of period T, the household would have

$$QR\left(1+i\right)^{T-1} + QR\left(1+i\right)^{T-2} + \cdots + QR\left(1+i\right)^{T-T} + S = QR \sum_{t=1}^{T} \left(1+i\right)^{T-t} + S$$

where S equals the salvage value or price of the farm assets at the end of year T. Now consider a particular value of QR, CC, such that

$$M_T = CC \sum_{t=1}^{T} \left(1+i\right)^{T-t} + S$$

[2]The student who is puzzled by the money needed for consumption might consider altering the model to allow for a fixed amount of each year's income going to consumption. The main point is that money for consumption would ordinarily be required under any investment option. Alternatively, we might simply assume that the family's consumption needs are met by a friend of the family.

where M_T is defined above. Rewriting, we have

$$CC = \frac{M_T - S}{\sum_{t=1}^{T} (1+i)^{T-t}} = \frac{M_T - S}{\left[\frac{(1+i)^T - 1}{i}\right]} \tag{1-22}$$

Equation (1-22) defines CC, the opportunity cost of capital tied up in the fixed plant. In general, CC equals the *user cost of capital* (or the implicit valuation of the flow of services provided by fixed factors over many periods or years). In real-world situations, it reflects the interest rate, depreciation, and possibly other factors as well. Aside from considerations of risk, an entrepreneur would (not) buy the plant if the resulting wealth at the end of year T were larger (less) than M_T.

Armed with this interpretation of the fixed cost of equity capital, we now turn to a household that is considering entering farming, that is, buying a farm and other necessary farm assets. Entry means the family labor will be fixed in the farm business. Taking family labor as the *residual claimant*, the return to family labor RFL, is defined as

$$RFL = TR - TVC - CC \tag{1-23}$$

We now define the household's *transfer earnings*, TE, as the minimum RFL that would prompt the household to enter farming. For a household that is already established in farming, TE is the minimum RFL that would keep them in farming. We imagine a process in which the household compares RFL with TE. If $RFL > TE$, the household enters (or stays in) farming. If $RFL < TE$, the household does not enter (or leaves) farming. If $RFL = TE$, the household is indifferent between farming and the next best alternative career. In this latter case, where $RFL = TE$, the farmer is said to be a *marginal farmer*.

On these arguments, profit is given by[3]

$$\begin{aligned} \pi &= TR - TVC - CC - TE \\ &= TR - TVC - TFC \end{aligned} \tag{1-24}$$

[3]Normal profit is sometimes defined as profit that includes a "normal" return on investment. In this book, profit is to be interpreted as pure profit. It is a residual over and above that which allows for a normal return on investment and/or transfer earnings to labor.

where *TFC* is total fixed cost, equaling the sum of *CC* and *TE*. If $\pi > 0$ (or, equivalently, if $RFL > TE$), the household enters the farm industry. The returns to the family's resources in farming exceed the returns in the nonfarm sector. If $\pi < 0$ (or, equivalently, if $RFL < TE$), the household stays out of farming. Although *CC* is based on opportunity cost of capital, *TE* is *not* based on the opportunity money cost of labor. Rather, *TE* is based on the opportunity cost of labor when both money and work satisfaction (or dissatisfaction) are taken into account.

Why the different treatment of the cost of capital and transfer earnings? Earning money from work means giving up leisure and performing tasks that may be more or less unpleasant. Being a life guard on a sunny California beach is more fun than collecting garbage in Newark, New Jersey. The members of a household might be content to receive $10,000 a year for its labor in farming, if they like farming, even though they might earn $15,000 in the next best career opportunity; perhaps, the father drives a truck and junior delivers newspapers. All you have to do to earn money from capital, on the other hand, is invest in a money market fund and put your feet up. (The wise investment of money in a world of uncertainty is an altogether different matter, however, as we shall see in Chapter 4.)

The typical diagram showing short-run cost curves gives curves for average fixed cost (*AFC*), average variable cost (*AVC*), average total cost (*ATC*), and marginal cost (*MC*). (To be internally consistent, the curves must be drawn in a particular way.) Because we have broken up total fixed cost into two components, we need to introduce two new average cost concepts: average cost of capital (*ACC*), defined as *CC* divided by output, and average transfer earnings (*ATE*), defined as *TE* divided by output. The familiar set of short-run cost curves, appropriately modified, is given in Fig. 1.2. The *AFC* curve is found by summing vertically *ACC* and *ATE*. As in basic theory, *ATC* is found by summing vertically *AVC* and *AFC*. Figure 1.2 is drawn for a farm that is a marginal farm when price equals P_0. At this price, total revenue (P_0q_0) equals the sum of total variable cost (AVC_0q_0), cost of capital (ACC_0q_0), and the transfer earnings (ATE_0q_0). Profit equals zero and the returns to family labor (*RFL*) equal *TE*.

We now consider the implications of our definition of a family farm for long-run cost relationships. In the usual treatment of competitive price theory, all the firm's inputs are assumed to be variable in the long run. In contrast, we assume that all inputs are variable except family labor, which is assumed fixed. Panel a of Fig. 1.3 provides a graphic representation of *ATE*. The long-run average total cost curve is given by *AC*, which equals the vertical sum of the *ATE* and *AVC*. The intersection of the demand and supply curves in panel b gives the equilibrium price P_c. At this price, the farm for whom the cost curves are given in panel a is a marginal farm in that profit equals zero and *RFL* equals $(ATE_c)q_c$ or, equivalently, $(P_c - AVC_c)q_c$. The firm would earn excess or pure profit for any price in excess of P_c. We assume that the household would either leave farming or not enter if price fell below P_c. For simplification, we assume that *AVC*, as illustrated in panel a, is the same for all farms, established or potential.

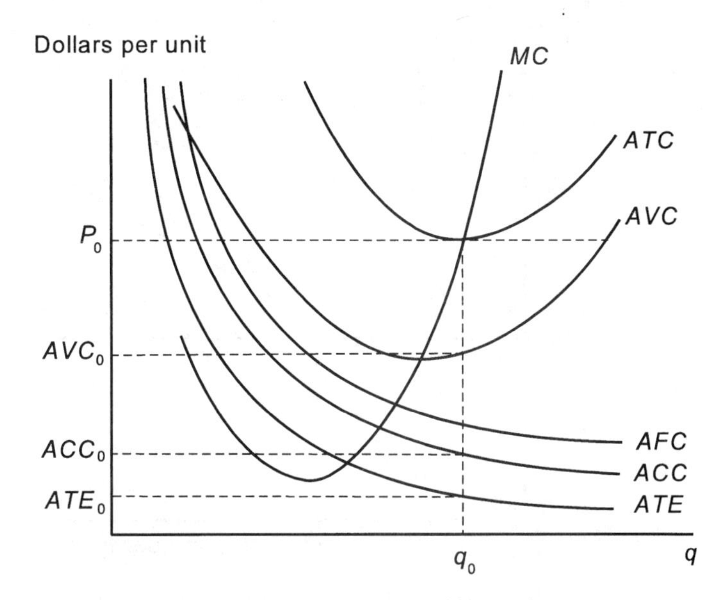

Figure 1.2 Short-run cost curves for a family farm.

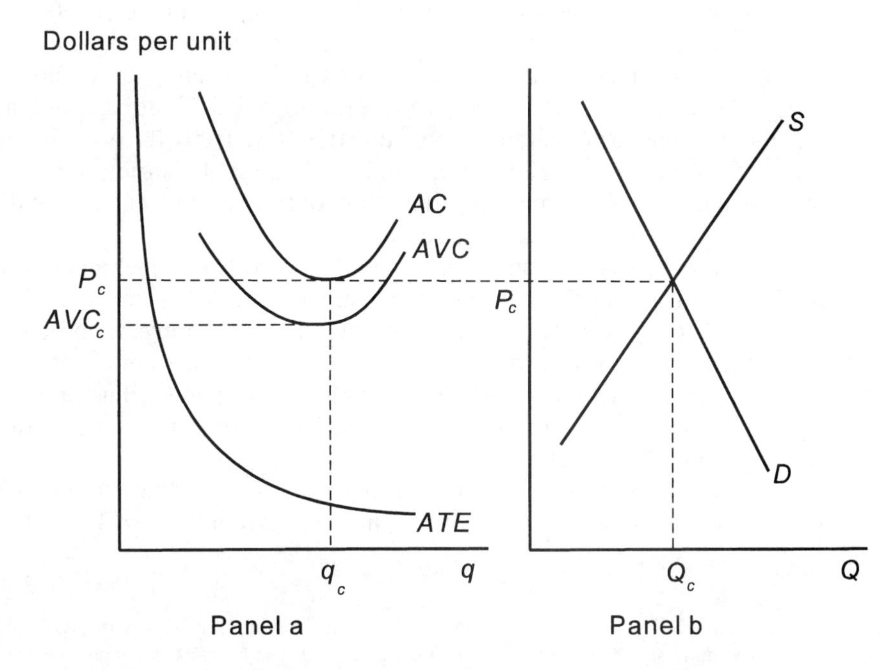

Figure 1.3 Long-run cost curves for a marginal family farm in long-run equilibrium.

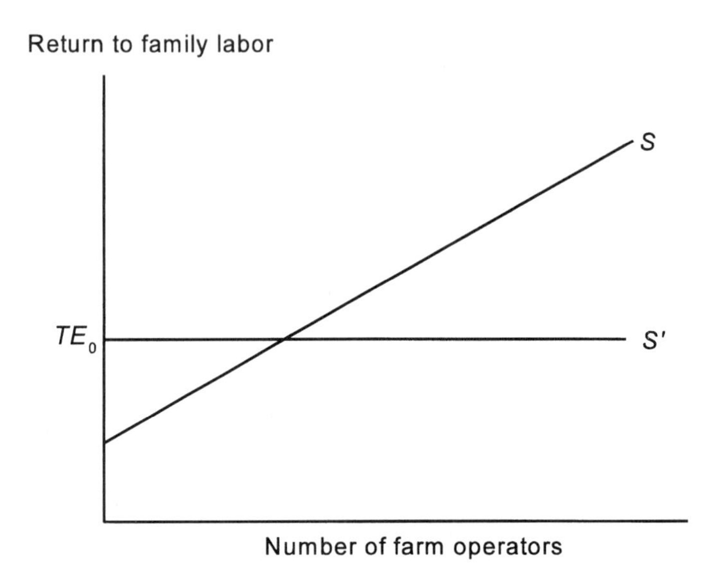

Figure 1.4 Supply curves for farm operators.

We next define the supply function for farm operators (or for farms or farmers) as a function that shows how many households would be willing in the long run to organize and operate farms (i.e., enter or stay in farming) at alternative levels of returns to family labor (RFL). If transfer earnings equal TE_0 for all households, as in Fig. 1.4, then the farm operator supply function is perfectly flat at S'. Importantly, if the supply functions for all farm inputs, including that for farm operators, are perfectly flat, then the long-run supply function for farm output will also be flat. We say that the industry is then a constant-cost industry.[4]

A more realistic assumption, however, is that transfer earnings vary among households. The transfer earnings for a movie star would presumably be much higher than for typical college professors; the transfer earnings for professors would likely exceed those with few skills. If transfer earnings vary among households, the long-run supply function for farm operators is upward sloping, illustrated by S in Fig. 1.4. In this case, the long-run supply function for farm output will be upward sloping as for any increasing cost industry.

Although family labor per farm is fixed, the aggregate input of family labor is variable in the long run because of the entry and exit of farmers. This is brought out

[4]A constant-cost industry is a competitive industry that buys small proportions of the available supplies of inputs such that input prices are unaffected by the industry's level of purchases. In the case of an increasing cost industry, at least one input price rises as more of it is purchased.

clearly by the derivation of the long-run product supply curve. In Fig. 1.5, we start with an initial position of long-run equilibrium at price P_0. We assume an upward sloping supply function for farm operators. The AVC curve in panel a is, by assumption, the same for all farmers. The AC curves, on the other hand, vary because transfer earnings vary among households. The transfer earnings for the ith farmer equals $(P_0 - AVC_0)q_0$. At price P_0, the return to this family's labor just equals its transfer earnings; its profit equals zero. At $P = P_0$, the ith farm is the marginal farmer. Suppose that demand rises from D_0 to D_1 in panel b. The returns to family labor rise with increases in price. The ith farmer now earns a pure profit and new households enter farming. A new long-run equilibrium position is illustrated for price equal to P_1. At this price the jth firm is the new marginal farm with transfer earnings equal to $(P_1 - AVC_1)q_1$. At $P = P_1$, the jth firm's profit equals zero.

Two further points may be noted. First, input prices other than the implicit price of family labor may rise with increased industry output. That is, some input prices may be jointly dependent with the levels of output and output price. Second, whereas labor in conventional competitive theory rises or falls because entrepreneurs hire or fire, in the present analysis, family labor varies strictly through the entry and exit of farms. Again we stress our intention to keep the analysis simple by not allowing for the off-farm employment of family labor.

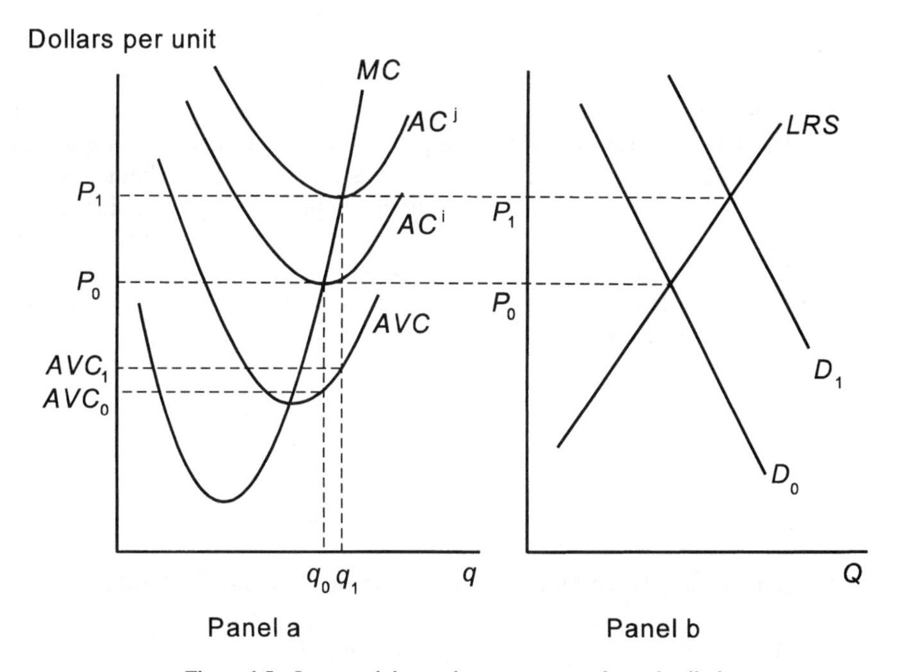

Figure 1.5 Increased demand attracts entry of new family farms.

1.4 CONCLUDING COMMENTS

This book centers on the determination of prices and other dimensions of performance in agricultural markets. We employ the time-honored technique of building structural models and analyzing their implications through the derivation and evaluation of the corresponding reduced form models. Of interest are stochastic models that include random variables as a means for dealing with uncertainty. Expectations and variances of random endogenous variables are themselves treated as endogenous. We do not, for example, believe we can gain a reasonably sophisticated understanding of agricultural pricing without analyzing the central role of price expectations.

Unlike major corporations, farms are usually family owned and managed. In addition, much of the labor input in agriculture is supplied by the farm family. We believe that understanding agricultural input pricing and the interdependency among input and output prices, particularly in the long run, requires modification of the definitions of the firm and entrepreneur as set forth in conventional theory. Accordingly, the conception of the family farm set forth here, together with the concept of the supply function for farm operators (or family labor), will often be fundamental to the long-run structural models developed in succeeding chapters.

APPENDIX 1.1: ELASTICITY

Elasticity, an important concept used frequently in the chapters that follow, is a characteristic of a functional relationship between two variables. Consider the demand $Q = D(P)$. Demand shifters do not appear explicitly because they are to be held constant. The elasticity of demand is defined as the percentage change in quantity demanded Q divided by the associated change in price P. Two formulas may be used to calculate elasticities depending on whether the changes are finite or infinitesimally small. Let (P_0, Q_0) and (P_1, Q_1) be two points in the same neighborhood of the demand function. The arc elasticity formula for calculating the elasticity of the quantity demanded with respect to price for finite changes is given by

$$\text{arc } \epsilon(Q \mid P) = \frac{\dfrac{Q_0 - Q_1}{(Q_0 + Q_1)/2}}{\dfrac{P_0 - P_1}{(P_0 + P_1)/2}} \tag{1-25}$$

The point elasticity formula is applicable in the case of infinitesimals:

$$\epsilon(Q \mid P) = \frac{\partial Q}{\partial P}\frac{P}{Q} \tag{1-26}$$

Take a simple case where $Q = 20 - 2P$ or, using the inverse form, $P = 10 - 0.5Q$. (See the D curve in the upper part of Fig. 1.6.) Consider the points (12, 4) and (14, 3). Arc elasticity equals –0.54. Point elasticity is given by

$$\epsilon(Q \mid P) = (-2)\frac{P}{Q}$$

Point elasticities for $Q = 12$, 13, and 14 are, respectively, –0.67, –0.54, and –0.43. Notice that the arc elasticity equals the point elasticity for the average of the two levels of output used in the arc elasticity formula. We may think of the arc elasticity as an approximate or representative measure of elasticity over a small range of output. Henceforth in this book, when we use the term elasticity, we will mean point elasticity. Since P is a function of Q, it is an easy matter to show in the present example that elasticity is a function of Q. Accordingly,

$$\epsilon(Q \mid P) = \frac{Q - 20}{Q}$$

For $Q = 10$, $\epsilon(Q|P) = -1.0$. For $Q < 10$, $\epsilon(Q|P) < -1.0$. For $Q > 10$, $\epsilon(Q|P) > -1.0$. If the elasticity is larger than –1 (or its absolute value is smaller than 1), demand is said to be inelastic. If the elasticity is less than –1 (or its absolute value is greater than 1), de-

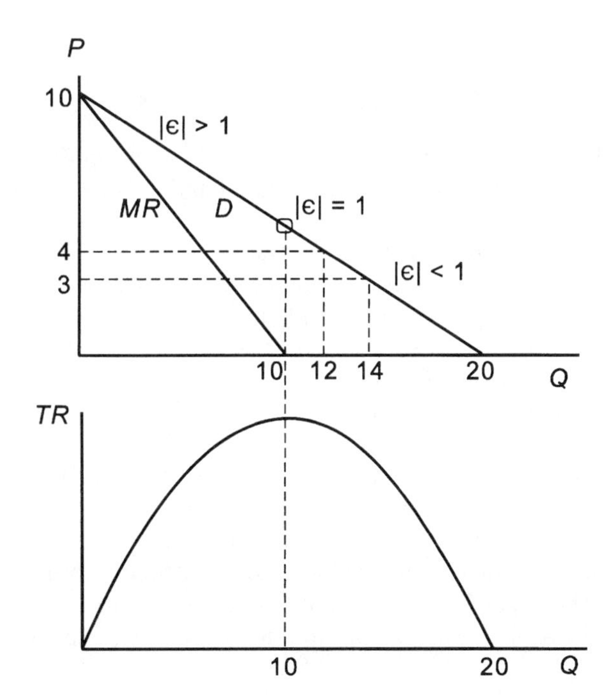

Figure 1.6 Demand, marginal revenue, and total revenue curves for various elasticities of demand.

mand is said to be elastic. If demand elasticity equals –1.0, demand is said to be unitarily elastic.

Why is the elasticity of demand $\epsilon(Q|P)$ an important concept? The answer to this question will emerge in due course, but Fig. 1.6 will get us off to a good start. Total revenue is given by the product PQ. Expressed as a function of Q, we have

$$TR = 10Q - 0.5Q^2$$

The graph of this quadratic total revenue function is given in the lower part of Fig. 1.6. Now picture a supply curve in the upper part of the diagram intersecting the demand curve in an inelastic neighborhood (where $|\epsilon| < 1$). A small shift in the supply curve to the right lowers total revenue to producers. Technological progress might lower producer income. If the demand–supply intersection were in an elastic neighborhood of demand, then technological progress would increase total revenue to producers. Producer income would increase. The analysis will be developed in greater detail in Chapter 4, but perhaps enough has been said to suggest that the economic effects of exogenous shocks (changes in exogenous variables) depend in a crucial way on whether demand is elastic or inelastic.

Although applied to the functional relationship between quantity demanded and price, the concept of elasticity can also be used to describe the relationship between quantity demanded and any other variable appearing on the right-hand side of demand. Let Z_i be the ith shifter of demand. Then

$$\epsilon\left(Q \mid Z_i\right) = \frac{\partial Q}{\partial Z_i} \frac{Z_i}{Q} \tag{1-27}$$

By definition, if Z_i equals income, then Eq. (1-27) defines the income elasticity of demand, which is positive for normal goods (ski lift tickets) and negative for inferior goods (lard).

If Z_i equals the price of a related product, then Eq. (1-27) defines cross-elasticity of demand. If the related product is a good substitute, as pork is for beef, then we expect the cross-elasticity to be positive. If the related product is a good complement, as ski boots are for skis, then we expect the cross-elasticity to be negative. (The advanced student will recognize that this brief discussion of cross-elasticity ignores subtleties associated with income and substitution effects.)

Although our discussion has centered on demand, it should be understood that the concept of elasticity can be used to characterize or describe the relationship between any two functionally related variables. Economists often apply the concept to supply, for example. The student should also recognize the difference between what might be called a structural elasticity and a reduced form elasticity, both involving the same two variables. The elasticity of the quantity demanded with respect to Z_1 for Eq. (1-9), a structural elasticity, differs from that for Eq. (1-12), a reduced form elasticity.

APPENDIX 1.2: AGGREGATION

Economic theories or models are ordinarily developed by showing first how the individual decision maker (consumer, producer, and input supplier) responds to changes in the environment (changes in price, say). The result is a behavioral relationship such as a demand or supply function. The next step is the aggregation or adding up of the behavioral relationships of all like decision makers. Our objective in this section is to show how this adding-up procedure is carried out graphically. The procedure is simple, but it is also important; it will be used repeatedly in the remainder of this book.

The easiest way to understand the procedure is to take an example. Let D_i and D_j be the demand curves for individual i and j, respectively, in panel a, Fig. 1.7. To obtain the total or aggregate demand, we find the quantities demanded for both consumers for each price in panel a and pair the sum of these quantities with the associated price in panel b. If price P equals P_0, individual i would be willing to buy q_{i0}; individual j would be willing to buy q_{j0}. To get one point on the aggregate demand, we plot the sum $(q_{i0} + q_{j0}) = Q_0$ in panel b for price P_0. This procedure can be followed for all possible prices. Notice that, for prices equal to or exceeding price P_1, individual j would not be willing to purchase anything. In this case the aggregate quantity Q equals q_i. The procedure used in aggregating D_i and D_j, in summing laterally D_i and D_j, can, of course, be applied to many consumers.

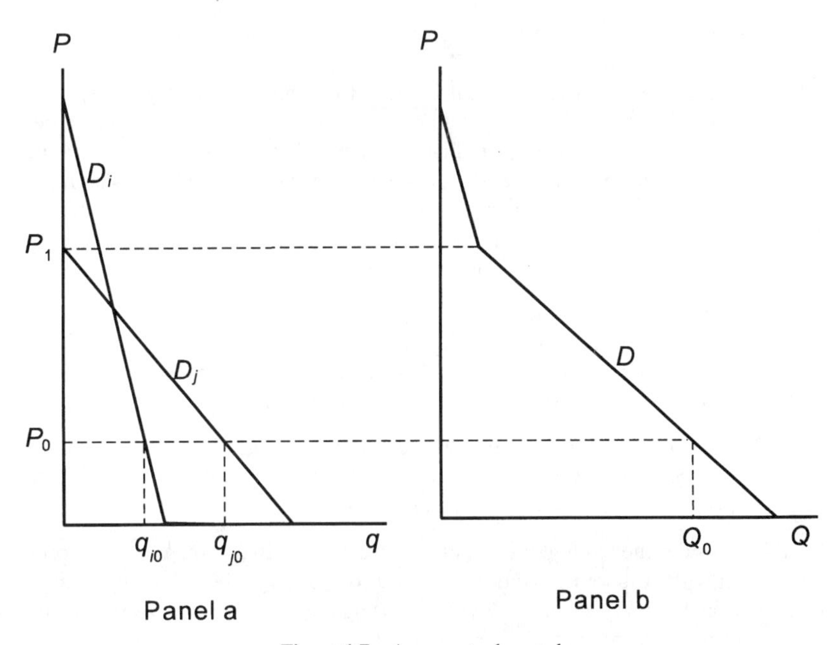

Figure 1.7 Aggregate demand.

The student should also think about subtracting laterally one curve from another. If D_i had been plotted in panel b together with D, we could subtract laterally D_i from D in panel b to get D_j in panel a.

One more thing. Suppose that we wanted to aggregate algebraically the functions for individual decision makers. Let $P = D_i^{-1}(q_i)$ and $P = D_j^{-1}(q_j)$ be the two inverse demand functions plotted in panel a. [In mathematical economics it is customary to express demand quantities as a function of price, $Q = D(P)$. If the equation is written with price as the dependent variable, the relationship is called the price-dependent or inverse demand function, $P = D^{-1}(Q)$.] To sum algebraically, we note that $Q = q_i + q_j$. Therefore, $Q = D_i(P) + D_j(P) = D(P)$. This is a handy result to keep in mind when solving problems.

PROBLEMS

1.1. You are given the following demand and supply functions:

$$P = 20 - 2Q + 0.8Z_1 \qquad \text{demand}$$

$$P = -1 + 0.5Q - 0.9Z_2 \qquad \text{supply}$$

a. Find the reduced form equations for P and Q. Find competitive equilibrium values for P and Q given $Z_1 = 1$ and $Z_2 = 4$.

b. Calculate the point elasticity of demand and supply in equilibrium.

c. Again letting $Z_1 = 1$ and $Z_2 = 4$, calculate the point structural elasticities $\epsilon_s(Q|Z_1)$ and $\epsilon_s(Q|Z_2)$ and the point reduced form elasticities $\epsilon_r(Q|Z_1)$, $\epsilon_r(Q|Z_2)$, $\epsilon_r(P|Z_1)$, and $\epsilon_r(P|Z_2)$. [*Note:* $\epsilon_s(Q|Z_1)$ and $\epsilon_s(Q|Z_2)$ are, respectively, the elasticities of demand and supply. The s subscript indicates structural elasticity; r indicates reduced form.]

d. Again letting $Z_1 = 1$ and $Z_2 = 4$, use the following two points along the demand curve to estimate an arc elasticity for demand: $Q_0 = 6.2$, $P_0 = 8.4$, $Q_1 = 4.2$, and $P_1 = 12.4$. Use the following two points to estimate an arc elasticity for supply: $Q_0 = 9.66$, $P_0 = 0.23$, $Q_1 = 10.66$, and $P_1 = 0.73$.

1.2. Demand is given by $Q = 1P^{-\beta}$.

a. Prove that demand elasticity equals $-\beta$ for all values of Q.

b. Suppose that Q is measured in bushels and P is measured in dollars per bushel. What is the unit of measurement for the coefficient 1?

1.3. Let X be a random variable that takes on the values of 2, 4, and 6 with the respective probabilities $\frac{1}{4}$, $\frac{1}{2}$, and $\frac{1}{4}$. Calculate $E(X)$, $V(X)$, and CX. Consider a new random variable Z, where $Z = X^2$. Calculate $E(Z)$, $V(Z)$, and CZ.

1.4. Consider the tossing of two unfair coins. For the first coin, $X = 0$ with probability $\frac{1}{4}$ and $X = 1$ with probability $\frac{3}{4}$. For the second coin, $Y = 1$ with probability $\frac{2}{3}$ and $Y = 2$ with probability $\frac{1}{3}$. Calculate $\text{Cov}(Y, X)$ using the definition of covariance given by Eq. (1-15). Then check your result by using Eq. (1-16).

1.5. Let $Y = a + bX$, where X is a random variable. Demonstrate that $\rho(Y, X)$, the correlation between Y and X, equals 1. (*Hint:* $\text{Cov}(Y, X) = \text{Cov}[(a + bX), X]$).

1.6. The demand for Christmas trees depends on how white Christmas is and is given by

$$P = 20 - 0.5Q + e$$

where Q equals the Christmas tree harvest. The tree harvest always equals 30. The random variable e takes on the values of +2, 0, and –2 with probabilities 1/4, 1/2, and 1/4, respectively. Also, $TR = PQ$.

a. Calculate $E(e)$, $V(e)$, $E(P)$, $V(P)$, $E(TR)$, and $V(TR)$.

b. Demonstrate that $\rho(TR, P) = 1$.

REFERENCES

DeGroot, Morris H., *Probability and Statistics*. Reading, Mass.: Addison-Wesley Publishing Co., 1975.

Helmberger, Peter G. *Economic Analysis of Farm Programs*. New York: McGraw-Hill, Inc., 1991.

2

Production Under Uncertainty

Understanding how farm outputs are determined is basic to understanding agricultural pricing. It cannot seriously be proposed that modeling the latter can proceed without modeling the former. Unfortunately, the concept of the supply function, together with its theoretical underpinnings, in the typical theory textbook is of limited applicability to agricultural production because of the failure to take account of production lags and uncertain market conditions. The objective of this chapter is to provide more realistic and convincing models of agricultural production and supply response, models that will be basic to the chapters that follow.

We first review briefly the model of a perfectly competitive market under certainty. Much of the attention centers on the derivation of the farmers' short-run supply function, a function that shows how much farmers in the aggregate would be willing to produce at alternative known prices. In Section 2.2 we take into account production lags when farmers do not know at planting time what the price will be at harvest time. This section abstracts from production uncertainty; actual output always equals planned output. Again, the emphasis is on short-run supply. It is shown that when farmers do not like risky prices (farmers are risk averse) the level of total output depends on expected price and the standard deviation of price. We also show how farmers' price expectations are formed by bringing the expected demand conditions into the picture. The student familiar with the theory of consumer behavior based on utility maximization may be surprised to learn that the theory of the firm under conditions of uncertainty is also based on utility maximization.

Section 2.3 allows for both price and production uncertainty. At planting time, farmers do not know what the price will be at harvest, and actual output need not equal planned output. A crucial concept is what we will call the gross return per unit of planned output. Total farm output will be shown to depend in the short run on the expectation and standard deviation of gross return.

Section 2.4 centers on the long-run supply response under conditions of price uncertainty. Two important ideas are developed that are not encountered in the usual treatment of the competitive model under certainty. The first pertains to how farmers, both established and potential, learn about and adjust to changes in uncertain market conditions. The second pertains to the speed of long-run adjustments to changes in exogenous variables and the likelihood that expansion of farm production capacity in response to demand growth occurs more quickly than the reduction of capacity in response to demand contraction.

2.1 PRODUCTION AND PRICING UNDER CERTAINTY

The short-run profit function for a firm under competitive conditions and in the absence of uncertainty may be expressed as follows:

$$\pi = Pq - C(q) - TFC \tag{2-1}$$

where π equals profit, P equals price, q equals output, and TFC equals total fixed cost. Total variable cost C is given by the function $C = C(q)$. We assume here that the farmer has already solved the cost minimization problem and knows the least-cost method of producing alternative levels of output.[1] Under competitive conditions, price P is viewed by the entrepreneur as a parameter or a given that does not vary with changes in the firm's output q. To investigate the implications of profit maximization and assuming that the optimal output is positive, we take the derivative of profit π with respect to output q and set it equal to zero:

$$\pi' = P - C'(q) = 0 \tag{2-2}$$

where $\pi' = \partial\pi/\partial q$ is the first derivative of π with respect to q and $C'(q) = \partial C(q)/\partial q$ is the first derivative of C also with respect to q. By definition, $C'(q)$ is the marginal

[1]The total variable cost function $C = C(q)$ for the two-variable input case is derived as follows: Total variable cost is defined as

$$C = V_1 x_1 + V_2 x_2$$

where, for this particular problem, V equals input price; x equals input level; and the subscript $i, i = 1, 2$, indicates input i. The production function is given by $q = q(x_1, x_2)$. Minimizing C subject to the production function (finding, graphically, the isocost lines that are tangent to isoquants) yields optimum input levels as functions of input prices (V_1 and V_2) and output (q) thus:

$$x_1 = f_1(V_1, V_2, q)$$
$$x_2 = f_2(V_1, V_2, q)$$

Making the appropriate substitutions and using the definition of total variable cost, we have

$$C = V_1 f_1(V_1, V_2, q) + V_2 f_2(V_1, V_2, q)$$

which, treating V_1 and V_2 as constants, is the total variable cost function $C = C(q)$.

cost of output, which is a function of q. We let MC equal marginal cost. Equation (2-2) asserts that to maximize profit the entrepreneur must set the first derivative equal to zero, thus choosing a level of output such that $P = MC$. This is an example of a first-order condition for an extremum position, whether a maximum or a minimum. To be sure that the entrepreneur has chosen the level of output q that maximizes profit, it must also be true that the second derivative of π with respect to q, π'', is less than zero. We note that $\pi'' = -C''(q)$. But $C''(q)$ is the derivative of the marginal cost function with respect to q. For any q, $C''(q)$ gives the slope of the MC function. Thus, if q is chosen such that $P = MC$ and the marginal cost function is upward sloping, we can be sure, well, almost sure, that profit has been maximized. To be 100 percent sure of a maximum, we must also insist that, at the point where $P = MC$, MC is not less than the average variable cost, $AVC = C(q)/q$. Otherwise, assuming that fixed cost cannot be avoided in the short run, it would pay the entrepreneur to shut down the plant and produce nothing at all. The quasi-rent, defined as total revenue minus total variable cost, is $Pq - C(q)$. The quasi-rent must be positive to give the entrepreneur the incentive to produce. It follows that $Pq - C(q) \geq 0$, or $P \geq C(q)/q = AVC$. This shows the desired result: Output price must be at least as large as the average variable cost. Otherwise, the firm will have no incentive to produce and will shut down operations.

Figure 2.1 allows a graphic analysis assuming the usual shaped MC and AVC curves. Average fixed cost is given by $AFC = TFC/q$. Letting TC equal total cost, that is, the sum of total variable cost $C(q)$ and total fixed cost TFC, average total cost is

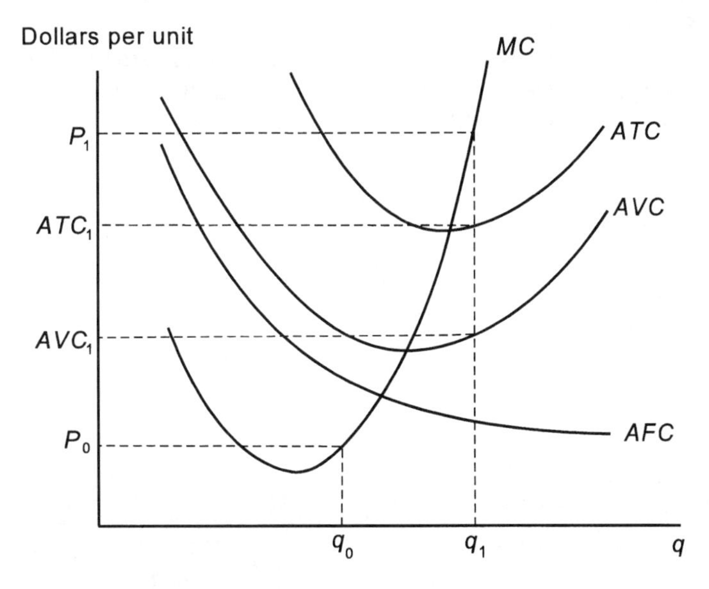

Figure 2.1 Short-run cost curves for a firm.

given by $ATC = TC/q$. Clearly, ATC is the sum of AVC and AFC. Graphic representations of the AFC and ATC functions are given by the AFC and ATC curves in Fig. 2.1. It may be noted that the ATC curve is the vertical sum of the AFC and AVC curves.

If price equals P_1, we look for the level of q such that $P = MC$. We also look to see if MC is upward sloping at that point and if $MC \geq AVC$. Thus it is clear that the level of output q_1 in Fig. 2.1 satisfies all the conditions; it yields the maximum profit. If $P = P_1$, then the maximum profit is given by $(P_1 - ATC_1)q_1$. Quasi-rent, defined as total revenue minus total variable cost, is positive and is given by $(P_1 - AVC_1)q_1$.

Still centering on Fig. 2.1, suppose that $P = P_0$. If the level of output q_0 is chosen, once again $P = MC$. The MC curve is upward sloping as well. The level of output q_0 does not, however, maximize profit. If q is set equal to q_0, the corresponding level of AVC exceeds P_0. Quasi-rent becomes negative. If price falls below the minimum of the AVC curve, the firm maximizes profit by shutting down and setting q equal to zero. The loss then equals TFC, but the loss would be even greater if the firm produced output with the total variable cost exceeding total revenue. That is, the loss would be even greater if quasi-rent were negative.

This discussion leads to an important result. The firm's supply curve is given by that part of the MC curve where $MC \geq AVC$. The supply curve is an example of a behavioral relationship in that it shows how much a profit-maximizing firm will produce at various alternative prices. To obtain the aggregate short-run supply curve for the industry, we need merely sum horizontally the individual supply curves of all the firms. (We are assuming here that all variable input prices are held constant.) Such an aggregate supply curve is given by SRS in Fig. 2.2. The intersection of the supply curve SRS and the demand curve D gives the equilibrium market price P_c and output Q_c. (Here and elsewhere, we will often use capital letters, in this case Q, to represent the aggregate of individual levels of a product or input, in this case q.) At P_c, consumers or buyers are able to buy as much as they desire and sellers are able to

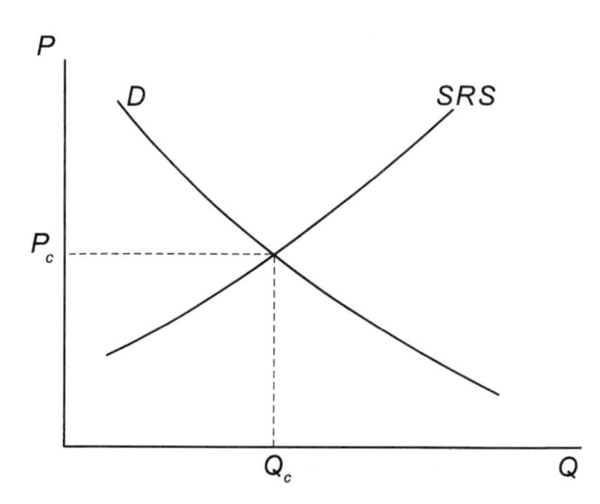

Figure 2.2 Demand and short-run supply curves.

sell as much as they desire. The intersection of demand and supply is an equilibrium because, at $P = P_c$, consumers are able to maximize utility subject to their budget constraints, and sellers are able to maximize profit, given production technology and input prices. Under competition, no other price allows the constrained maximization of the objectives of all market participants.

As noted in Chapter 1, the notion of market equilibrium is important not so much for its own sake, but because it allows comparative static analysis. Once we allow for shifts in the demand and supply curves, we are often able to frame hypotheses about how exogenous shocks affect price and output. We could easily compile a long list of hypotheses with regard to the price and output effects of shifts in demand and supply. This is why the theory of supply and demand is of fundamental importance in economics.

This completes our brief review of one of the most important models in microeconomic theory, a model that purports to explain how a competitive market works and how market performance (price and output) responds to exogenous shocks, that is, changes in exogenous variables. As argued in Chapter 1, however, the competitive market as envisaged in economic theory often does not correspond very well to markets observed in the real world. A prime objective in agricultural economics is to modify the theory of competitive markets as an important step in trying to understand why farm commodity markets perform the way they do. Take an example. According to theory, an increase in demand causes output to rise. But suppose that in January the demand for corn increases dramatically because, say, the Russians decide to increase corn imports. The available supply of corn is fixed, at least until after the next harvest. With bad weather the next harvest might be less than normal; the supply of corn might then be less than before the price hike. What happens to the prediction that an increase in demand causes output to expand? Clearly, the standard theory of supply and demand cries out for modification if it is to be of much use in understanding how the corn market works in practice, and much the same holds for other farm markets as well.

2.2 MARKET PERFORMANCE UNDER PRICE UNCERTAINTY

One distinguishing feature of agricultural production is that it always involves biological processes. This has many important implications for the application of economic theory to the study of agricultural production, marketing, and pricing. Agricultural production, crop production particularly, requires environmental inputs such as sunlight, water, heat, and time that are often beyond the control of farmers. The control of production quality and yields is often far from complete. In addition, crop production requires land, so the production activities are spread thinly over millions of acres like butter on a slice of bread. Moreover, production activities involve the growth of animals and plants, and the length of time between the applications of inputs and the garnering of output might take anywhere from six weeks in the case of broilers to several years in the case of tree fruits. Crop production also typically takes place on an annual cycle. The result is that agricultural production involves heterogeneous outputs produced in small lots, often on a seasonal basis, scattered hither and yon over wide geographic areas.

Although we will discuss the nature of agricultural production and its implications for pricing in many of the chapters that follow, the aspect of production of central interest here is the time lag between the application of inputs and the completion of output. If a sufficient number of carpenters is put to the task, a house can be built in a week. No number of farmers can produce a corn crop in less than several months. Because of lags, it is possible that farmers may need to make production decisions (how and how much to produce) without knowing what price will be received for their outputs. The production plans may also go awry because of bad weather and other factors beyond the farmer's control. In short, we will in this chapter face up to the fact that farmers must make important production decisions in a world beset by uncertainty and where time itself is an important factor of production.

Modeling the Farmer's Decisions

To take account of production lags, we assume that the farmer's production function is given by

$$q_{t+1} = q(a_t, k_t, h_t)$$

where q_{t+1} equals output in period $t + 1$ and a_t, k_t, and h_t equal, respectively, the levels of land, producer goods (capital), and labor committed to production in period t, at planting time, say. Producer goods include such inputs as fertilizer, diesel fuel, and the services or use of machines such as tractors and combines. We abstract for the time being from random elements such as the weather, outbreaks of disease, or other factors that could disrupt production plans. In other words, we are not concerned here with what might be called production uncertainty, which is a complication taken up in the section that follows.

The profit function for the farmer is given by

$$\pi_{t+1} = P_{t+1}q_{t+1} - C(q_{t+1}) - TFC_{t+1} \tag{2-3}$$

where the notation is the same as before except that we have added the subscript $t + 1$ to all the variables, indicating that all variables are defined for period $t + 1$. (Bear in mind, however, that both variable and fixed cost are incurred at planting time in period t.) Since TFC_{t+1} is fixed, we will simply drop the subscript from TFC in what follows.

Although the farmer does not know in period t what P_{t+1} will be for sure, we suppose that he or she views price as a random variable, that is, as a variable that can take any one of many possible values, generating an outcome that is not known ahead of time. The uncertainty of the outcome can be quantified by using probabilities, and we assume that the farmer makes a guess as to what the probability distribution (or density) of the random variable is. It is to be emphasized that, since price is random, profit is random as well. We suppose for the moment that the perceived probability distribution for profit is consistent with that for price, leaving until later the question how price expectations are formed. How does the farmer choose the level of output in this situation? How does the choice of output vary with changes in the farmer's perception of the probability distribution of price? These are the questions we shall try to answer.

The analysis of price determination under certainty assumed that the farmer strived to maximize profit. Why not utility? It comes to the same thing, of course, if utility u is expressed as an increasing function of certain profit, that is, if $u = u(\pi)$ and if $u' = \partial u/\partial\pi > 0$. Clearly, the higher is π in these circumstances, the higher is u; maximizing one maximizes the other. If profit is random, however, profit maximization is not possible. The farmer cannot choose the value of output that maximizes profit because the latter is determined in part by price, which the farmer neither knows nor controls at the time of the production decisions.

In this situation, it might be supposed that the farmer will strive to make output choices that, on average, yield the highest level of profit, that is, maximize expected profit. As a first approximation, this assumption might yield useful theoretical results, but there is a problem. As shown later, it is likely in a wide range of circumstances that the farmer's expected profit and variance of profit are positively related. Increasing expected profit increases variance of profit. Why might this matter? The answer comes down to the likelihood that most people enjoy a stable pattern of income and consumption over time.

Unlike the lucky bear who feasts during the happy days of summer and dreams away the harsh days of winter, living off fat, people want food, clothing, and housing on a continuous basis. We are not able to store up the blessings of life in our bodies and minds during the days of plenty to be drawn down during days of scarcity. To the extent that a farmer's consumption and financial viability depend on income, this suggests that he or she will, holding average or expected income constant, prefer a stable income stream to an unstable one. This leads us to believe that farmers might be interested in more than average profits. They might be interested in the dispersion or variance of profit as well.

In Chapter 3, we will consider complications associated with the existence of a banking system that facilitates borrowing and saving and allows the farmer to stabilize consumption to some extent even though income is unstable. We will indeed explore a menu of options open to the farmer who wishes to avoid unstable income and consumption streams through time. For the moment, however, we seek to analyze the behavior of the farmer who must make a trade-off between expected income and income stability.

At a very general level, we might assume that utility is a function of the perceived probability distribution for profit in all its detail and then seek to analyze the implications of utility maximization. We take a more pedestrian approach that is both fruitful and much simpler by assuming that the farmer's utility is a function of expected income and the standard deviation of income.[2] Expressed mathematically, we have

$$u_t = u\{E_t\,(\pi_{t+1}), [V_t(\pi_{t+1})]^{1/2}\} \tag{2-4}$$

where u_t equals the farmer's level of utility in period t, $E_t(\pi_{t+1})$ equals the farmer's perception at time t of the expected value of profit π_{t+1}, $[V_t(\pi_{t+1})] \geq 0$ equals the farmer's perception at time t of the variance of profit, and $[V_t(\pi_{t+1})]^{1/2}$ is the standard deviation of profit. (See Chapter 1 for a discussion of expectations and variances.) At this juncture it is necessary to simplify notation and for the student to take the time to become familiar with a few new symbols. We will often make use of the following conventions:

Random Profit, π_{t+1}	Random Variable, Z_{t+1}
$E_t(\pi_{t+1}) = \Theta$	$E_t(Z_{t+1}) = \Theta_z$
$V_t(\pi_{t+1}) = \sigma^2$	$V_t(Z_{t+1}) = \sigma_z^2$
$[V_t(\pi_{t+1})]^{1/2} = \sigma$	$[V_t(Z_{t+1})]^{1/2} = \sigma_z$

Using this notation, Eq. (2-4) may be written as $u_t = u(\Theta, \sigma)$. Also, if price P_{t+1} is a random variable, then $E_t(P_{t+1}) = \Theta_p$ and $V_t(P_{t+1}) = \sigma_p^2$. In general, many parameters may be required to characterize in detail a probability distribution, but we assume that only two are relevant. One, Θ, measures the mean or central tendency of the random variable π_{t+1}, and the other, σ, is the standard deviation measuring its spread or dispersion around its expected value. We may suppose for the moment, without inquiring into the particulars, that the farmer is able to estimate (or guesses) values for Θ and σ.

Figure 2.3 provides alternative graphic representations of the utility function given by Eq. (2-4). In all three panels, indifference curves are given by the curves labeled u_{t0}, u_{t1}, and u_{t2}, where it is assumed that $u_{t2} > u_{t1} > u_{t0}$. If the farmer has up-

[2]An alternative approach commonly found in the literature assumes that the decision maker maximizes the expected utility of profit, $E_t u(\pi_t)$, $u(\pi_t)$ being called a von Neumann–Morgenstern utility function representing risk preferences. As shown by Meyer (1987), there are close relationships between the mean-standard deviation approach we use here and the expected utility approach. However, there is some controversy on whether the expected utility approach provides an accurate representation of individual behavior. There is empirical evidence that risk preferences that are linear in the probabilities (as assumed in the expected utility model) do not provide an appropriate representation of individual preferences. (See Machina, 1987.) To the extent that the mean-standard deviation approach holds in situations for which risk preferences are *not* linear in the probabilities, our approach can be interpreted to be more general than the expected utility approach. It also has the advantage of being relatively simple.

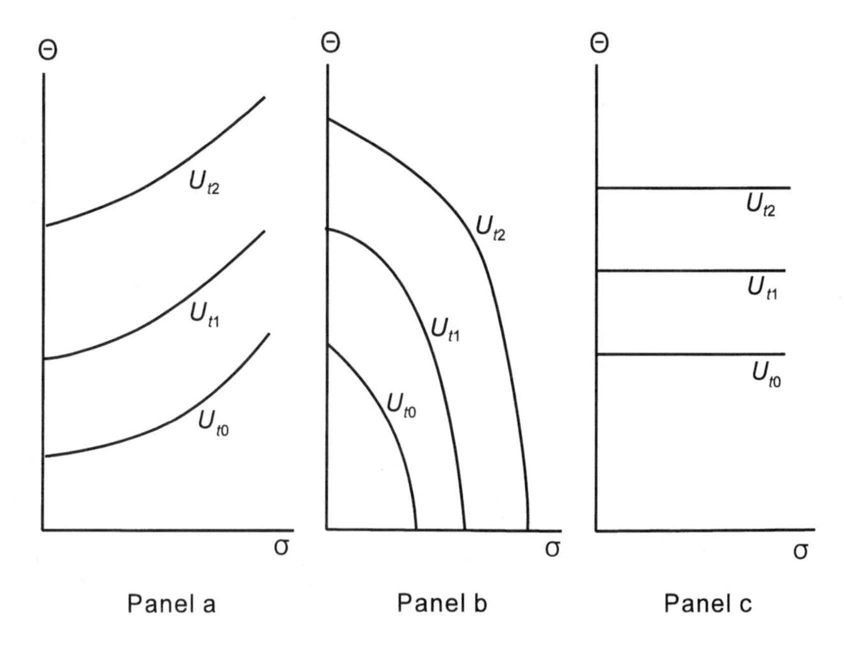

Figure 2.3 Indifference curve maps for risk-averse, risk-inclined, and risk-neutral farmers.

ward sloping indifference curves, as in panel a, we say the farmer is *risk averse*. Risk-averse farmers dislike dispersion of profits. Given an expected value for π_{t+1}, risk-averse farmers would be willing to pay money in order to reduce its dispersion. Alternatively stated, if the risk-averse farmer is expected to incur greater dispersion of profit, he or she must receive compensation in the form of higher expected profit if utility is to be kept at the same level. The farmer's marginal utility for expected profit is given by the first partial derivative of u with respect to Θ, that is, by $u'_{\Theta} = \partial u / \partial \Theta$. We might think of u'_{Θ} as the change in utility associated with a unit change in Θ. We assume that u'_{Θ} is always positive. The farmer's marginal utility for the dispersion of profit, as measured by the standard deviation, is given by the first partial derivative of u with respect to σ, that is, by $u'_{\sigma} = \partial u / \partial \sigma$. For the risk-averse farmer, u'_{σ} is always negative. To see the implications of this for the nature of the farmer's indifference curve map, we take the total differential of $u_t = u(\Theta, \sigma)$:

$$du_t = u'_{\Theta}\, d\Theta + u'_{\sigma} d\sigma \qquad (2\text{-}5)$$

The student may think of du_t as a small change in utility associated with small increments in Θ and σ. If we are interested in the movement along an indifference curve, with utility held constant, then $du_t = 0$ and Eq. (2-5) yields

$$\frac{d\Theta}{d\sigma} = -\frac{u'_\sigma}{u'_\Theta} \tag{2-6}$$

which is positive if $u'_\Theta > 0$ and $u'_\sigma < 0$. Equation (2-6) gives the marginal rate of substitution between Θ and σ, holding utility constant: It measures the slopes of the indifference curves in Fig. 2.3. For a *risk-averse* farmer, $u'_\sigma < 0$, and the slope of the indifference curve in panel a is $d\Theta/d\sigma = -u'_\sigma/u'_\Theta > 0$.

If the farmer's indifference curve map is as in panel b of Fig. 2.3, the farmer is said to be *risk inclined*. Here, the farmer, like a bigtime gambler, enjoys risk. More particularly, although $u'_\Theta > 0$, it is also true that $u'_\sigma > 0$. All indifference curves are then downward sloping.

Panel c depicts the utility function for a farmer who is said to be *risk neutral*. Whereas $u'_\Theta > 0$, $u'_\sigma = 0$. Utility depends on Θ, but not on σ. In this restrictive but nonetheless important case, maximizing utility is equivalent to maximizing expected profit. In this book, we will often assume that farmers or other economic agents are risk neutral to simplify analysis.

Two further definitions will prove useful in what follows. Define the *risk premium* as the maximum amount of money the farmer is willing to pay in order to eliminate income dispersion, driving σ equal to zero. This can be written as the sure amount of money X that implicitly satisfies

$$u(\Theta, \sigma) = u[(\Theta - \mathrm{X}), 0] \tag{2-7}$$

The risk premium can be interpreted as a measure of the farmer's willingness to pay money to avoid risk. It is a monetary value of the implicit cost of private risk bearing. The quantity of money given by $(\Theta - X)$ is called the *certainty equivalent*; it involves no risk ($\sigma = 0$) and yet yields the same level of utility as does $u(\Theta, \sigma)$.

Graphic representations will further clarify the meanings of the risk premium and the certainty equivalent. Figure 2.4 gives the graph of a single indifference curve for a risk-averse farmer with utility held constant at u_{t1}. Choose any point along this curve, such as the point (Θ_1, σ_1). The length of the line segment $(\Theta_1 - \Theta_0)$ measures the risk premium. The certainty equivalent, on the other hand, is given by Θ_0. Notice that the farmer is indifferent between $u(\Theta_1, \sigma_1)$ and $u(\Theta_0, 0)$. The certainty equivalent Θ_0 is the same for all points along the indifference curve u_{t1}. The risk premium increases, at least in the case of Fig. 2.4, as we move along the indifference curve u_{t1} in a northeasterly direction. It is clear from Fig. 2.4 that the risk premium is always positive for risk-averse farmers. Alternatively, the risk premium equals zero and is negative, respectively, for farmers who are risk neutral and risk inclined.

To aid in understanding the utility concepts just introduced, we now introduce a particular utility function, which, although somewhat restrictive, is relatively easy and instructive to analyze:

$$u_t = \phi_0 + \phi_1\Theta - \phi_2\sigma^2 \tag{2-8}$$

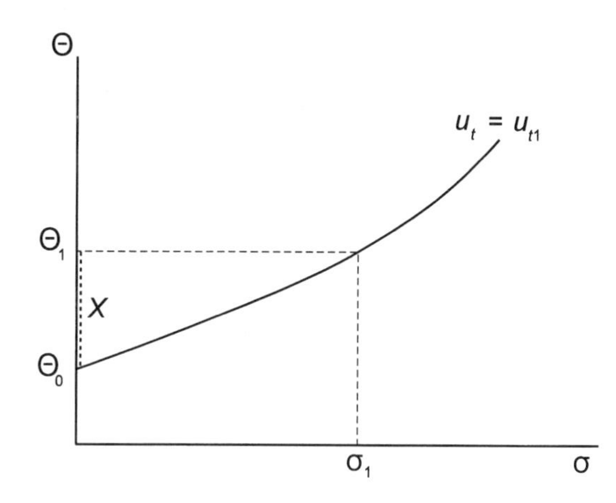

Figure 2.4 Risk premium and certainty equivalent.

where ϕ_0, ϕ_1, and ϕ_2 are nonnegative constants. (For a more general treatment, see Appendix B.) Note that $\phi_2 > 0$ implies a risk-averse decision maker. Consider a specific example:

$$u_t = 4 + 8\Theta - 2\sigma^2$$

Letting $u_t = 52$, the equation for the corresponding indifference curve is $\Theta = 6 + (1/4)\sigma^2$. The slope of the indifference curve for $u_t = 52$ equals $d\Theta/d\sigma = \sigma/2$, which is the marginal rate of substitution between Θ and σ. Notice that the second derivative of Θ with respect to σ is positive. The slope increases with increases in σ. Of the many combinations of values for Θ and σ that yield $u_t = 52$, consider $\Theta = 10$ and $\sigma = 4$. To calculate the risk premium associated with this point, we use Eq. (2-7) and solve the following equation for X:

$$52 = 4 + 8(10 - X) - 2(0)$$

The risk premium X equals 4. The certainty equivalent equals $(10 - X) = 6$.

More generally, with the utility function given by Eq. (2-8) and the definition of risk premium given by Eq. (2-7), we have

$$\phi_0 + \phi_1\Theta - \phi_2\sigma^2 = \phi_0 + \phi_1(\Theta - X) - \phi_2 0$$

Solving for X, the risk premium, we have $X = \phi_2\sigma^2/\phi_1$.

The conventional wisdom in economics is that risk aversion is likely the most useful assumption to make in the development of theories of choice in uncertain or risky environments. Put yourself in a gambling casino. A fair coin is to be tossed once. If it comes up heads, you must pay the house \$10,000. If it comes up tails, you receive \$10,100. Your expected gain from the game is \$50. How much would you be willing to pay for the privilege of playing this game? A risk-inclined person would be willing to pay more than $50 for the opportunity. Risk-neutral people would be willing to pay exactly $50. We believe that most people confronted with this game would be risk averse, however, and would not play if they had to pay $50. It seems likely many people would not play the game even if it cost them nothing at all. In other words, we believe most people are averse to risk when significant sums of money are involved. To be sure, we would all like to receive $10,100, but a $10,000 loss would be a calamity. (The well-to-do reader might up the sums to $100,000 and $100,100!)

We now assume, returning to Eq. (2-3), that the risk-averse farmer's objective is to maximize u_t through choosing the optimal value for q_{t+1}. If π_{t+1} is given by Eq. (2-3), we then have

$$\Theta = \Theta_p q_{t+1} - C(q_{t+1}) - TFC \tag{2-9}$$

where Θ equals the perceived expected value for profit, as noted, and Θ_p is the perceived expected value for price. In addition,

$$\sigma^2 = q_{t+1}^2 \sigma_p^2 \tag{2-10}$$

where σ^2 equals the perceived variance of profit and σ_p^2 equals the perceived variance of price. Hence we have $q_{t+1} = \sigma/\sigma_p$. This allows expressing Θ as a function of σ, as follows:

$$\begin{aligned} \Theta &= \Theta_p \frac{\sigma}{\sigma_p} - C\left(\frac{\sigma}{\sigma_p}\right) - TFC \\ &= f(\sigma) \end{aligned} \tag{2-11}$$

[For the moment we will not consider variations in Θ_p, σ_p, and TFC. These arguments do not, therefore, appear in $f(\sigma)$.] Equation (2-11) is an important result that gives the feasible relationship between expected profit Θ and the standard deviation for profit σ. Taking the derivative of Θ with respect to σ, we have[3]

$$\frac{\partial \Theta}{\partial \sigma} = \frac{1}{\sigma_p}\left[\Theta_p - C'(q_{t+1})\right] \tag{2-12}$$

Given $\sigma_p > 0$, the function $f(\sigma)$ is positively sloped when the expected price Θ_p exceeds marginal cost $C'(q_{t+1})$. The function is negatively sloped when the expected

[3]Here we use the function of a function rule of differentiation. Suppose that we have $Y = f(X)$ and $X = f(Z)$. Then

$$\frac{dY}{dZ} = \frac{dY}{dX}\frac{dX}{dZ}$$

price $\Theta_p < C'(q_{t+1})$. Obviously, $\partial\Theta/\partial\sigma = 0$ if $\Theta_p = C'(q_{t+1})$, that is, if expected price equals marginal cost.

Taking the second derivative of Θ with respect to σ yields

$$\frac{\partial^2\Theta}{\partial\sigma^2} = -\frac{C''(q_{t+1})}{\sigma_p^2} \tag{2-13}$$

But $\partial^2\Theta/\partial\sigma^2 < 0$ if the marginal cost function $C''(q_{t+1})$ is upward sloping in the relevant region, which we assume it is. Thus $\Theta = f(\sigma)$ is shaped like a hill, as in Fig. 2.5. The indifference curve map for a risk-averse farmer is also given in Fig. 2.5, and the student will quickly see how the farmer can choose σ and Θ in order to maximize utility. Start at $\sigma = \sigma_0$, for example, with $\Theta = 0$, and consider small increments in σ. As we move to larger and larger levels of σ along the function $f(\sigma)$, Θ rises, and as it does we achieve successively higher indifference curves and higher levels of utility. At point A, for example, we are better off than at $\sigma = \sigma_0$. As the student of economics will quickly realize, the point of utility maximization is given at $\sigma = \sigma_1$ and $\Theta = \Theta_1$, where $f(\sigma)$ is tangent to the indifference curve labeled u_1. Utility will not be maximized at any other point along the curve $f(\sigma)$.

At the risk of belaboring the point, consider point A, where, say, the marginal utility of Θ equals 1 util and that for $\sigma = -½$ util. Therefore, the slope of the indiffer-

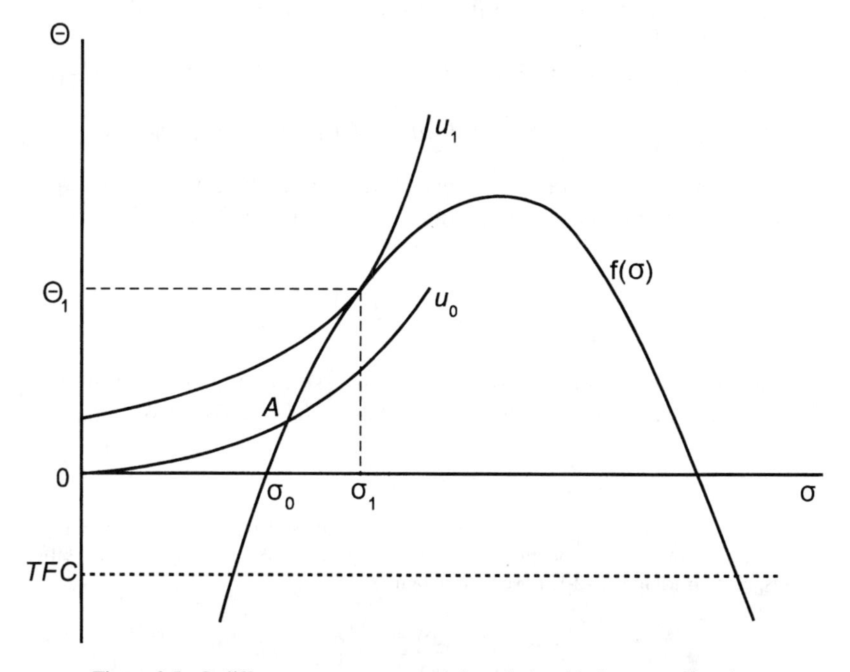

Figure 2.5 Indifference curve map and the relationship between Θ and σ.

ence curve u_1 equals +½ at this point. Let the slope of the $f(\sigma)$ curve equal 2.0, where a 1-unit increase in σ is associated with a 2-unit increase in Θ. To see that such a point is not optimal, suppose that the farmer does in fact increase σ by 1 unit, which is associated with an increase in Θ of 2 units. The 1-unit increase in σ causes utility to fall by −½ util, but the 2-unit increase in Θ causes utility to rise by 2 utils. The result is a net increase in utility of 1.5 utils. Point *A* cannot be optimal. Following similar steps, it can be shown that any point besides (Θ_1, σ_1) on the curve $f(\sigma)$ is not optimal either. Thus the only point that is optimal is the point (Θ_1, σ_1) in Fig. 2.5.

We have stressed the importance of the tangency condition for utility optimization because of the role it will now play in showing how the farmer's output varies with Θ_p and σ_p. Our problem is this: The farmer seeks to maximize utility $u_t = u(\Theta, \sigma)$ subject to the constraint $\Theta = f(\sigma)$. Both Θ and σ are functions of q_{t+1}, and the problem may be framed by noting that

$$u_t = u\left[(\Theta_p q_{t+1} - C(q_{t+1}) - TFC), \sigma_p q_{t+1}\right] \tag{2-14}$$

To find the level of q_{t+1} that maximizes u_t, we set $\partial u/\partial q_{t+1} = 0$ and insist that $\partial^2 u/\partial q_{t+1}^2 < 0$. To keep things simple, we will make use of the particular utility function introduced before:

$$u_t = \phi_0 + \phi_1\Theta - \phi_2\sigma^2 \tag{2-15}$$

where ϕ_0, ϕ_1, and ϕ_2 are positive numbers. Again we stress that $\phi_2 > 0$ implies a risk-averse decision maker.

We make substitutions for Θ and σ^2 using Eqs. (2-9) and (2-10), which yields:

$$u_t = \phi_0 + \phi_1[\Theta_p q_{t+1} - C(q_{t+1}) - TFC] - \phi_2\sigma_p^2 q_{t+1}^2 \tag{2-16}$$

To maximize u_t, we take the derivative of u_t with respect to q_{t+1} and set it equal to zero. Dividing by ϕ_1 and rearranging the terms, we have

$$\Theta_p - C'(q_{t+1}) = \frac{2\phi_2\sigma_P^2}{\phi_1} q_{t+1} \tag{2-17}$$

It is left to the student to show that this result can also be obtained by equating the slope of an indifference curve to the slope of $\Theta = f(\sigma)$.[4] We also insist that the sec-

[4]*Hint*: Let $u_t = u_{t0}$ be a particular level of utility. Solve for Θ as a function of σ using Eq. (2-15). Notice that the slope of the indifference curve is given by $\partial\Theta/\partial\sigma = 2\phi_2\sigma/\phi_1$. Now differentiate Θ with respect to σ using Eq. (2-11). We then obtain

$$\frac{\partial\Theta}{\partial\sigma} = 1/\sigma_p\left[\Theta_p - C'(q_{t+1})\right]$$

Don't forget that $\sigma = q_{t+1}\sigma_p$.

ond derivative of u_t with respect to q_{t+1} is less than zero, which it is if the marginal cost curve slopes upward in the relevant region, which it does.

Importantly, if ϕ_2 equals zero in Eqs. (2-15) and (2-17), then the farmer simply equates expected price to marginal cost. This is, of course, the case of risk neutrality, for which output decisions are based on the expected price, but not on the standard deviation of price. We will often make use of this simple case in the chapters that follow.

To help fix ideas, we build on the previous numerical example where $\phi_0 = 4$, $\phi_1 = 8$, and $\phi_2 = 2$. Now suppose that $C(q_{t+1}) = q_{t+1}^2$ and $TFC = 1$. Then

$$u_t = 4 + 8(\Theta_p q_{t+1} - q_{t+1}^2 - 1) - 2\sigma_p^2 q_{t+1}^2$$

Letting $\partial u_t / \partial q_{t+1} = 0$, we have

$$q_{t+1} = \frac{8\Theta_P}{16 + 4\sigma_p^2}$$

Clearly, q_{t+1} increases with ceteris paribus increases in Θ_p; q_{t+1} decreases with ceteris paribus increases in σ_p^2. Also, notice that $(\partial u_t^2 / \partial q_{t+1}^2) < 0$, thus assuring that we have found a maximum and not a minimum.

If the utility function is given by Eq. (2-8), the risk premium is given by $X = \phi_2 \sigma^2 / \phi_1$, as we have seen. Since $\sigma^2 = q_{t+1}^2\, \sigma_p^2$, we have

$$X = \frac{\phi_2 \sigma_p^2}{\phi_1} q_{t+1}^2 \tag{2-18}$$

The risk premium reflects the influence of both risk (σ_p^2) and risk aversion (ϕ_2) on the welfare of the entrepreneur. The *marginal risk premium MX* is simply the derivative of the risk premium Eq. (2-18) with respect to q_{t+1}. More specifically, we have

$$MX = \frac{\partial X}{\partial q_{t+1}} = \frac{2\phi_2 \sigma_p^2}{\phi_1} q_{t+1} \tag{2-19}$$

Notice that the expression to the right of the equality sign in Eq. (2-17) is the marginal risk premium. Clearly, the marginal risk premium increases linearly with q_{t+1}, given our assumed utility function.

We now define the *marginal cost of operation MCO* as the sum of the marginal risk premium and the marginal cost of production. The *MX* curve in Fig. 2.6 shows the marginal risk premium. The *MC* curve shows marginal cost of production. Adding vertically the *MX* and *MC* curves gives the *MCO* curve, which shows the marginal cost of operation. Notice, in the spirit of a controlled experiment, that if we lowered σ_p toward zero *MX* would disappear. The same thing would happen if the farmer's utility function changed such that the parameter ϕ_2 fell toward zero (corresponding to a risk-neutral decision maker).

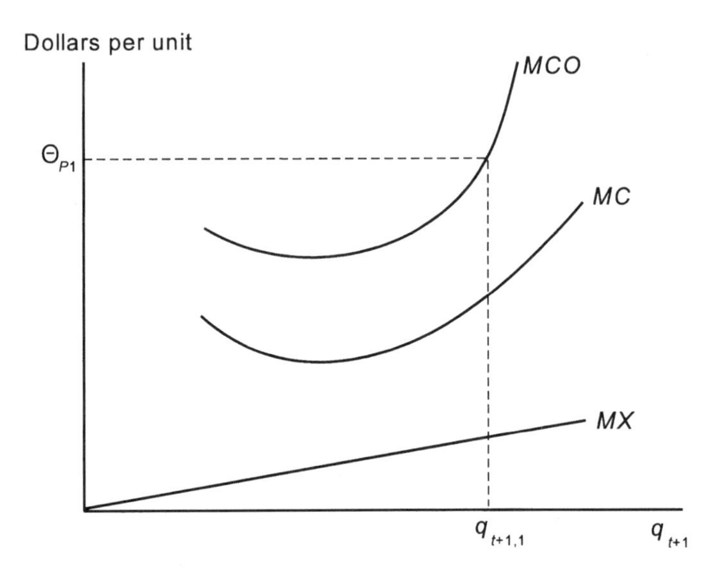

Figure 2.6 Marginal cost curves.

From Eq. (2-17), utility maximization clearly implies that the expected price Θ_p must be equated to the sum of the marginal cost of production plus the marginal risk premium. In other words, utility maximization implies that the expected price Θ_p equals the marginal cost of operation *MCO.*

But there is more to the story than this. Returning to Fig. 2.6, we may say provisionally that if Θ_p equals Θ_{p1} then the farmer will choose to produce $q_{t+1,1}$. Why must the conclusion be accepted only provisionally? Recall from our discussion of the case of certainty that the firm equates price to marginal cost only if marginal cost is not less than average variable cost. A similar complication arises in the study of uncertainty.

Since the total cost of risk is given by the risk premium X in Eq. (2-18), the average cost of risk is given by the *average risk premium AX*:

$$AX = \frac{X}{q_{t+1}} = \frac{\phi_2 \sigma_p^2}{\phi_1} q_{t+1} \tag{2-20}$$

Figure 2.7 draws together the several strands of the argument and allows some interesting and informative comparative static analysis, as shown next. First, we copy *MC*, *MX*, and *MCO* curves from Fig. 2.6. The *AX* curve is the graphic representation of Eq. (2-20). Notice that the *AX* curve is linear and rises half as fast as does *MX* with increases in q_{t+1}.

We next define the *total variable cost of operation* as the sum of the risk premium and the total variable cost of production. If we divide this total by output, we

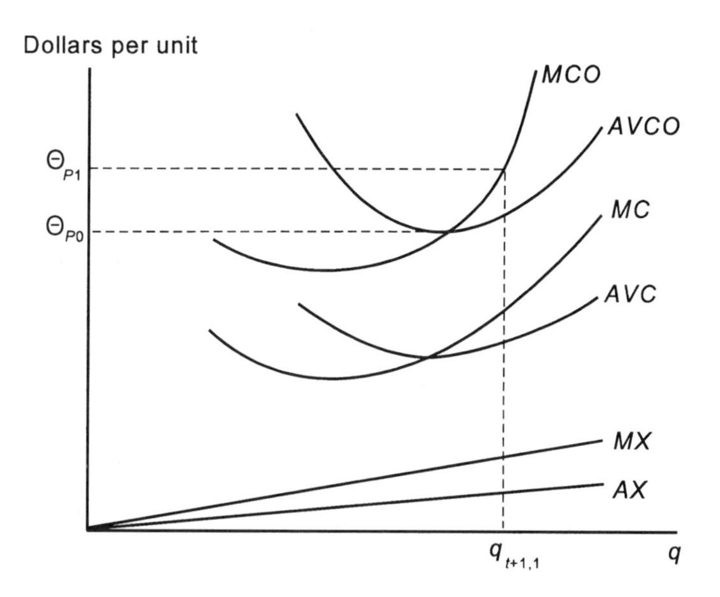

Figure 2.7 Short-run cost curves for a risk-averse farmer.

get the *average variable cost of operation*, *AVCO*. The *AVCO* curve in Fig. 2.7 shows *AVCO* for various levels of output. It can easily be derived by simply adding vertically the *AX* and *AVC* curves. (To avoid clutter, we will dispense with the average fixed cost curve.)

We may now set forth the conditions that must be satisfied if utility is to be maximized. To maximize utility, the farmer must choose a level of output such that the expected price Θ_p equals the marginal cost of operation *MCO*, but at this level of output, the *MCO* curve must be upward sloping *and* the expected price must not be less than the minimum of the *AVCO* curve. In Fig. 2.7, let expected price equal Θ_{p1}, the same as in Fig. 2.6. The level of output $q_{t+1,1}$ satisfies all the optimality conditions; it maximizes utility. If the expected price were less than Θ_{p0}, on the other hand, the farmer would produce nothing at all. It is not enough that expected price covers the average variable cost of production; it must also compensate the farmer for incurring the risk associated with operations.

Again, we note that in the risk-neutral cases the *MX* and *AX* curves do not appear. The farmer equates expected price to the marginal cost of production. The supply curve for the risk-neutral farmer consists of that part of the *MC* curve that does not lie below the *AVC* curve.

A digression provides further insights on decisions as to whether to shut down operations and leads to an interesting result. Consider Fig. 2.8, where total fixed cost is set initially at TFC_0 and where the relationship between Θ and σ is given by $f_0(\sigma)$. Expected profit is always negative because expected price is less than the average

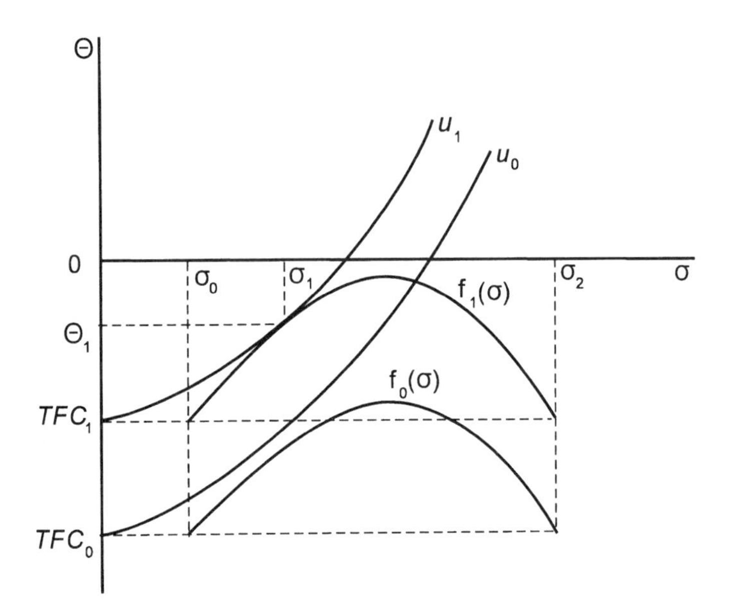

Figure 2.8 Indifference curve map and the relationship between the expectation and standard deviation of profit for two levels of total fixed cost.

cost of operation. Notice that between σ_0 and σ_2 the expected price must exceed the average variable cost of operation. Why? Because the expected loss associated with any σ in this range is less than total fixed cost. The farmer still chooses to produce nothing. Maximum utility is given by u_0; the loss equals TFC_0. The expected loss could be decreased by producing output only if the farmer incurs risk that he or she is unwilling to accept. The expected price does *not* exceed the average variable cost of operation given by *AVCO* in Fig. 2.7. If total fixed cost is lowered from TFC_0 to TFC_1, however, as shown in Fig. 2.8, such that $f(\sigma)$ shifts upward to the $f_1(\sigma)$ curve, the farmer chooses to produce output such that $\sigma = \sigma_1$ and $\Theta = \Theta_1$. Utility rises from u_0 to u_1. Lower fixed cost encourages the farmer to take chances! Conclusion? Although exogenous changes in total fixed cost have no impact on optimal output under certainty, the same does not hold under uncertainty.

Turning now to comparative static analysis, we go back to Fig. 2.7. That part of the *MCO* curve that is not below *AVCO* is the farmer's supply curve; it shows how much the farmer would be willing to produce at alternative expected prices, holding the standard deviation of price constant. Clearly, output expands as the expected price rises. Suppose, on the other hand, that the expected price Θ_p is held constant and the standard deviation of price σ_p decreases. A decrease in σ_p causes the *MX* curve (and the *AX* curve for that matter) to fall. Clearly, *MCO* falls as well and, with a constant Θ_p, output expands. This is, of course, what we would have expected. Indeed, as σ_p tends toward zero, price uncertainty tends to disappear and the farmer simply equates a certain price and the marginal cost of production. It is left to the student to analyze the effects of changes in the farmer's utility function, the production function, and input prices.

The student should be aware that these results could be obtained even though the utility function does not take the convenient form given by Eq. (2-15). We could, for example, assume that the marginal risk premium is a function of q_{t+1} and σ_p, increasing in both q_{t+1} and σ_p. These specifications are very reasonable for risk-averse farmers. The marginal risk premium need not be a linear function of q_{t+1} as in Eq. (2-19).

Some of the preceding results can also be obtained mathematically.[5] Go back to Eq. (2-17). Solving the first-order condition, Eq. (2-17), for q_{t+1} gives the farmer's supply function $q_{t+1} = s(\Theta_p, \sigma_p)$, indicating how much a utility-maximizing farmer would produce when facing an output price with mean Θ_p and standard deviation σ_p. The properties of the farmer's supply function can be obtained by using the implicit function rule.[6] We have

$$\frac{\partial q_{t+1}}{\partial \Theta_p} = -\frac{\phi_1}{-\phi_1 C''(q_{t+1}) - 2\phi_2 \sigma_p^2} > 0 \qquad (2\text{-}21)$$

Since the marginal cost curve is upward sloping in the relevant region, $C''(q_{t+1}) > 0$, and the denominator of the right-hand side is negative. As Θ_p rises, so does q_{t+1}. Again applying the implicit function rule, we have

$$\frac{\partial q_{t+1}}{\partial \sigma_p} = -\frac{-4\phi_2 \sigma_p q_{t+1}}{-\phi_1 C''(q_{t+1}) - 2\phi_2 \sigma_p^2} < 0 \qquad (2\text{-}22)$$

As σ_p rises, q_{t+1} falls. (These results hold if the expected price exceeds the average variable cost of operation.)

[5]A rigorous treatment of general comparative static analysis is presented in Appendix A. An application to the analysis of production decisions under risk can be found in Appendix B. For the more advanced student, these appendixes provide a formal derivation of the results discussed in the text.

[6]Briefly, suppose that $f(X, Y, Z) = 0$. It might be difficult to solve explicitly for any one of the three variables as a function of the other two. We can nonetheless obtain the partial derivative of one variable, Y, say, with respect to another, X, using the implicit function rule:

$$\frac{\partial Y}{\partial X} = -\frac{\dfrac{\partial f(\cdot)}{\partial X}}{\dfrac{\partial f(\cdot)}{\partial Y}}$$

Aggregative Analysis with Rational Expectations

Assuming that all farmers have the same cost curves and that all farmers have the same perceptions with regard to Θ_p and σ_p, we may then proceed with aggregative analysis. With N_t identical farmers, we have

$$\begin{aligned} Q_{t+1} &= N_t q_{t+1} \\ &= N_t s(\Theta_p, \sigma_p) \\ &= S(\Theta_p, \sigma_p) \end{aligned} \tag{2-23}$$

where $q_{t+1} = s(\Theta_p, \sigma_p)$ is a representative farmer's supply function and $S(\Theta_p, \sigma_p)$ is the aggregate supply function. The farmer's supply curve in Fig. 2.7 is the graphic representation of $q_{t+1} = s(\Theta_p, \sigma_p)$ with σ_p held constant.

The aggregative market equilibrium model then becomes

$$Q_{t+1} = S(\Theta_p, \sigma_p) \qquad \text{aggregate supply} \tag{2-24a}$$

$$Q_{t+1} = D(P_{t+1}, e_{t+1}) \qquad \text{aggregate demand} \tag{2-24b}$$

where e_{t+1} is an exogenous stochastic or random term that shifts demand and where the demand function is downward sloping ($\partial D/\partial P_{t+1} < 0$). It is assumed that the expected value of e_{t+1}, as formed in period t, equals zero; the variance of e_{t+1} is given by σ_e^2. In this formulation, the quantity supplied equals the quantity demanded for consumption under market equilibrium, which means that commodity storage is disallowed. (Storage will be included in a more advanced model set forth in Chapter 6.) The system given by Eq. (2-24) is a closed system if we take Θ_p and σ_p as exogenously determined; we then have two equations with two unknowns, Q_{t+1} and P_{t+1}. The values for Θ_p and σ_p in period t determine production plans that determine output in period $t + 1$. Available output in period $t + 1$ is predetermined by the production decisions made at time t. In other words, the supply curve at harvest time $t + 1$ is perfectly inelastic. The intersection of the demand curve and a perfectly inelastic supply curve determines the price. This is an example of the recursive economic systems that are of great importance in agricultural economics. In this model, Θ_p and σ_p determine Q_{t+1}, which then determines P_{t+1}.

This model, although of use as a steppingstone to more advanced models, is of little value in its own right. The assumptions are too restrictive, even aside from omitting commodity storage. In the absence of any explanation of how Θ_p and σ_p are determined, the explanation of how production is determined is absurdly superficial. We need to make Θ_p and σ_p endogenous variables. To do so will require additional equations to make the model complete. How to proceed?

One is reminded of questions often asked of the politician suspected of covering up wrongdoing. What did he know and when did he know it? What should we assume farmers know and when should we assume that they know it?

One way to proceed is to invoke the assumptions of the cobweb model. More particularly, we could assume that all farmers are risk neutral, equating expected price to marginal cost to find optimal output. The standard deviation of price then becomes irrelevant. Assume further that in period t, when production decisions are made, $\Theta_p = P_t$. In other words, assume that all farmers expect that next year's price equals this year's price, which is known. Adding the equation $\Theta_p = P_t$ to Eq. (2-23) and dropping σ_p completes the model.

We will not pause here to develop the implications of cobweb theory because the theory leads to hypotheses that are not very satisfactory. It can be shown, for example, that if a linear supply curve is more steeply inclined than is a linear demand curve, ignoring the negativity of the latter's slope, then price gyrations will become ever larger with the passage of time. Such markets are unstable, and although we are accustomed to seeing considerable variability in farm prices, we never observe the kind of madness that cobweb theory, at least in its linear form, predicts. It is also of interest to note that, according to cobweb theory, all farmers are hopelessly uneducable; they never learn from experience that a relatively high price in one year is likely to be followed by a relatively low price the next. We must look elsewhere for assumptions that will admit of a proper theory of expectations.

This brings us to Muth's famous, although sometimes disputed, hypothesis of *rational expectations*. Far from assuming that farmers are stupid, Muth (1961) goes to what would seem, in a world of uncertainty, to be the opposite extreme. In period $t +$ 1, assume that the aggregate demand, given by Eq. (2-24b), takes the form

$$P_{t+1} = \beta_0 - \beta_1 Q_{t+1} + e_{t+1} \tag{2-25}$$

where $\beta_0 > 0$, $\beta_1 > 0$, and demand is linear, with price being the dependent variable. Recalling that $E_t(e_{t+1}) = 0$ and $V_t(e_{t+1}) = \sigma_e^2$, we next take both the expectation and variance of P_{t+1} thus:

$$E_t(P_{t+1}) = \beta_0 - \beta_1 Q_{t+1} \tag{2-26}$$

$$V_t(P_{t+1}) = \sigma_e^2 \tag{2-27}$$

Note that production plans are made in period t so that, in period t, Q_{t+1} is known.

Now go back to Eq. (2-23), which asserts that the levels of q_{t+1} and Q_{t+1} are both functions of expected price [recalling that $\Theta_p = E_t(P_{t+1})$] and the standard deviation of price (recalling that $\sigma_p = [V_t(P_{t+1})]^{1/2}$). Following Muth, we propose using Eq. (2-26) to make Θ_p endogenous and, using Eq. (2-27), to make σ_p endogenous. If we do this, our aggregative market model becomes

$$
\begin{aligned}
P_{t+1} &= \beta_0 - \beta_1 Q_{t+1} + e_{t+1} && \text{demand} \\
Q_{t+1} &= \gamma_0 + \gamma_1 \Theta_p - \gamma_2 \sigma_p && \text{supply} \\
\Theta_p &= \beta_0 - \beta_1 Q_{t+1} && \text{expected price function} \\
\sigma_p &= \sigma_e && \text{standard deviation of price function}
\end{aligned}
\tag{2-28}
$$

where a linear supply function is assumed for simplicity. The model has four equations. The four unknowns are P_{t+1}, Q_{t+1}, Θ_p, and σ_p. Although the third equation is expected demand, we nevertheless refer to it as the expected price function. The reason for this is that in more realistic models, models that take commodity storage into account, for example, expected price functions will appear that are more complicated than expected demands. It might also be noted that the standard deviation of price function is exceedingly simple in this formulation, but this, too, will not be the case for all models of interest, as we shall see.

For a numerical example, suppose that

$$P_{t+1} = 10 - 0.5\, Q_{t+1} + e_{t+1}$$

$$Q_{t+1} = 4.2 + 6\Theta_p - 0.2\sigma_p$$

where $e_{t+1} = +1$ half the time and -1 half the time. Therefore, $E_t(e_{t+1}) = 0$ and $V_t(e_{t+1}) = 1$. Assuming rational expectations, $\Theta_p = E_t(P_{t+1})$ and $\sigma_p^2 = V_t(P_{t+1})$. Solving $\Theta_p = 10 - 0.5\, Q_{t+1}$ and $Q_{t+1} = 4.2 + 6\Theta_p - 0.2$ for the two unknowns, we have $\Theta_p = 2$ and $Q_{t+1} = 16$. If $e_{t+1} = -1$, $P_{t+1} = 1$. If $e_{t+1} = +1$, $P_{t+1} = 3$. On average, $P_{t+1} = 2$, and this is what farmers suppose when they make their production plans.

Figure 2.9 allows for graphic analysis. The expected price (demand) function is given by the D_Θ curve corresponding to $\Theta_p = \beta_0 - \beta_1 Q_{t+1}$. The supply function is given by the S_Θ curve corresponding to $Q_{t+1} = \gamma_0 + \gamma_1\Theta_p - \gamma_2\sigma_e$, recognizing that $\sigma_p = \sigma_e$. The intersection of D_Θ and S_Θ determines the equilibrium values for expected price $E_t'(P_{t+1})$ and actual output Q_{t+1}'. The intersection of actual demand D ($P_{t+1} = \beta_0 - \beta_1 Q_{t+1} + e_{t+1}$) and predetermined supply S (equals Q_{t+1}') determines equilibrium price P_{t+1}'. Actual demand could, of course, lie below D_Θ, in which case the actual price would be less than expected.

It was emphasized in Chapter 1 that the theory of demand and supply is a powerful engine for generating hypotheses as to the effects of exogenous shocks on market performance. The worth of the present model must also derive from its effectiveness in generating hypotheses. Two examples will suffice to show that much of what was learned in theory courses can now be brought to bear in understanding how commodity markets work, given lags in the production process. Suppose that at some point in time a new crop variety is introduced, which, for given levels of inputs, increases crop yields. In Fig. 2.9, the supply S_Θ shifts to the right, lowering both ex-

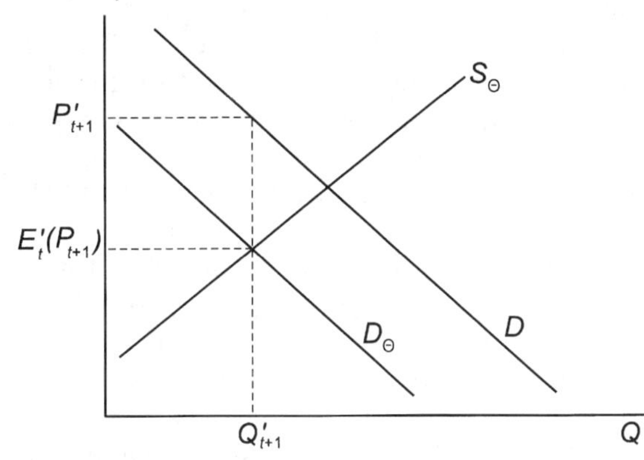

Figure 2.9 Expected price, supply, and demand curves under price risk.

pected price and increasing output. Under real-world conditions we might suppose that farmers are at first surprised by prices that are lower than expected. They would, however, eventually learn and appreciate how the technological change is affecting crop prices. For the second example, suppose that export demand for the commodity shifts to the right because foreign countries decrease tariffs on U.S. exports. In this case, expected demand D_Θ shifts to the right. Expected price and production both increase. Shifts in supply and permanent shifts in demand affect both actual prices and expected prices.

The model given by Eqs. (2-28) can also be solved algebraically. When this is done, the solutions or reduced form equations for the endogenous variables are

$$
\begin{aligned}
Q_{t+1} &= \frac{\gamma_0 + \gamma_1\beta_0 - \gamma_2\sigma_e}{1+\gamma_1\beta_1} \\
P_{t+1} &= \frac{\beta_0 - \beta_1\gamma_0 + \beta_1\gamma_2\sigma_e}{1+\gamma_1\beta_1} + e_{t+1} \\
\Theta_p &= \frac{\beta_0 - \beta_1\gamma_0 + \beta_1\gamma_2\sigma_e}{1+\gamma_1\beta_1} \\
\sigma_p &= \sigma_e
\end{aligned}
\tag{2-29}
$$

Notice that if σ_e increases, keeping $E_t(e_{t+1})$ equal to zero, then Q_{t+1} falls and P_{t+1} and Θ_p rise. The student should also consider the effects of changes in β_0 and γ_0 as examples of shifts in demand and supply.

We now pause to consider what has been accomplished. Muth's hypothesis is called the rational expectations hypothesis because if the expected price and the standard deviation of price are given by Eqs. (2-26) and (2-27), and this is what the model asserts, then how can it be maintained that farmers develop their expectations on some other basis? As rational market participants, the argument goes, farmers should understand how the market works, and they should take full advantage of all current relevant information. If, for example, price in period t is high because e_t happens to be large, rational farmers should not suppose that price in period $t + 1$ is necessarily going to be high, as cobweb theory might suggest. If e_t and e_{t+1} are independent random variables, for example, then high values for e_t and P_t must not be viewed as good indicators of what is likely to happen in period $t + 1$.

At this point the skeptic might argue, first, that the model as it stands is much too simple to explain how any real-world markets work and, second, that even if the model is a plausible approximation to some real-world market, no one knows what the parameters of the model equal. The first objection is serious enough, even if it is maintained that the model given by Eq. (2-28) is to be viewed mainly as a stepping-stone for the building of more realistic formulations. It must be admitted that the adequacy of any theoretical model in explaining real-world markets is always subject to debate and controversy.

Regarding the second argument, we note that the economic agents in virtually all economic theory are highly idealized relative to their real-world counterparts. At this point, the reader might take advantage of introspection. When you go into a supermarket, do you fill your shopping cart by maximizing a utility function subject to a budget constraint? Do you know what your utility function is? Are you familiar with the thousands of items in the typical supermarket? If the idealized decision making of theoretical consumers helps us understand how retail markets work, why should we reject "rational" or idealized behavior on the part of farmers who need to make production plans on the basis of price expectations? In a fundamental way, Muth's hypothesis of rational expectations is merely an extension of standard theoretical procedures from a world of certainty to a world of uncertainty.

Expectations of Exogenous Variables

There is, however, one problem with the preceding formulation that we have glossed over, a problem that we now propose to meet head on. The problem pertains to the explicit introduction of exogenous shifters of demand and supply. To see the problem and to get started on finding the solution, suppose that demand in period $t + 1$ is given by

$$P_{t+1} = \beta_0 - \beta_1 Q_{t+1} + \beta_2 Z_{t+1} + e_{t+1} \tag{2-30}$$

where Z_{t+1} is a demand shifter such as per capita income. Since Q_{t+1} is determined in period t and $\Theta_e = 0$, we have

$$\Theta_p = E_t(P_{t+1}) = \beta_0 - \beta_1 Q_{t+1} + \beta_2 E_t(Z_{t+1}) \tag{2-31}$$

The analysis to this point has centered on how to make $E_t(P_{t+1})$ endogenous, that is, to provide a tentative explanation of how $E_t(P_{t+1})$ gets determined. What about the determination of $E_t(Z_{t+1}) = \Theta_z$?

The unfortunate fact is that the rational expectations hypothesis says very little about how $E_t(Z_{t+1})$ is determined. It is best to interpret the hypothesis as providing a tentative explanation for the determination of the expectations of endogenous variables, recognizing that it offers no explanation for the determination of the expected values of exogenous variables. It may be argued that a variety of techniques could be used by farmers to estimate $E_t(Z_{t+1})$. An example is useful. Farmers might find that the best predictor of $E_t(Z_{t+1})$ is Z_t. Today's value, in other words, is the best predictor of tomorrow's value. Equation (2-31) may be modified accordingly. Alternatively, suppose that Z_t follows very closely an upward trend. That is,

$$Z_{t+1} = a + b(t + 1) + u_{t+1}$$

where $E_t(u_{t+1}) = 0$ and a and b are parameters. Then $E_t(Z_{t+1}) = a + b(t + 1)$. If $b = 2$, for example, then $E_t(Z_{t+1})$ rises every year by 2. Through substitution, we see that

$$E_t(P_{t+1}) = \beta_0 - \beta_1 Q_{t+1} - \beta_2[a + b(t + 1)] \tag{2-32}$$

Deviations from the trend may be used to estimate σ_z^2, the variance of Z. The latter is important because

$$V_t(P_{t+1}) = \beta_2^2\sigma_z^2 + \sigma_e^2 \tag{2-33}$$

This assumes that Z_{t+1} and e_{t+1} are independent random variables, in which case we need not worry about covariance. Using results given by Eqs. (2-32) and (2-33) allows modification of the system given by Eqs. (2-28). The bottom line is that models such as that given by Eqs. (2-28) can be fleshed out by adding exogenous shifters and through specifying plausible processes that farmers could use in estimating the expectations and variances of those shifters. This discussion calls attention to Muth's suggestion that farmers in predicting Θ_p and σ_p should not only understand how the market works, as noted previously, but that they should take full advantage of all relevant information presently available in predicting what is going to happen in the future. They should learn from past experience.

2.3 MARKET PERFORMANCE UNDER PRICE AND PRODUCTION UNCERTAINTY

The preceding model assumes that planned production equals actual production, an assumption that may be realistic for animal or irrigated crop production. For the major field crops, however, such as feedgrains, wheat, soybeans, and cotton, the assumption appears to be unduly restrictive, and we now set about generalizing the model analyzed in the previous section.

To do this, we center our attention on crop production: The farmer plants acreage and applies variable inputs such as fertilizer in one period and harvests a crop in the next. Yield per acre is subject to the random influence of weather. We assume that the production function may be written

$$q_{t+1} = q(a_t, k_t, h_t)v_{t+1} \tag{2-34}$$

where, as before, a_t, k_t, and h_t equal, respectively, the levels of land, capital, and labor committed to production in period t, and where v_{t+1} measures weather effects (e.g., deviations from average rainfall). Let $E_t(v_{t+1}) = 1$. Then

$$E_t(q_{t+1}) = q(a_t, k_t, h_t) \tag{2-35}$$

and $q_{t+1} = \Theta_q v_{t+1}$ where $E_t(q_{t+1}) = \Theta_q$. In Chapter 4 we will explore in more detail an example of a production function such as this. Here we merely assume that to maximize utility the farmer first derives a cost function $C = C(\Theta_q)$ that shows the minimized total variable cost incurred in period t associated with alternative levels of expected or planned output.

Modeling the Farmer's Decisions

The farmer's profit function is given by

$$\pi_{t+1} = P_{t+1}q_{t+1} - C(\Theta_q) - TFC \tag{2-36}$$

Since $q_{t+1} = \Theta_q v_{t+1}$, we have

$$\pi_{t+1} = G_{t+1}\Theta_q - C(\Theta_q) - TFC \tag{2-37}$$

where G_{t+1} is defined as the product of P_{t+1} and v_{t+1} ($G_{t+1} = P_{t+1}\, v_{t+1}$). The variable G_{t+1} is the gross return per unit of planned output, but in what follows we will often simply refer to G_{t+1} as the gross return. Now compare this with the profit function given by Eq. (2-3) for the case involving price but not production uncertainty. In the previous case, the choice variable was actual output q_{t+1} chosen in period t. In the present case, the choice variable is planned output Θ_q, also chosen in period t. In the previous case, the only random variable was price P_{t+1}. Now the only random variable is gross return G_{t+1}. All that we have accomplished up to this point is that we have changed the notation; but at the level of the individual farmer, this is about all that is needed. The previous analysis of the firm based on utility maximization applies straightway. The counterpart to Eq. (2-17) is

$$\Theta_G - C'(\Theta_q) = \frac{2\phi_2\sigma_G^2}{\phi_1}\Theta_q \tag{2-38}$$

where $E_t(G_{t+1}) = \Theta_G$ is the farmer's perceived expected gross return and $V_t(G_{t+1}) = \sigma_G^2$ is the corresponding perceived variance. Solving the first-order condition for Θ_q yields the farmer's planned output supply function $\Theta_q = s(\Theta_G, \sigma_G)$. Following the analysis in the previous section, the farmer's planned output responds positively to increases in the expected gross return and negatively to increases in the standard deviation of expected gross return. With N_t identical farmers, we have $\Theta_Q = N_t\Theta_q$; the aggregate planned output supply function becomes $\Theta_Q = N_t s(\Theta_G, \sigma_G) = S(\Theta_G, \sigma_G)$.

Aggregative Analysis

We may now assemble part of the model we soon hope to complete as follows:

$$\Theta_Q = N_t\Theta_q, \qquad \text{aggregation identity}$$

$$\Theta_Q = S(\Theta_G, \sigma_G), \qquad \text{aggregate expected supply}$$

-- (2-39)

$$Q_{t+1} = N_t q_{t+1}, \qquad \text{aggregation identity}$$

$$Q_{t+1} = N_t\Theta_q v_{t+1}, \qquad \text{aggregate production function}$$

$$Q_{t+1} = D(P_{t+1}, e_{t+1}), \qquad \text{aggregate demand}$$

This five equation model has five endogenous variables or unknowns, Θ_Q, Θ_q, Q_{t+1}, q_{t+1}, and P_{t+1}, *providing* we take Θ_G and σ_G as determined exogenously. On the arguments advanced in the previous case of price uncertainty only, such a procedure is unacceptable. We need to treat Θ_G and σ_G as endogenous variables, but to do so will require finding two additional equations to complete the model. How to find them? The answer, of course, is to make use of the assumption of rational expectations. What we will show by way of example is that the expectation and standard deviation of gross return are both functions of expected output. We have

$$\Theta_G = f_1(\Theta_q), \qquad \text{expected gross return function} \tag{2-40a}$$

$$\sigma_G = f_2(\Theta_q), \qquad \text{standard deviation of gross return function} \tag{2-40b}$$

Adding these two functions to Eqs. (2-39) completes the model. We will return to the model shortly, but first the example.

Let demand in period t be given by $P_{t+1} = 10 - Q_{t+1} + e_{t+1}$, where e_{t+1} takes on the values of +1 and –1 with equal probabilities. Also, assume that v_{t+1} takes on the values +1.25 and +0.75 with equal probabilities. Since $Q_{t+1} = N_t\Theta_q v_{t+1}$, we have

$$
\begin{aligned}
G_{t+1} &= P_{t+1}v_{t+1} \\
&= 10v_{t+1} - N_t\Theta_q v_{t+1}^2 + e_{t+1}v_{t+1}
\end{aligned}
\tag{2-41}
$$

Taking the expectation, we have

$$E_t(G_{t+1}) = 10 - 1.0625N_t\Theta_q, \qquad \text{expected gross return function} \tag{2-42}$$

where $E_t(v_{t+1}^2) = 1.0625$. This assumes that e_{t+1} and v_{t+1} are independent random variables. [With independence, $E_t(e_{t+1}v_{t+1}) = E(e_{t+1})E(v_{t+1}) = 0$.] The student will notice that the expected gross return Θ_G does not equal the simple product of expected v_{t+1} and expected price Θ_p. Clearly, v_{t+1} and price are not independently distributed random variables.

To compute the variance of the gross return G_{t+1}, we use the basic definition of variance, taking advantage of the following:

e_{t+1}	v_{t+1}	$P_{t+1}v_{t+1}$	$P_{t+1}v_{t+1} - E_t(P_{t+1}v_{t+1})$
+1	+1.25	$(13.75 - 1.5625N_t\Theta_q)$	$(3.75 - 0.5N_t\Theta_q)$
–1	+1.25	$(11.25 - 1.5625N_t\Theta_q)$	$(1.25 - 0.5N_t\Theta_q)$
+1	+0.75	$(8.25 - 0.5625N_t\Theta_q)$	$(-1.75 + 0.5N_t\Theta_q)$
–1	+0.75	$(6.75 - 0.5625N_t\Theta_q)$	$(-3.25 + 0.5N_t\Theta_q)$

Next, square the four deviations given in the last column and add the resulting sum of squares. Recognizing that all four cases are equally likely, we multiply the resulting sum by 0.25. This yields

$$V_t(G_{t+1}) = 7.3125 - 2.5N_t\Theta_q + 0.25N_t^2\Theta_q^2, \qquad \text{variance of gross return function} \tag{2-43}$$

Taking the square root of both sides yields the standard deviation of the gross return function. We have thus derived the functions $\Theta_G = f_1(\Theta_q)$ and $\sigma_G = f_2(\Theta_q)$. It is clear from the example, however, that we could have chosen other demand functions and more realistic probability distributions for e_{t+1} and v_{t+1}. The algebra might become messy, but the principles would remain the same.

Let us again pause to take stock of what has been accomplished. Adding Eqs. (2-40a) and (2-40b) to those given above the dashed line in Eqs. (2-39) yields four equations and the four unknowns Θ_Q, Θ_q, Θ_G, and σ_G. In period t, farmers plan output and commit production inputs based on the expected value and the standard de-

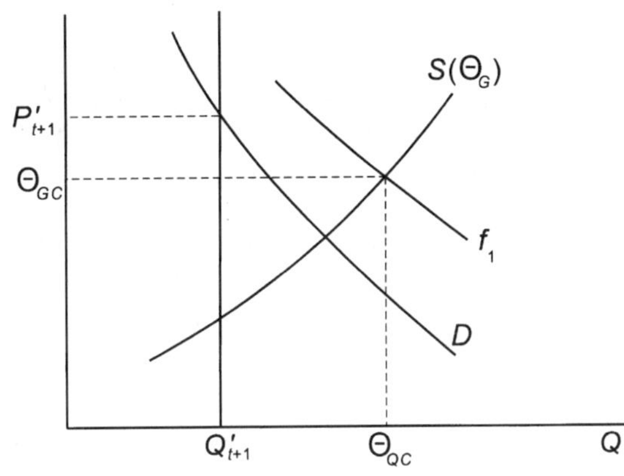

Figure 2.10 Expected gross return, expected supply, actual demand, and actual supply curves under price and production risk.

viation of the gross return per unit of planned output given by G_{t+1}. The farmer's perceptions of these strategic parameters are formed rationally, reflecting as they do the manner in which the market works. Planned output is determined in period t, but the values of the random variables e_{t+1} and v_{t+1} cannot be known until period $t + 1$. Given aggregate planned output, predetermined in period t, v_{t+1} determines actual output in period $t + 1$. Actual output and actual demand determine price in period $t + 1$.

Figure 2.10 allows a graphic analysis on the assumption that the standard deviation of the gross return is held constant.[7] The $S(\Theta_G)$ curve shows how much farmers would be willing to produce at alternative levels of expected gross return per unit of planned output. The f_1 curve shows what the expected gross return per unit would equal for various levels of planned output. In other words, the f_1 curve is a graphic representation of Eq. (2-40a). The intersection of these two curves gives equilibrium levels for planned output Θ_{Qc} and expected gross return Θ_{Gc}. Actual demand in period $t + 1$ is given by the D curve. Because of a drought, say, actual output Q'_{t+1} is much less than planned output. Price in period $t + 1$ is given by P'_{t+1}. At this juncture the student should give some thought to the effects of exogenous shifts in the various curves that determine both planned and actual market performance. Importantly, the usual supply–demand diagram is replaced with a diagram similar to Fig. 2.10 for markets characterized by price and production uncertainty.

[7]Alternatively, since $\sigma_G = f_2(\Theta_q)$, we have for the individual farmer $\Theta_q = s[\Theta_G, f_2(\Theta_q)]$. We solve for $\Theta_q = s'(\Theta_G)$. Then $\Theta_Q = N_t\Theta_q = N_t s'(\Theta_G) = S(\Theta_G)$. On this interpretation, the $S(\Theta_G)$ curve in Fig. 2.10 shows how much farmers plan to produce for alternative levels of expected gross return, allowing for the effects of endogenous changes in the standard deviation of the gross return.

2.4 LONG-RUN ADJUSTMENTS TO SHIFTS IN DEMAND

The preceding models of production and pricing have been short-run models; each farmer was assumed to have a fixed plant, and the number of farmers N_t was taken as given. In what follows, we take up the long-run response of a farm industry to changes in demand, allowing for changes in the number of farm producers. Of central interest are (1) the role of an upward sloping supply function for farm operators or families, (2) the role of expectations in the long run, and (3) the lengths of time required for the expansion and contraction of farm numbers. It may be noted parenthetically that the subject of long-run supply will be considered in more detail in Chapter 4, but with much less attention given to the lengths of time required by long-run adjustments.

For the present objectives, we hark back to the model of short-run pricing given in Section 2.2, where price but not production was assumed to be uncertain. We allow for an exogenous shifter of demand Z_{t+1}, which might be per capita income, population, or any one of several possible shifters. The variable Z_{t+1} is assumed to be random, and we further assume that Z_{t+1} and e_{t+1} are independently distributed. [See Eq. (2-30).] With these specifications, the short-run model is given by

$$\begin{aligned} P_{t+1} &= \beta_0 - \beta_1 Q_{t+1} + \beta_2 Z_{t+1} + e_{t+1}, && \text{demand} \\ Q_{t+1} &= S[E_t(P_{t+1}), (V_t(P_{t+1}))^{1/2}], && \text{short-run supply} \\ E_t(P_{t+1}) &= \beta_0 - \beta_1 Q_{t+1} + \beta_2 E_t(Z_{t+1}), && \text{expected price function} \\ V_t(P_{t+1}) &= V_t(Z_{t+1}) + V_t(e_{t+1}), && \text{variance of price function} \end{aligned} \quad (2\text{-}44)$$

The model has as many equations as unknowns if both the expectation and the variance of Z_{t+1} are treated as parameters that are determined exogenously. More on this assumption in a moment.

A graphic representation of the expected price (demand) function for period $t + 1$, with the expectation formed in period t, is given by D'_Θ in panel b of Fig. 2.11; the short-run supply is given by SRS'_Θ. In panel a, the average variable cost of operation curve and the marginal cost of operation curve are given by $AVCO$ and MCO. These curves, assumed to be the same for all established and potential producers, are defined to include the average cost of risk (the average risk premium), as explained in Section 2.2. We further suppose that each farmer is free to vary all inputs except that family labor is fixed. Thus the number of farms is variable, but the family labor per farm is not.

Equilibrium aggregate output and expected price are given by Q'_{t+1} and $E'_t(P_{t+1})$, but we want now to set forth conditions that make this short-run equilibrium a long-run equilibrium as well. For this purpose, note that the area given by $q'_{t+1}\,[E'_t(P_{t+1}) - AVCO']$ in panel a equals the marginal farm's quasi-rent, defined as total revenue minus total variable cost of operations. The quasi-rent is the remuner-

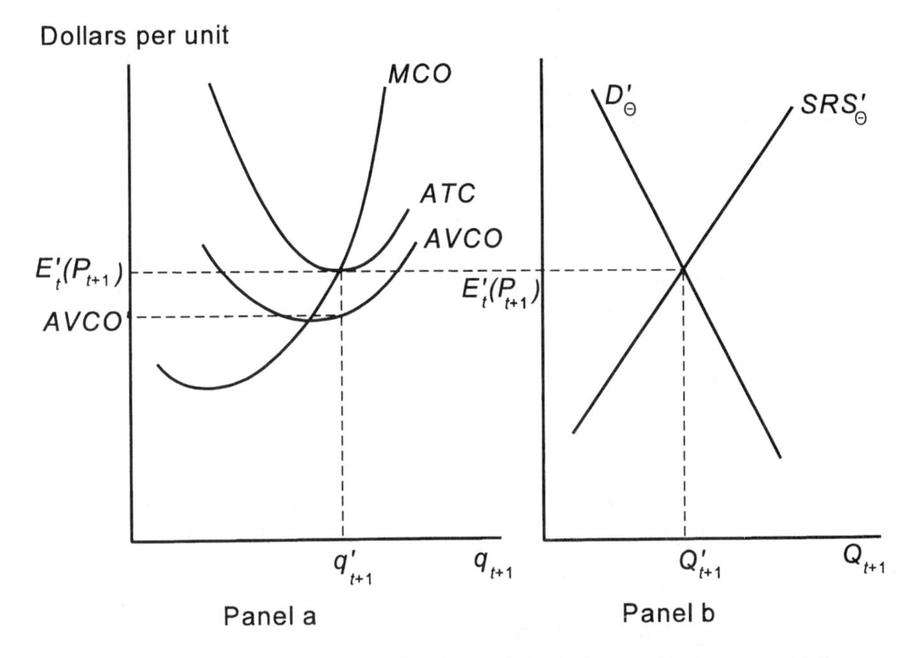

Figure 2.11 Cost curves for the marginal farmer in both short- and long-run equilibrium.

ation of the fixed factors (family labor in this case). As explained in Chapter 1, define transfer earnings as the minimum return to family labor required to keep the household in farming. Then a marginal farm is a farm for which quasi-rent just equals its transfer earnings, that is, where the farmer is indifferent between using family labor in farming versus some other nonfarm activity. Now suppose that, for every potential entrant, current transfer earnings exceed his or her potential quasi-rent as a farmer and that no established farmer has transfer earnings less than his or her current quasi-rent. On these assumed conditions, there is no incentive for entry or exit in farming, which corresponds to a *long-run equilibrium*; the short-run equilibrium in Fig. 2.11 is also a long-run equilibrium. The farmer whose cost curves are given in panel a is the marginal farmer since transfer earnings exactly equal quasi-rent. The total receipts of the marginal farmer just cover the sum of (1) the cost of all purchased inputs, (2) the risk premium, and (3) the farmer's transfer earnings. Other established farmers using the same technology, with transfer earnings that are less than that for the marginal farmer, will enjoy excess profit. With this setup, actual output in period $t + 1$ equals Q'_{t+1} in Fig. 2.11, but price will depend on the actual demand in period $t + 1$, not on expected demand.

As long as all the parameters in Eq. (2-44) remain the same, including, importantly, the expectations and variances of e_{t+1} and Z_{t+1}, aggregate output and the number of farmers will remain the same through time. The industry will remain in the initial position of long-run equilibrium. What we want to consider are the consequences of a ceteris paribus change in $E_t(Z_{t+1}) = \Theta_Z$. To do this, it is instructive to examine the processes that determine Z_{t+1} and what farmers know about those processes.

We have previously considered the case where Z followed an upward trend and suggested how farmers might take advantage of the past trend in predicting the future. Figure 2.12 shows an alternative pattern of observed values of Z for the current and previous years: $t, t-1, t-2, \ldots, t-n$. The figure also shows what the future pattern of values for Z will be, a pattern that farmers cannot know in period t. The average or expected value of Z for the current and previous years is given by Θ_{Z1}. The expected value for future years (unknown in year t) is given by Θ_{Z2}. What we are supposing here is a very simple case where Z fluctuates around one expected value for many years and where, of a sudden, a new pattern emerges and Z fluctuates around a new, higher mean. The magnitudes of the deviations of the actual values of Z around the means tend to be the same for both means, indicating that it is the expectation of Z, not its variance, that has changed. The student should recognize that the hypothetical example given in Fig. 2.12 is merely suggestive. In real-world markets there will typically be a great number of exogenous variables, each determined by a process far more complex than that given in Fig. 2.12.

How do farmers respond to this upward shift of demand? We propose to respond to this question in an informal manner. That is, we eschew the development of a rigorous model that assumes farmers and potential farmers are willing to spend money and time collecting and analyzing data in order to make sophisticated guesses.

In the years following the upward shift in expected demand, farmers (both established and potential) will observe a new pattern of Z's that differs from the old pattern. Both output price and profits will tend to be higher than anticipated. This

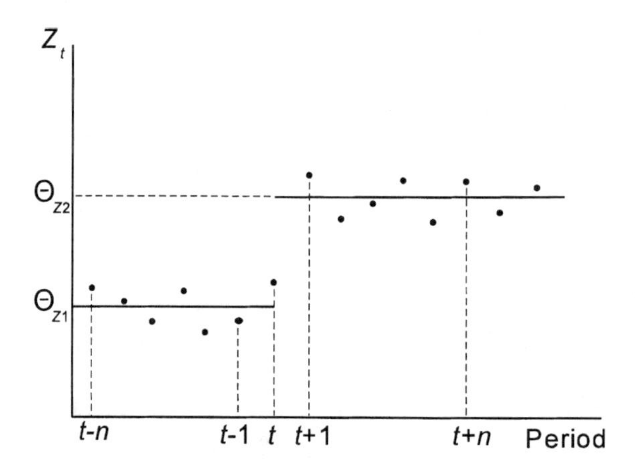

Figure 2.12 Upward shift in the expected value of an exogenous variable.

leads eventually, although not necessarily immediately, to the new entry of farmers. With entry, aggregate production and aggregate input use increase. Increased production causes expected prices to decline. Increased levels of inputs cause the prices of some inputs (those with upward sloping supply functions) to rise. Eventually, a new long-run equilibrium will be reached such that the last entrant earns zero expected profit. The returns to this family's labor just equal the family's transfer earnings.

The essential story can be told with the aid of Fig. 2.13. In the initial long-run equilibrium, output equals Q' and expected price equals P'. At first, before farmers realize that expected demand has shifted from D'_{Θ} to D''_{Θ}, output stays at Q', but the average or expected price rises to P^{+}. Once farmers learn of the increase in demand, they expand output along the short-run supply curve SRS'_{Θ} to Q^{*}, and average price falls to P^{*}. Potential farm families come to see that excess profits can be earned in farming, that the returns to their family labor used in farming exceed their current transfer earnings. They enter farming. The short-run supply curve drifts to the right, to SRS''_{Θ}, where the profit of the last entrant disappears. In the new long-run equilibrium, output equals Q''; expected price equals P''.

Having derived two points of long-run equilibrium, other points can be derived in similar fashion, including points associated with decreases in expected demand. The locus of many such points is the long-run supply curve given by *LRS* in Fig. 2.13. The student will have noticed, of course, that no commitment has been made as to the lengths of time, the number of periods, required before output increases from Q' to Q^{*} and from Q^{*} to Q'' in response to the increase in expected demand. Models can be constructed that predict how quickly such responses will be made. We prefer to leave the door open as to the length of time required for adjustments to take place. It might take 5 years or 20 years depending, for example, on whether the farm output in question is soybeans or oranges.

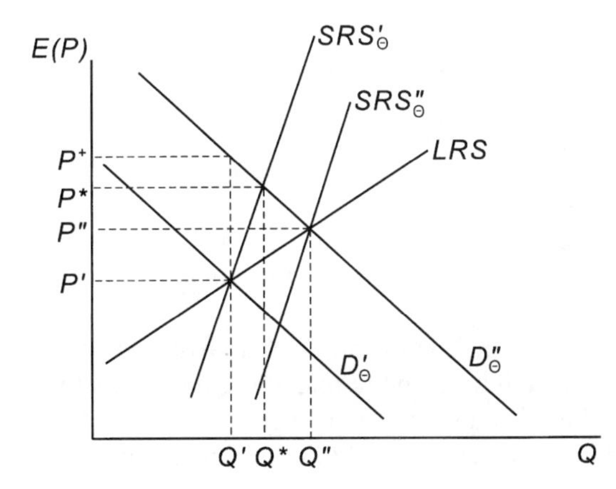

Figure 2.13 Demand curves, short-run supply curves, and long-run supply curve.

There is, however, one issue regarding the period of adjustment that is of great importance in applied economics. The time required for the long-run expansion of output in response to a 10 percent increase, say, in demand [$E_t(Z_{t+1})$ rises by 10 percent] is very likely much shorter than that for a 10 percent decrease. Why is that? Take an example. A barn that lasts 100 years can be built in a few months. In response to a perceived demand increase, a household can within a few short years buy or rent land, have buildings constructed, and buy machines. Land that had been abandoned in previous years can be brought back into production. (We are considering here new capacity, not a change in the ownership of established capacity, as an established farmer retires, say, and sells his or her plant to an entrant.)

But now consider a perceived decline in demand. At any particular point in time, some capital assets, such as buildings, machines, and breeding animals, may be more or less used up. That is, they may be fully depreciated and/or obsolete. Such capacity can exit overnight, and the resulting adjustment may be sufficient to restore the farm industry to long-run equilibrium if only the decrease in demand is sufficiently small.

For large decreases in perceived demand, however, exit may be both more painful and slower to occur. Some established producers will find that, in the short run, quasi-rents are positive but profits are negative. A farmer might sell the physical plant to an entrant, but we are not interested in the identities of those farmers who exit in this manner. The important point is that a relatively large decrease in demand, starting in long-run equilibrium, causes capital losses for all farmers except the lucky few with fully depreciated assets. As assets become used up (i.e., fully depreciated), they will be sold for scrap or simply abandoned. In the case where assets are highly specialized or where the transaction costs of selling the asset are high, the scrap value of an asset may be very low. In such situations, the disengagement of productive assets may be very slow: It may take many years for the industry to reach a new equilibrium point. This, in a nutshell, is why many agricultural economists believe that long-run downward adjustments in physical plant take longer, perhaps much longer, than long-run upward adjustments.

Finally, our analysis is subject to several qualifications. First, our attention centered on a shift in the expected value of a demand shifter that exhibits neither an upward nor a downward trend. Analysis of upward or downward trends is messy, but the basic ideas set forth here apply nonetheless. With a downward trend in output demand, for example, current excess profit may be wholly consistent with the exit of physical assets as farmers anticipate hard times to come. Here what matters is whether past trends are an accurate guide to the future. Shifts in the trend line become the unexpected exogenous shocks that may call for either a speeded up rate of exit or entry.

Second, our model abstracts from complications that arise out of adjustments to the fixed plants that occur through time. In the real world it appears that virtually all farmers make changes in their fixed plants as time goes by. In good times, farmers make new investments; in bad times, they tighten their belts and wait for better times to come. For some, the better times come only when they exit farming, salvaging what equity capital they can and seeking employment in the urban sector.

Third, it is important to understand what determines the salvage value of a fixed farm plant. We have supposed that a fixed plant is employed to the point where most remnants of it must be sold for junk. This conception may be useful if our model is assumed to be aggregative in nature, pertaining to the whole of agriculture, with food being the output. Most farmland, machines, and service buildings are of little or no value in the nonfarm sector. Alternatively, however, we might think of more narrowly defined industries, such as the milk production industry, in which case the best alternative use of a fixed plant might be in another farm industry. If milk prices fall far enough, with little hope of improvement in the future, a farmer may decide to stop milking cows in favor of developing a corn–soybean operation. In this case the salvage value of the farm may consist of the "junk value" of assets highly specialized to the dairy industry, used milking machines, for example, plus the discounted flow of future quasi-rents to the initial land, family labor, buildings, and machines that are suitable for crop production. Additional investment of equity capital may be needed to switch from dairying to crop production.

If the farm industry under consideration is narrowly defined around a single commodity or two, we must be mindful of closely competing uses for the land and other assets, including farm family labor. In such cases the supply curve for farm families for the farm industry in question could conceivably be very flat. In addition, the time required for making long-run adjustments to changes in expected demand might consist of a year or two, even in the case of a demand decrease. The quantity of production might be substantially and quickly responsive to changes in expected prices. This probably goes a long way toward explaining why the prices of farm commodities tend to move together. To some degree the fixed plant of the farm sector consists of land, labor, and machines that can be switched more or less readily from one commodity to another, depending on relative expected prices. Much of the fixed plant may not be readily switched to the nonfarm sector, however, when food prices fall relative to prices in the other sectors of the economy.

Fourth, although our analysis centers attention on long-run equilibrium, we should not suppose that farm industries often achieve such blessed states of stability. Rather, it is likely that in the real world, where markets are constantly bombarded by exogenous shocks of many kinds, farms barely begin adjusting to one situation when a new one emerges. Farm industries adjust continually toward moving long-run equilibria. The dynamic process hypothesized in the model is at least as important as is the notion of equilibrium itself.

Finally, our analysis dealt with a single exogenous shock, a change in expected product demand, that was capable of altering the position of long-run equilibrium. Many other shocks can be conceived that would have similar effects. More particularly, changes in technology and in input supply functions will alter the output supply function. The factors affecting the supply function of farm operators, such as the nonfarm wage rate, might also be mentioned. In Chapter 4 we will take up in more detail the analysis of the long-run effects of changes in exogenous variables, including shifts in expected product demand. In the process we will abstract from many of the dynamic issues that have been of interest here.

PROBLEMS

2.1. You are given the following utility functions for farmers A and B:

$$u_t = 10\Theta - 5\sigma^2, \qquad \text{farmer } A$$

$$u_t = \phi_0 + \phi_1\Theta - \phi_2\sigma, \qquad \text{farmer } B$$

a. For each farmer, find the marginal utilities for expected profit Θ and for the standard deviation of profit σ. Also find the expressions for the marginal rates of substitution between Θ and σ. Are the farmers risk averse? Explain.

b. Let farmer A's utility equal 20. What does the certainty equivalent equal? If $\sigma = 4$, what does the risk premium X equal?

c. Find the algebraic expression for the risk premium X for both farmer A and B.

2.2. The profit function for farmer A (see Problem 2.1) is given by $\pi_{t+1} = P_{t+1}q_{t+1} - 2q_{t+1}$, where we assume a long-run case with $TFC = 0$. Price is random, but output is not. Find the optimal level of output. Show that the marginal risk premium rises linearly with q_{t+1}.

2.3. The profit function for farmer B (see Problem 2.1) is given by

$$\pi_{t+1} = P_{t+1}q_{t+1} - aq^2_{t+1} - TFC$$

where a is a parameter. Price is random, but output is not. Find the optimal output. Show that the marginal risk premium is a constant.

2.4. Suppose that all farmers are exactly the same as farmer B (see Problems 2.1 and 2.3). Let $\phi_1 = 2, a = 1$, and $\phi_2 = 1$. There are 100 farmers. Demand in period $t + 1$ is given by

$$P_{t+1} = 400 - Q_{t+1} + e_{t+1}$$

where $\Theta_e = 0$ and $\sigma_e^2 = 16$. Find the equilibrium levels for Q_{t+1}, Θ_p, and σ_p^2. If $e_{t+1} = 4$, what will P_{t+1} equal?

2.5. The data for this problem are the same as for Problem 2.4 except that

$$P_{t+1} = 400 - Q_{t+1} + 2Z_{t+1} + e_{t+1}$$

where Z_{t+1} is a demand shifter. Suppose that $Z_{t+1} = 10 + 0.01\,(t + 1) + u_{t+1}$, where the year $t + 1$ is 1981, and $E_t(u_{t+1}) = 0$ and $V_t(u_{t+1}) = 9$. The covariance between Z_{t+1} and e_{t+1} equals zero. Find the equilibrium levels for Q_{t+1}, Θ_p, and σ_p^2. If $e_{t+1} = -8$ and $u_{t+1} = 5$, what will P_{t+1} equal?

2.6. Suppose that in a short-run situation 100 identical farmers have the utility function given by

$$u_t = 4 + 4\Theta - \sigma^2$$

The production function for the represented farmer is given by $q_{t+1} = \Theta_q v_{t+1}$. Also, v_{t+1} equals 1.2 and 0.8 with equal probabilities. The total variable cost function is $C = \Theta_q^2$. Aggregate demand is given by

$$P_{t+1} = 300 - Q_{t+1} + e_{t+1}$$

where e_{t+1} equals +10 and –10 with equal probabilities.

a. Derive the aggregate expected supply function.

b. Show that both Θ_G and σ_G^2 are functions of Θ_q.

c. Derive an equation system (the system is nonlinear) that, if solved, would yield competitive equilibrium.

REFERENCES

Anderson, Jock R., John L. Dillon, and Brian Hardaker, *Agricultural Decision Analysis*. Ames, Iowa: Iowa State University Press, 1976.

Chavas, J. P., R. D. Pope, and H. Leathers, "Competitive Industry Equilibrium under Uncertainty and Free Entry," *Economic Inquiry*, 26 (1988): 331–344.

Machina, Mark J., "Choice under Uncertainty: Problems Solved and Unsolved," *Journal of Economic Perspective* (Summer 1987), 121–154.

Meyer, Jack, "Two-moment Decision Models and Expected Utility Maximization," *American Economic Review*, 77, no. 3 (June 1987), 421–430.

Muth, John F., "Rational Expectations and the Theory of Price Movements," *Econometrica*, 29, no. 3 (July 1961), 315-335.

Sandmo, Agnar, "On the Theory of the Competitive Firms under Price Uncertainty," *American Economic Review*, 61, no. 1 (March 1971), 65–73.

Sheffrin, Steven M., *Rational Expectations*. New York: Cambridge University Press, 1983.

von Neumann, John, and Oskar Morgenstern, *Theory of Games and Economic Behavior*. Princeton, N.J.: Princeton University Press, 1944.

3

Coping with Risk: Financial Transactions, Diversification, and Hedging

To maximize utility, risk-averse farmers take both the expected profit and the standard deviation of profit into account when choosing levels of output (planned output). It was shown in Chapter 2 that, ceteris paribus, the higher the marginal cost of risk, the lower the levels of inputs and output. In short, risk-averse farmers respond to risk by decreasing output. Other strategies are possible, however, and in the first three sections of this chapter we consider the following: (1) use financial transactions to stabilize consumption, (2) diversify operations by producing more than one kind of output, and (3) hedge production decisions by taking advantage of futures markets. In Section 3.4, we consider as an empirical example of farm output supply analysis an acreage response equation for U.S. soybeans. In Section 3.5, we summarize various farm output supply functions that can be derived depending on the choice of the farm production model.

Before analyzing the three risk avoidance strategies cited above, we note that other means for coping with risk exist, including sharecropping, purchasing crop insurance, and engaging in collective action to secure government assistance. The latter approach will be considered at some length in Chapters 9 and 10. The point the student should bear in mind is that the farmer has a panoply of strategies that afford some protection from risk.

3.1 STABILIZING CONSUMPTION THROUGH FINANCIAL TRANSACTIONS

Central to the analysis of Chapter 2 was the assumption that the farmer's utility is a function of the expected level and standard deviation of profit. Alternatively, we may suppose that the farmer's utility in period t is a function of the expected level and standard deviation of consumption in period $t + 1$. Let κ_{t+1} denote consumption in period $t + 1$. Also, let Θ_κ equal $E_t(\kappa_{t+1})$, the expected value of κ_{t+1}, and σ_κ^2 equal $V_t(\kappa_{t+1})$, the variance of κ_{t+1}. Then $u_t = u(\Theta_\kappa, \sigma_\kappa)$. As in Chapter 2, π_{t+1}, Θ, and σ, are, respectively, profit, expected profit, and the standard deviation of profit.

Define a financial transaction as the amount of money B_t that is saved (if $B_t > 0$) or borrowed from (if $B_t < 0$) a financial institution such as a bank. Assume that the choice of B_t is made according to the following rule:

$$B_{t+1} = \Psi(\pi_{t+1} - \Theta) \tag{3-1}$$

where $(\pi_{t+1} - \Theta)$ is the deviation of actual profit from its expected value and $0 \leq \Psi \leq 1$ is a choice variable reflecting the proportion of $(\pi_{t+1} - \Theta)$ that is saved (or borrowed). B_{t+1} may be positive if $(\pi_{t+1} - \Theta) > 0$, meaning that the farmer adds to savings (or pays off loans). Alternatively, B_{t+1} may be negative, if $(\pi_{t+1} - \Theta) < 0$, meaning that the farmer draws down savings (or borrows from a bank). It is commonplace for people to keep liquid reserves to maintain consumption levels in the face of adversity. Idle money means relinquishing interest income, however, and long-term savings accounts pay higher rates of interest than do short-term accounts. Maintaining liquidity costs money.

The financial transaction rule given by Eq. (3-1) may be interpreted as follows: The farmer keeps a money reserve in either a demand deposit or a short-term, low-interest savings account. This reserve can be used to stabilize consumption over time. In the years when profit is high, the farmer can contribute to building the reserve through saving ($B > 0$). Alternatively, in the years when profit is low, consumption can be financed (in part) from drawing down the money reserve ($B < 0$). If the reserve is empty in some years, following a run of low profits, the farmer secures an equity line of credit from a financial institution that allows him or her to borrow money ($B < 0$).

Before proceeding with analysis, recall from Chapter 1 that total fixed cost TFC_{t+1} may be interpreted as the minimum amount of money the farmer must place in a money market account annually over the life of the fixed plant to pay for the initial investment in the fixed plant. Broadly speaking, the farmer's quasi-rent may be used to both support consumption and maintain equity capital, but the amount of money set aside to maintain equity capital need not be the same every year. In lean years, for example, quasi-rent may be used entirely for consumption. It need not be supposed, moreover, that the farmer is seeking to preserve equity capital. The farmer might elect to consume all or a portion of his or her equity capital over the life of the fixed plant. Of course, loans used to finance part of the fixed plant necessitate fixed annual payments (interest charges plus loan repayment) that cannot be varied over time to stabilize consumption.

The financial transactions B_t are typically not costless. We will assume that the higher Ψ is in Eq. (3-1), the higher the cost to the farmer in terms of interest payments foregone and/or interest payments on borrowed funds. This is fairly intuitive since $\Psi = 0$ implies no financial transaction ($B = 0$), while $\Psi > 0$ means that the farmer will face the cost of transactions according to Eq. (3-1). Let C_B equal the total cost of financial transactions used to stabilize consumption. Then $C_B = C_B(\Psi)$, where $dC_B/d\Psi > 0$ and $C_B(0) = 0$.

On these arguments, the amount of money available for the farmer's consumption in period $t + 1$, κ_{t+1}, is given by profit π_{t+1}, minus investment, B_{t+1}, minus the cost of financial transactions, C_B. Using Eq. (3-1), this gives

$$\kappa_{t+1} = \pi_{t+1} - \Psi(\pi_{t+1} - \Theta) - C_B(\Psi) \tag{3-2}$$

Therefore,

$$\Theta_\kappa = E_t(\kappa_{t+1}) = \Theta - C_B(\Psi) \tag{3-2a}$$

$$\sigma_\kappa^2 = V_t(\kappa_{t+1}) = (1 - \Psi)^2\sigma^2 \tag{3-2b}$$

But, as we saw in Chapter 2, under price uncertainty [see Eq. (2-11)] we have

$$\Theta = \Theta_P \frac{\sigma}{\sigma_P} - C\left(\frac{\sigma}{\sigma_P}\right) - TFC$$

Therefore, noting that $\sigma = \sigma_\kappa/(1 - \Psi)$ from Eq. (3-2b), Eq. (3-2a) may be written as follows:

$$\Theta_\kappa = \frac{\Theta_P \sigma_\kappa}{(1-\psi)\sigma_P} - C\left[\frac{\sigma_\kappa}{(1-\psi)\sigma_P}\right] - C_B(\psi) - TFC \tag{3-3}$$

Alternatively, under price and production uncertainty, we have

$$\Theta = \Theta_G \frac{\sigma}{\sigma_G} - C\left(\frac{\sigma}{\sigma_G}\right) - TFC$$

where Θ_G is the expectation of G_{t+1}, the gross return per unit of planned output, and σ_G is its standard deviation. [To get this result, go back to Eq. (2-37) and take both the expectation and variance of π_{t+1}. Substituting σ/σ_G for planned output Θ_q in the expression for expected profit yields the above result.] Notice that to allow for both price and production uncertainty in the present analysis we need merely revise Eq. (3-3) through replacing Θ_P with Θ_G and σ_P with σ_G.

The farmer's choice problem is to maximize $u_t = u(\Theta_\kappa, \sigma_\kappa)$ subject to the constraint given by Eq. (3-3). Optimal values are to be chosen for σ_κ and Ψ. A graphic analysis is instructive. Figure 3.1 gives the graphic relationship of $\Theta_\kappa = f_B(\sigma_\kappa)$ for three alternative values of Ψ. From Eq. (3-3), it is clear that raising the value of Ψ

both lowers the maximum of and flattens $f_B(\sigma_\kappa)$. If $\Psi = 0$, then consumption always equals income. Accordingly, the function $f_B(\sigma_\kappa|\Psi = 0)$ is the same as $f(\sigma)$ from Chapter 2. If $\Psi = 1$, then consumption becomes a constant, equaling $\Theta_{\kappa 1}$ in Fig. 3.1. A graphic representation of $f_B(\sigma_\kappa)$ is also given for $\Psi = \frac{1}{2}$. If the farmer is risk averse, with indifference curves given by u_0, u_1, and u_2, and if Ψ is arbitrarily set at zero, point *A* yields the maximum utility. Likewise, if $\Psi = \frac{1}{2}$, point *B* gives maximum utility. Given the setup in Fig. 3.1, however, the farmer maximizes utility by setting Ψ equal to 1. The solution is a corner solution with Θ_κ and σ_κ set equal to $\Theta_{\kappa 1}$ and 0, respectively.

Aside from corner solutions, maximizing utility may be viewed as a two-step procedure: First, choose Θ_κ and σ_κ for alternative values for Ψ; second, choose the optimal Ψ. For each value of Ψ in the first step, the farmer finds a tangency position that maximizes utility, as in the case of points *A* and *B*. The farmer then chooses in the second step the tangency solution associated with the highest indifference curve. Once the optimal values for Θ_κ and σ_κ are determined, Eqs. (3-2a) and (3-2b) provide the optimal values for Θ and σ. The latter are linked to the optimum value for q_{t+1} (or Θ_q) as in Chapter 2. (Also, see the first of the problems given at the end of this chapter.)

Going back to Eq. (3-3), we note that the higher $C_B(\Psi)$ is, that is, the more expensive the holding of liquid reserves is, the greater is the sacrifice of expected con-

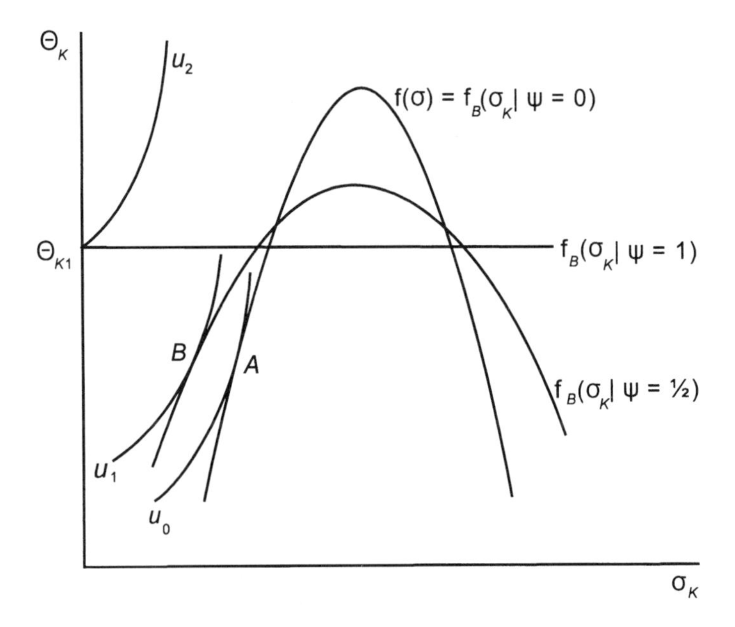

Figure 3.1 Indifference curve map and the relationship between the expectation and standard deviation of consumption for various levels of liquidity.

sumption required to maintain consumption variability at a given level. Importantly, the existence of financial institutions that facilitate saving and borrowing affords the farmer a very important means for avoiding risk and its associated cost. But liquidity is costly to maintain, as noted, and there is no reason to suppose that Ψ will be set equal to 1. If the optimal value of Ψ is less than 1, the farmer is stuck with making a trade-off between expected profit and the standard deviation of profit, as in Chapter 2.

Even so, the models of Chapter 2 need to be reconsidered in light of dissociating consumption and profit through financial transactions. Factors that determine the cost of holding liquid reserves, such as the difference between short- and long-term interest rates, for example, must be seen as possible shifters of farm supply functions. Also, as noted in Chapter 2, assuming that farmers are risk neutral greatly simplifies analysis, and we will frequently employ such an assumption in the remainder of this book. The existence of modern financial markets almost certainly lessens the cost of risk taking, lending support for the view that simple models based on risk neutrality might provide good first approximations.

It might also be noted, as an aside, that in less developed countries the extended family substitutes, to some extent at least, for modern financial markets. The extended family or kinship group often allows the nuclear family (married couple with children) that has suffered a setback in food production to seek food from relatives who live over the hill, so to speak, in a different microclimate. This also tends to sever the connection between food production (profit) and food consumption.

3.2 PRODUCT DIVERSIFICATION

Product diversification is still another means for coping with uncertainty and risk. The old aphorism regarding the carrying of eggs—don't put them all in one basket—has been given solid theoretical support in the literature on the optimum design of sets of investments called portfolios. Maximizing an investor's utility under risk and risk aversion might involve spreading his or her total investment over stocks and bonds representing different companies and over a variety of assets, such as real estate, oil tankers, and precious oil paintings.

The basic idea can most easily be seen by considering the following two games. In the first, a coin is tossed. If it comes up heads, the player receives a dollar; tails means the player gets nothing. The expected payoff equals 50 cents. The variance of the payoff equals 25. In the second game, *two* coins are tossed simultaneously. A head yields 50 cents; a tail yields nothing. The expected payoff again equals 50 cents, given by (¼) \$1.00 + (¼) 0 + (½) \$0.50. The variance of the payoff of the second game equals 12.5. Diversification cut the variance in half. The reason for this is that the outcome of the toss of one coin is independent of or uncorrelated with the outcome of the toss of the other. To see why this is crucial, suppose that a poltergeist, whom we met in Chapter 1, casts a magnetic spell on the two coins such that both either come up heads or tails with perfect correlation. In this event, diversification does not reduce variance. On the other hand, suppose that the spell causes the two coins never to turn up the same; one and only one head appears every time. In this case the variance of the payoff is reduced to zero. The player always receives 50 cents. A perfect negative correlation drives variance to zero. This little example suggests that a farmer who proposes to reduce risk through planting more than one crop ought to be very concerned with how closely the crop prices are correlated.

To see this more clearly and in more detail, we now consider a model constructed for the express purpose of examining conditions that are conducive (or inimical) to using diversification as a means of reducing risk. (See Appendix B for a general and more formal treatment of these ideas.) A farmer with a fixed acreage $\overline{a}$ can produce either corn or soybeans or both. The net returns per acre of corn, revenue per acre minus all variable cost per acre, is given by G_c. That for soybeans is given by G_s. We treat G_c and G_s as random variables, thus allowing for both price and production risk. Neglecting fixed cost and assuming that the farmer plants both crops, profit, expected profit, and variance of profit are given by the following three equations:

$$\pi_{t+1} = G_c a_c + G_s a_s$$

$$\Theta = \Theta_{Gc} a_c + \Theta_{Gs} a_s \tag{3-4}$$

$$\sigma^2 = a_c^2 \sigma_{Gc}^2 + a_s^2 \sigma_{Gs}^2 + 2 a_c a_s \operatorname{Cov}(G_c, G_s)$$

where Θ_{Gc} and Θ_{Gs} equal, respectively, expected net cash returns to corn and soybean acreage, σ_{Gc}^2 and σ_{Gs}^2 are the respective variances of the net cash returns, and $\operatorname{Cov}(G_c, G_s)$ equals the covariance between G_c and G_s. Drawing on Chapter 1, we know that

$$\operatorname{Cov}(G_c, G_s) = \sigma_{Gc} \sigma_{Gs} \rho(G_c, G_s)$$

where $\rho = \rho(G_c, G_s)$ is the correlation between G_c and G_s satisfying $-1 \le \rho \le 1$. Covariance and correlation are closely related. They may be negative, positive, or zero.

We assume, as before, that the farmer's utility function is given by

$$u_t = \phi_0 + \phi_1 \Theta - \phi_2 \sigma^2$$

with $\phi_1 > 0$ and $\phi_2 > 0$, corresponding to risk aversion. We may then proceed to maximize utility as follows: Make substitutions for Θ and σ^2 using the last two of Eqs. (3-4). Write covariance in terms of correlation and eliminate the variable a_s using the equation $a_s = \bar{a} - a_c$.[1] Assuming that the optimal a_c is positive, differentiate both u_t and $\partial u_t/\partial a_c$ with respect to a_c in order to obtain the first- and second-order conditions for a maximum, as follows:

$$a_c = \frac{\phi_1\left(\Theta_{Gc} - \Theta_{Gs}\right) + 2\phi_2\bar{a}\left(\sigma_{Gs}^2 - \sigma_{Gc}\sigma_{Gs}\rho\right)}{2\phi_2\left(\sigma_{Gc}^2 + \sigma_{Gs}^2 - 2\sigma_{Gc}\sigma_{Gs}\rho\right)} \tag{3-5}$$

$$2\sigma_{Gc}\sigma_{Gs}\rho - \sigma_{Gc}^2 - \sigma_{Gs}^2 < 0 \tag{3-6}$$

It is immediately apparent that, holding variances constant, a_c increases with increases in Θ_{GC} and falls with increases in Θ_{GS}.

One of the most important lessons that can be learned from these results is that diversification can decrease risk under some conditions but not under others, much as in the case of the coin-tossing game considered previously. Perhaps the easiest way to see this is to assume that Θ_{Gc} equals Θ_{Gs}. Then Eq. (3-5) becomes

$$a_c = \bar{a}\sigma_{Gs}\frac{\sigma_{Gs} - \sigma_{Gc}\rho}{\sigma_{Gc}^2 + \sigma_{Gs}^2 - 2\sigma_{Gc}\sigma_{Gs}\rho} \geq 0$$

On this assumption we consider, in turn, three values for ρ: 0, –1, and +1. If $\rho = 0$, then the optimal value of a_c equals $\bar{a}$ times the ratio $\sigma_{Gs}^2/(\sigma_{Gc}^2 + \sigma_{Gs}^2)$. As σ_{Gc}^2 tends toward zero, the way to minimize variance of profit is to plant a lot of corn. Alternatively, as σ_{Gc}^2 tends toward zero, the farmer plants a lot of soybeans. This is what we would expect. If the variances are the same, then half the acreage goes to corn and half to soybeans.

The second case of interest is when $\rho = +1.0$. Here, diversification fails to reduce risk. Assuming again that $\Theta_{Gc} = \Theta_{Gs}$, we have

$$a_c = \bar{a}\frac{\sigma_{Gs}\left(\sigma_{Gs} - \sigma_{Gc}\right)}{\left(\sigma_{Gs} - \sigma_{Gc}\right)^2} \geq 0$$

from Eq. (3-5). Experimentation with arithmetic examples will convince the student that, if $\rho = +1$ and $\sigma_{Gs}^2 < \sigma_{Gc}^2$, then the farmer will specialize in soybean production. (Suppose, for example, that $\sigma_{Gs}^2 = 4$ and $\sigma_{Gc}^2 = 9$.) If $\rho = +1$ and $\sigma_{Gs}^2 > \sigma_{Gc}^2$, on the

[1]These steps lead to the following expression for u_t:

$$u_t = \phi_0 + \phi_1[\Theta_{Gc}a_c + \Theta_{Gs}(\bar{a} - a_c)] - \phi_2[a_c^2\sigma_{Gc}^2 + (\bar{a} - a_c)^2\sigma_{Gs}^2 + 2a_c(\bar{a} - a_c)\sigma_{Gc}\sigma_{Gs}\rho(G_c, G_s)]$$

Differentiating this expression with respect to a_c and setting the resulting derivative equal to zero yields Eq. (3-5).

other hand, the farmer will specialize in corn production. In either event, if $\rho = +1$, the farmer will not diversify to decrease risk. (Reduction in the cost of production might be a motivation for diversification quite aside from risk considerations.)

In the third case, where $\rho = -1.0$, diversification works like the devil. If σ^2_{Gs} is smaller than σ^2_{Gc}, the farmer reduces risk by planting more soybeans than corn and contrariwise if σ^2_{Gs} is larger than σ^2_{Gc}. Moreover, profit risk can be completely eliminated. Indeed, given $\rho = -1$, the variance of profit σ^2 in Eq. (3-4) can be written as

$$\sigma^2 = (a_c\sigma_{Gc} - a_s\sigma_{Gs})^2$$

which can be set equal to zero by choosing $a_c = (\sigma_{Gs}/\sigma_{Gc})a_s$.

Suppose, however, that $\Theta_{Gc} \neq \Theta_{Gs}$. Examples can be used to show that diversification can still reduce the variance of profit and increase utility. In general, a more negative correlation between G_c and G_s means that utility maximization provides an incentive to diversify while taking into consideration the trade-offs between expected profit and variance of profit.

If we consider only those sets of parameter values that are conducive to risk reduction through diversification, we may then determine how a_c is affected by changes in various parameters through differentiation of Eq. (3-5). That a_c is increasing in Θ_{Gc} and decreasing in Θ_{Gs} has already been noted. Some of the other effects are downright tricky. How does acreage planted to corn respond to increased risk aversion, that is, to an increase in the parameter ϕ_2 in the utility function? It can be shown that $(\partial a_c/\partial \phi_2) < 0$ if $\Theta_{Gc} > \Theta_{Gs}$. If the farmer becomes more averse to risk, he or she will plant more soybeans when soybeans is on the average less profitable than corn. The reason for this is that the risk benefit associated with soybean production increases with the degree of risk aversion of the decision maker. Furthermore, letting the correlation between G_c and G_s, $\sigma(G_c, G_s)$, move from +1 to –1, ceteris paribus, increases the effectiveness of diversification. This tends to cause the optimal values for both expected profit and the variance of profit to fall. The reason for this is that, as $\rho(G_c, G_s)$ moves from +1 to –1, the risk benefits from lowering the variance of profit increase. In a sense, buying economic stability becomes a better deal as $\rho(G_c, G_s)$ falls. Finally, an increase in σ^2_{Gc}, ceteris paribus, tends to cause corn acreage to fall. Similarly, as σ^2_{Gs} rises, soybean acreage tends to fall. This illustrates that, under risk aversion, an increase in risk for one crop tends to have a significant effect on the acreage planted of the corresponding crop. The student's understanding of these effects will be greatly increased by carefully working through Problem 3.3.

3.3 THE THEORY OF FUTURES MARKETS AND HEDGING

In pioneering articles that appeared at about the same time, Holthausen (1979) and Feder, Just, and Schmitz (1980) showed that in a world of price uncertainty and in the presence of a futures market, a producer maximizes utility by choosing a level of output such that the marginal cost of production equals the futures price. If the futures price is a good approximation of the expected price (both prices were taken as exogenous in the above articles), then the producer in effect equates expected price to marginal cost, as does the risk-neutral producer. Because risk taking is shifted to speculators in the futures market, agricultural production is no longer impeded by price uncertainty. This result is of fundamental importance in agricultural economics.

In what follows, we derive this result and extend the analysis to include both price and production uncertainty. We also make the futures price endogenous in a simple, highly idealized model of a futures market viewed as an estimator of expected prices. We approach futures markets from a point of view that is distinct from that found in many basic textbooks in agricultural economics. Several excellent treatments [e.g., Leuthold, Junkus, and Cordier (1989)] are available in which writers describe in some detail futures markets, futures contracts, history and evolution of futures contracting, and the manner in which and the reasons why country and terminal elevators, exporters, users of raw material (flour millers, for example), farmers, and other agents in farm commodity and input markets buy and sell futures contracts. Building on the analysis given in Chapter 2, our objective here is to incorporate the market for futures into theory of the firm and to assess the impact of futures on market prices of farm commodities.

To set the stage for students unfamiliar with futures markets, we begin by providing some descriptive information on futures contracts, options, and cash forward contracting, leaving more detailed treatments to the end-of-chapter references. A futures contract is a highly standardized and detailed agreement to buy or sell a commodity for delivery at some future date at a price set at the time of the agreement. Futures contracts for corn, soybeans, wheat, live beef, hogs, cotton, orange juice, sugar, and other commodities are bought and sold on formally organized exchanges such as the Chicago Board of Trade or the New York Coffee, Sugar, and Cocoa Exchange. An example helps to explain what is involved. You call a commodity broker on May 1 and sell one corn futures contract for December delivery. You pay a brokerage fee and make a margin deposit to cover the possibility of an adverse price change, thus protecting the buyer against your default on the agreement. With this contract you agree to sell 5,000 bushels of no. 2 yellow corn to be delivered at a specific location in December at, say, \$2.67 per bushel. Actual delivery in December would normally take the form of a warehouse receipt that indicates where the corn is being held; you would not be expected to dump the stuff on someone's driveway. In point of fact, however, relatively few sellers of futures contracts expect to make deliveries. At some time prior to the end of the delivery month, November 1, for example, you would likely decide to buy a futures contract for December delivery at the November 1 price. If the price of the December futures has fallen to \$2.40, you lose. If the price rises to \$3.00, you make a profit.

Importantly, arbitrage between the cash and futures markets means that in December the futures price per bushel will equal (at least roughly) the cash price. This is crucial to the practice of hedging, which can best be understood through an example of a merchant storing corn. Suppose that on November 1, the cash price of corn is $2.60 per bushel and the price of futures for May delivery in $3.00. A merchant figures the cost per bushel of storing corn until May (including a normal return on investment) is 38 cents per bushel. The merchant buys corn at $2.60 and sells corn futures at $3.00 per bushel. Suppose that by next May the price of corn has risen only to $2.80. The merchant both buys corn futures and sells corn at $2.80. Speculation in the storage operation results in a loss of 18 cents per bushel. Speculation in the futures market, however, results in a profit of 20 cents per bushel. The reader can readily calculate the outcomes if the price of corn in May equals $2.50 or $3.25. Does it make sense for the merchant to hedge in this manner, buying corn and at the same time selling corn futures and then, later, selling corn and buying corn futures? This question will be taken up in Chapter 6 where the theory of commodity storage is considered in some detail.

Farm producers can establish a minimum price for their outputs by purchasing put options, available for such commodities as corn, soybeans, cattle, and sugar. Such an option gives the buyer the right, but not the obligation, to sell a futures contract, for the duration of the option, at a specified price called the strike price. Like futures contracts, commodity options are bought and sold by brokers on an exchange. The buyer of a put option pays the person granting the option through the exchange a premium that is not refundable. Commodity options may be exercised at any time prior to expiration. Exercising a put option is tantamount to selling a futures contract at the strike price. A profit accrues to the buyer if the futures price falls below the strike price. The option may be allowed to lapse if the futures price rises above the put option's strike price. In effect, the purchase of a put option provides protection against a fall in price, but preserves the opportunity to gain from a price increase. Obviously, such protection can only be obtained at a cost.

A cash forward contract is an agreement between a farmer and a buyer to deliver a given amount of commodity in exchange for payment at a later date. The agreement specifies the grade to be delivered and sets the price in advance. Since production is subject to uncertainty, farmers are rarely advised to sell forward their entire expected crop. Most country elevators offer cash forward contracts, either buying futures contracts or put options as protection from price risk. Although farmers make minimal direct use of futures contracts and put options, they do make considerable use of forward contracts, with up to 25 to 50 percent of the U.S. soybeans and corn sold through such contracts in some years. Cash forward sales are apparently less widely used by U.S. wheat growers than by corn–soybean farmers, perhaps reflecting the greater weather uncertainty involved in wheat production.

How does cash forward contracting affect the pricing of farm commodities? One way to approach this question is to construct a sector model that includes a futures market, a system of grain elevators, and a farm production industry. To limit the complexity and scope of the model, we propose instead to analyze a market in which the farmer has the option of hedging through planting a crop and speculating in an idealized futures market.

A Model of the Futures Market

We begin by constructing a model of a futures market seen here as a special kind of gambling casino. Put yourself in the shoes of a customer. Enter the front door of the casino in period t and immediately you see a neon sign that gives what we will call the stated price $F_t(P_{t+1})$. (The student may think of this as the current price, at time t, of a futures contract for delivery of one unit of the commodity in period $t + 1$.) If you think P_{t+1} will actually be less than the stated price $F_t(P_{t+1})$, you are given the opportunity to put your money where your opinion is, to bet that in the second period $P_{t+1} < F_t(P_{t+1})$. Gambling casinos must establish rules of the game and so it is here. To place a bet, you may go to what is called the *short* desk and enter a selling contract. According to the contract, the casino is obligated to send you either a bill or a check in period $t + 1$. Your profit from the contract may be expressed thus:

$$\pi_{t+1} = [F_t(P_{t+1}) - P_{t+1}]g_s \tag{3-7}$$

where g_s is the quantity that you choose in your selling contract in period t. If P is the price of a ton of onions, for a hypothetical example, g_s may be stated in terms of tons of onions. In the parlance of the literature on futures markets, you would be said to have sold a futures contract or to have sold onion futures. Your contract is similar in some respects to agreeing in period t to sell onions in period $t + 1$ at a price $F_t(P_{t+1})$ with the understanding that you will therefore need to buy onions in period $t + 1$ at the price P_{t+1}. This explains the s subscript in Eq. (3-7), indicating seller. You will not be expected, of course, to actually buy or sell onions in period $t + 1$. Now then, if you guess right you get a check from the casino in period $t + 1$. If you guess wrong, alas, you get a bill to pay.

Your utility is a function of expected profit Θ and the standard deviation of profit σ. Your problem is to maximize utility through making the proper choice of g_s. (The timid, the virtuous, and maybe even the wise have the option of setting g_s equal to zero.) Following previous analysis, it is clear that the optimal choice of g_s is a function of three variables: your subjective evaluation of the expected price, Θ_p; your subjective evaluation of the standard deviation of price, σ_p; and $F_t(P_{t+1})$, the stated price over which you have no control. The optimal choice of g_s depends, of course, on your utility function. Assuming risk aversion, it can be shown that g_s rises with increases in $F_t(P_{t+1})$, but falls with increases in Θ_p or in σ_p.[2] If we let $g_{si} = g_{si}[\Theta_{Pi}, \sigma_{Pi}, F_t(P_{t+1})]$ be the ith seller's supply function for onion futures, then the aggregate supply function becomes

$$G_s = \sum_i^{N_s} g_{si}\left[\Theta_{pi}, \sigma_{pi}, F_t\left(P_{t+1}\right)\right] \tag{3-8}$$

where N_s equals the number of sellers (of onion futures), which will be viewed as an endogenous variable. The higher $F_t(P_{t+1})$ is, for example, the larger the number of customers who would choose to become sellers instead of buyers. The higher $F_t(P_{t+1})$ is, the higher G_s becomes. The aggregate supply function for futures contracts is upward sloping.

Suppose that upon first seeing the value of $F_t(P_{t+1})$ you decide it is too low, that in all likelihood P_{t+1} will exceed $F_t(P_{t+1})$. On this assumption you may decide to become a buyer of onion futures, in which case you go to what might be called the *long* desk. Your profit from the bet you place is figured thus:

$$\pi_{t+1} = [P_{t+1} - F_t(P_{t+1})]g_d \tag{3-9}$$

where g_d equals the tons of onions you plan to buy. If you place a bet, you will receive either a check or a bill in period $t + 1$ depending on whether π_{t+1} in Eq. (3-9) is positive or negative. In this case you would be called a buyer of futures. In a sense you are agreeing to buy in period $t + 1$ onions at the price $F_t(P_{t+1})$ with the idea of selling them at P_{t+1}. Following the analysis for the sellers, we may simply postulate an aggregate demand function for buyers of onion futures thus:

$$G_d = \sum_j^{N_d} g_{dj}\left[\Theta_{pj}, \sigma_{pj}, F_t\left(P_{t+1}\right)\right] \tag{3-10}$$

where N_d is the numbers of buyers, an endogenous variable. The higher $F_t(P_{t+1})$ is, the lower the optimal values for g_{dj}, assuming risk aversion. Also, the higher $F_t(P_{t+1})$ is, the lower the number of buyers N_d.

[2]An explicit treatment is left to the student as a useful exercise. (*Hint:* Derive the algebraic expressions for Θ and σ. Suppose that your utility function is given by

$$u_t = \phi_0 + \phi_1\Theta - \phi_2\sigma^2$$

Write u_t as a function of g_s. Then maximize u_t with respect to the choice variable g_s. The optimal value of g_s is given by

$$g_s = \frac{\phi_1}{2\phi_2\sigma_P^2}\left[F_t\left(P_{t+1}\right) - \Theta_p\right]$$

If $\Theta_P = F_t(P_{t+1})$, you may retire to the lounge and have a beer.)

We next take up the question of how $F_t(P_{t+1})$ is determined by the casino. It is agreed by the parties involved that all bets are off unless G_d equals G_s. The casino and all the customers therefore experiment with a large set of values for $F_t(P_{t+1})$ until a particular value is obtained, call it the equilibrium futures price $F_{tc}(P_{t+1})$, such that $G_d = G_s$. All contracts made at this market clearing price are binding. Obviously, we are making use of demand and supply theory. The number of sellers of futures and the amount each is willing to sell rises with increases in $F_t(P_{t+1})$. The number of buyers of futures and the amount each is willing to buy fall with increases in $F_t(P_{t+1})$. In fact, if each market participant filed his or her schedule of intentions to buy or sell for all alternative values of $F_t(P_{t+1})$, the operator of the casino or futures market could simply announce $F_{tc}(P_{t+1})$ at the outset.

We are assuming that all customers have utility functions, each with his or her own ideas about the expected price and the standard deviation of price. The customers are not betting against the house; they bet against each other. This is a game of wits in a world of chance. We abstract from a cover charge or any commissions or any fees customers must pay the casino for the services that it renders.

How do gamblers (speculators) form their estimates of Θ_P and σ_P? One way of proceeding is to invoke the assumption of rational expectations, as in Chapter 2, assuming that all market participants know how the market works, how prices in period $t + 1$ will be determined, and how best to estimate the expectations and variances of all exogenous and endogenous variables. Although we believe this approach is fruitful in the modeling of many markets, it provides no insight as to the performance of a futures market. Why is this? If all market participants, including the patrons of our imaginary casino, have the same rational estimates of $E_t(P_{t+1})$ and $V_t(P_{t+1})$, there will be neither challenge nor fun; there will be no speculation. In our imaginary casino the object is to outguess your rivals.

We will not dwell here on how flesh and blood speculators go about forming their expectations. Perhaps they make elaborate charts, assemble econometric models, gaze longingly into the eyes of their lovers; who knows? It is important to realize, however, that the people who make the best estimates of what P_{t+1} will equal, particularly if they are well financed, are the people who will make money at this game. But surely those who best understand how markets work and have the best information for predicting future events, in this case what second-period demand will likely be and the production plans of producers in period t, will be the best estimators of the second-period price. The ignorant and the foolish will lose money at this game and will sooner or later wise up and quit; alternatively, they may be assumed to run out of money. What this means is that large speculators who survive through time, perhaps outcompeting a crowd of amateurs who come and go, are informed and astute.

Based in part on the preceding considerations, but also recognizing the requirements of building a model, we will often assume in what follows that on average the equilibrium stated price in the casino, the price that equates G_d and G_s, equals the expected price. Mathematically, we have

$$F_t(P_{t+1}) = E_t(P_{t+1}) \quad (3\text{-}11)$$

What this means is that on average the speculators get it right. Indeed, we might argue plausibly that, although futures markets exist for the pleasure of speculators, the ability of speculators to make shrewd (rational) guesses as to prices that will prevail in the future improves the performance of farm commodity markets, including, as we will see in Chapter 6, the markets for commodity storage. We might indeed view the futures market process as an estimation procedure. Suppose that you were given a blank check by the government and asked to supply the nation with short-range predictions of future spot prices. Most people would likely think immediately of the need to organize a team of researchers—a bevy of economists, statisticians, computer programmers, and clerks. But a futures market involves many such teams of researchers (speculators), each team trying to outsmart all the others and trying to make a profit in the process. What better estimation procedure could be devised by the human mind?

We now consider a market situation in which the farmer who makes output decisions in period t has the option of speculating in a futures market as well. Two cases are considered in turn, the first involving price risk, but no production risk, and the second involving both price and production risk. Analysis of both cases requires modification of the models considered in Chapter 2.

Hedging under Price Risk

The profit function for a farmer who faces price risk only and sells futures is as follows:

$$\pi_{t+1} = P_{t+1}q_{t+1} - C(q_{t+1}) - TFC + [F_t(P_{t+1})g_s - P_{t+1}\, g_s] \quad (3\text{-}12)$$

where the notation is the same as in Chapter 2 except that we have now added the sales of futures in brackets. Recall that the value for q_{t+1} is chosen in the first period. If q is measured in bushels of wheat, for example, we may think of g_s as the quantity of futures measured in bushels of wheat as well. It is left to the student to write out the expressions for expected profit Θ and the variance of profit σ^2. We assume that the farmer wishes to maximize utility, where $u_t = \phi_0 + \phi_1\Theta - \phi_2\sigma^2$. Maximizing u with respect to the choice variables q_{t+1} and g_s yields the first-order conditions for a maximum as follows:

$$\phi_1[\Theta_p - C'(q_{t+1})] - 2\phi_2\sigma_p^2(q_{t+1} - g_s) = 0$$

$$\phi_1[F_t(P_{t+1}) - \Theta_p] + 2\phi_2\sigma_p^2(q_{t+1} - g_s) = 0$$

Solving the latter expression for g_s and using the result to rid the first expression of g_s through substitution yields

$$F_t(P_{t+1}) = C'(q_{t+1}) \tag{3-13a}$$

$$g_s = \frac{\phi_1[F_t(P_{t+1}) - \Theta_p] + 2\phi_2\sigma_p^2 q^*}{2\phi_2\sigma_p^2} \tag{3-13b}$$

where q^* in Eq. (3-13b) equals the optimum level of output. Given the option of selling futures, the farmer chooses the level of output that equates the marginal cost of production with the price of futures. This is an important result: *Under optimal use of futures contracts, the cost of risk no longer impedes production. The existence of future trading and hedging, under the assumed conditions, increases output and lowers the expected market price.*

The model given by Eq. (2-28) must be changed for commodities with contracts traded in futures markets. With futures trading and hedging, aggregate farm output is a function of the futures price, not of the expectation and standard deviation of market price (of Θ_P and σ_P). In fact, the standard deviation of market price may be dropped from the model. Equation (3-11) may be added to make the futures price an endogenous variable. A useful exercise for the interested student is to write the new equation system and identify the new set of endogenous variables.

What of the extent of speculation or gambling in the market for futures? Going back to Eq. (3-13b), we see that if the farmer's subjective price expectation Θ_p equals the futures price $F_t(P_{t+1})$, then the farmer's sales of futures g_s equal his or her optimal level of production q^*. The farmer evades all risk through hedging. This is important in that, in real life, many farmers might not have much confidence in their ability to estimate expected price, relying instead on the futures price as the best estimate. If, on the other hand, $F_t(P_{t+1}) > \Theta_p$, the farmer's sales of futures exceeds q^* and contrariwise, if $F_t(P_{t+1}) < \Theta_p$. Is it possible that the farmer might actually buy futures, that is, carry out a so-called "Texas" hedge? (Hedging is usually defined as taking opposite positions in commodity and futures markets.) The answer is yes if Θ_p exceeds $F_t(P_{t+1})$ by a sufficient margin and/or if σ_p^2 is very small because the farmer, who perhaps has been out in the sun too long, "knows" what the price will be. Other comparative static results are left to the student.

Hedging under Price and Production Risk We now take up the more challenging case when the farmer is confronted with both production and price risk. As in the Chapter 2 model, we let the farmer's production function be given by

$$q_{t+1} = q(a_t, k_t, h_t)v_{t+1}$$

Recall also that $G_{t+1} = P_{t+1}v_{t+1}$. The profit function for the farmer who sells futures may be expressed as follows:

$$\pi_{t+1} = G_{t+1}\Theta_q - C(\Theta_q) - TFC + [F_t(P_{t+1})g_s - P_{t+1}g_s] \tag{3-14}$$

where G_{t+1} equals the gross return per unit of planned output, $P_{t+1}v_{t+1}$, and Θ_q equals planned or expected output. The expectation and variance of profit are given by

$$\Theta = \Theta_G \Theta_q - C(\Theta_q) - TFC + F_t(P_{t+1})g_s - \Theta_p g_s \tag{3-15a}$$

$$\sigma^2 = \Theta_q^2 \sigma_G^2 + g_s^2 \sigma_p^2 - 2\Theta_q\, g_s\, \sigma_G\, \sigma_p\, \rho(G_{t+1}, P_{t+1}) \tag{3-15b}$$

where the correlation between G_{t+1} and P_{t+1} is given by $\rho = \rho(G_{t+1}, P_{t+1})$ and where the covariance between G_{t+1} and P_{t+1} is given by $\sigma_G\sigma_p\rho(G_{t+1}, P_{t+1})$. These expressions for Θ and σ^2 may be inserted in the farmer's utility function, which is again given by $u_t = \phi_0 + \phi_1\Theta - \phi_2\sigma^2$. To simplify further, we assume that the farmer takes the futures price $F_t(P_{t+1})$ as his or her estimate of the expected price Θ_p. Given $\Theta_P = F_t(P_{t+1})$, maximizing u_t with respect to the choice variables Θ_q and g_s yields the following first-order conditions:

$$\begin{aligned} \phi_1[\Theta_G - C'(\Theta_q)] - 2\phi_2[\Theta_q\, \sigma_G^2 - g_s\, \sigma_G\, \sigma_p\, \rho] &= 0 \\ -2\phi_2[g_s\, \sigma_p^2 - \Theta_q \sigma_G\, \sigma_p\, \rho] &= 0 \end{aligned} \tag{3-16}$$

The latter of these two expressions allows solving for g_s in terms of Θ_q. This, in turn, allows eliminating g_s from the first expression through substitution. When these steps are taken, we have

$$\Theta_q = \frac{\phi_1\left[\Theta_G - C'\left(\Theta_q\right)\right]}{2\phi_2\sigma_G^2\left(1-\rho^2\right)} \tag{3-17a}$$

$$\begin{aligned} g_s &= \left(\sigma_G\, \rho / \sigma_p\right)\Theta_q \\ &= \frac{\phi_1\rho\left[\Theta_G - C'\left(\Theta_q\right)\right]}{2\phi_2\sigma_G\sigma_p\left(1-\rho^2\right)} \end{aligned} \tag{3-17b}$$

In this formulation, the farmer's planned output is, according to Eq. (3-17a), a function of the expected gross returns, Θ_G; the variance (or standard deviation) of gross returns, σ_G^2; and the correlation between the gross returns and price, $\rho(G_{t+1}, P_{t+1})$. By comparing Eq. (3-17a) with Eq. (2-38), the student will quickly see an important result. If $0 < |\,\rho\,(G_{t+1}, P_{t+1})\,| \leq 1$, then planned output is larger with futures trading and hedging than without. Again, the futures market is seen as a potential means for decreasing the cost of risk, expanding planned output, and lowering the expected market price.

Some further discussion of these results is in order. Earlier in this chapter we saw that, when the farmer can plant two crops, diversification can be used to reduce risk, particularly if the correlation between crop prices is negative and close or equal to –1. What is crucial here is the correlation ρ between the price the farmer receives and the gross return per unit of planned output. This correlation can be zero, positive, or negative. It may be zero if the farmer's yield (per acre or per unit of planned output) is uncorrelated with national yield. This could happen if climatic conditions are variable greatly across space. In the case of uniform climatic conditions, however, the farmer's yield will likely be correlated with national yield. If the latter condition holds, then elastic demand is conducive to negative correlation, $\rho < 0$, because a decrease in yield causes a relatively small *increase* in price, such that gross returns per unit of output *decline*. Alternatively, inelastic demand is conducive to a positive correlation, $\rho > 0$, because a decrease in yield causes a relatively large *increase* in price, such that gross returns per unit of output *rise*.[3] In the intermediate case, when demand is unitarily elastic, the correlation ρ will equal zero without regard to whether the farmer's yield is correlated with the national yield.[4]

In any case, the correlation coefficient ρ influences both production and hedging decisions. As the absolute value of ρ approaches 1, the farmer equates the expected gross return to the marginal cost of planned output. Hedging is highly effective in that the cost of risk no longer impedes expected production. The farmer's supply of output is a function of the expected gross return Θ_G, and the standard deviation of gross returns σ_G no longer appears in the supply function. If ρ is positive, the farmer hedges through selling futures. If ρ is negative, the farmer hedges through buying futures. As the ratio $\sigma_G\rho/\sigma_p$ approaches +1 (or –1), the farmer hedges fully in that $|g_s| = \Theta_q$. If ρ equals zero, and providing σ_P does not equal zero as well, then the farmer does not participate in the futures market. The production decisions are made without regard to the goings on in the market for futures.[5]

It should be re-emphasized that, although farmers at present make little direct use of the markets for futures, they do take advantage of cash forward contracting

[3]For intuitive insight, take the case when the aggregate supply of land is perfectly inelastic. Also, hold planned output constant. If demand is elastic, total revenue falls when yield is low, even though price rises; total revenue rises when yield is high, even though price falls. Therefore, the gross return per unit of planned output is negatively correlated with price. If demand is inelastic, on the other hand, total revenue rises along with price when yield is low; total revenue falls along with price when yield is high. Here the gross return per unit of planned output is positively correlated with price.

[4]Farmers using cash forward contracts are often advised to sell forward no more than half their expected output. This advice seems to suggest that $0 \le \sigma_G\,\rho/\sigma_q \le 0.5$. See Eq. (3.17b).

[5]The model given by Eqs. (2-39) and (2-40) for the cases of price and production uncertainty must be changed if futures trading and hedging exist. As noted, aggregate planned output Θ_Q becomes, in general, a function of Θ_G, σ_G, and $\rho(G_{t+1}, P_{t+1})$. If $|\rho|$ equals 1, Θ_Q is a function of Θ_G only. If ρ equals zero, Θ_Q is a function of Θ_G and σ_G, as in Chapter 2.

on a large scale. From the point of view of modeling farm output decisions, the futures price referred to in the preceding analysis might just as well be thought of as the set price in cash forward contracts.

There are several reasons why farmers make little direct use of futures markets, not the least of which is that cash forward contracting is better suited to their purpose. Futures contracts are highly standardized and involve minimum quantities that are often very large relative to a farmer's expected output. The unit of output for the futures contract for soybeans is 5000 bushels, for example, which exceeds the normal production of most farmers. Adverse changes in the basis may also expose farmers to risk if they use futures markets. Basis is defined as the difference between the price for a futures contract and the price for the same or a similar commodity for actual delivery at a particular location. In general, basis risk (i.e., unpredictable changes in basis over time) provides a disincentive for hedging. (For further detail on forward contracting, the reader is referred to the references at the end of the chapter.)

3.4 SOYBEAN ACREAGE RESPONSE: AN EMPIRICAL EXAMPLE

For an empirical example of agricultural supply, consider the following estimated acreage response equation for U.S. soybeans.[6]

$$\begin{aligned} A_t = 3.2269 &+ \underset{(6.21)}{28.8638F_t(PS)} - \underset{(4.29)}{45.4827F_t(PC)} \\ &+ \underset{(19.99)}{0.8983A_{t-1}} - \underset{(4.24)}{0.1166I_t} + \underset{(3.66)}{5.6696B_t} \end{aligned} \tag{3-18}$$

where

A_t = acreage planted to soybeans (million acres) in year t

$F_t(PS)$ = the average price of soybean futures (dollars per bushel) for November delivery during the months of March, April, and May, divided by an index of farm input prices

[6] This equation was estimated using multiple regression analysis and time-series data for 1961–1991. The numbers given in parentheses below the estimated coefficients are t-ratios. The coefficient of multiple determination (R-squared) equaled 0.98.

$F_t(PC)$ = the average price of corn futures (dollars per bushel) for December delivery during the months of March, April, and May, divided by an index of farm input prices

A_{t-1} = acreage planted to soybeans (million acres) in year $t-1$

I_t = acreage diverted under farm programs for feed grains, wheat, and cotton (million acres)

B_t = a binary variable that equals zero when commodity programs were not in effect (1974–1977 and 1980–1981) and one when they were in effect

For an explanation of this equation, we note that acres planted to soybeans, A_t, may be viewed as a proxy for planned or expected output. This ignores the steps a farmer might take to vary expected yield in response to changes in the expected price, such as altering the input of fertilizer. Under the assumption that the futures price is a good proxy for the expected cash price, futures prices at planting time for delivery at harvest time are used as explanatory variables in the acreage response equation. More specifically, we have included the price of soybean futures for November delivery, $F_t\ (PS)$, as a proxy for the expected soybean price. To take account of changes in input prices, the expected price was divided by the index of farm input prices used in production (1910 – 1914 = 100). In the preceding analysis of diversification, we saw how acreage planted to one crop varies inversely with the expected price (gross return) for a competing crop. The price of corn futures for December delivery, $F_t\ (PC)$, is included in the equation on the hypothesis that corn and soybeans are competing crops. Acres planted to soybeans, lagged one year, is included to characterize possible slow acreage adjustments over time. Indeed, the full adjustment of soybean acreage to a change in the expected price might take place over several years, not just in a single season. Inserting lagged acreage in the equation allows for this possibility and provides the basis for estimating both the short-run and long-run elasticities of acreage response. Finally, the two variables I_t and B_t are included in the equation to separate out the effects of farm programs for feed grains, wheat, and cotton, a subject that will be explored in Chapters 9 and 10.

Drawing on economic theory and previous research, we believe the estimated coefficients appearing in the above equation have the correct signs; they are also statistically significant. The sample means for $F_t\ (PS)$ and A_t equal, respectively, 0.8415 and 52.1516. The estimated short-run elasticity of soybean acreage response is 0.47, which equals approximately (28.8638)(0.8415)/52.1516. The estimated long-run acreage response elasticity is 4.58, which appears to be a high estimate relative to those appearing in the literature.[7] According to these estimates, a 10 percent increase in the expected price of soybeans would increase acreage planted by 4.7 percent in the short run and by 45.8 percent in the long run. Importantly, these estimates are based on holding the expected price of corn constant. We would expect, however, that as acres were switched from corn to soybeans, the expected price of corn would begin to rise, which would tend to choke off the increase in soybean acreage. We note, for example, that a 10 percent increase in the expected price of corn would decrease planted soybean acreage by 3.2 percent. [The sample mean for $F_t\,(PC)$ equals 0.3643.] Interestingly, a simultaneous 10 percent increase in both the expected prices of soybeans and corn would have a limited effect on soybean acreage, which indicates that soybeans and corn compete vigorously for U.S. cropland. This simple example of an estimated acreage response equation illustrates the importance of expectations and the role of competing outputs in the farm sector. It also calls attention to the importance of addressing substitution issues among multiple products in agricultural economics.

3.5 PRODUCTION MODELS AND SUPPLY FUNCTIONS

Taken together, Chapters 2 and 3 contain a small menu of production and pricing models under risk, with each model or subset of models having its own distinctive farm supply function. It is important for the student to understand the various production models, to know what supply function goes with what model, and to recognize that real-world applications involve identifying the "right" or most appropriate model for whatever commodity market is of interest. Table 3.1 is included as a handy reference for this purpose. The student should recognize that supply function (1), which is associated with price uncertainty and risk neutrality, can also be associated with other production models. Indeed, if $\Theta_p = F_t(P_{t+1})$, then the supply functions (1) and (2) in Table 3.1 are the same. In this case, neglecting the effects of price variance

[7] Suppose $A_t^* = \Upsilon_0 + \Upsilon_1 F_t\,(PS)$, where A_t^* equals acreage planted in long-run equilibrium. According to the partial adjustment hypothesis,

$$A_t - A_{t-1} = \Upsilon_2(A_t^* - A_{t-1})$$

where Υ_2 measures the percentage of desired adjustment $(A_t^* - A_{t-1})$, actually taking place from period $t-1$ to period $t\,(A_t - A_{t-1})$. Also, $0 < \Upsilon_2 \leq 1$. The closer is Υ_2 to unity, for example, the greater the adjustment in period t. Using the last equation to get rid of A_t^* in the first equation yields

$$A_t = \Upsilon_2\Upsilon_0 + \Upsilon_2\Upsilon_1 F_t\,(PS) + (1 - \Upsilon_2)A_{t-1}$$

Taking advantage of Eq. (3-18), we see that the estimates for Υ_2 and Υ_1 are, respectively, 0.1017 (equals 1 – 0.8983) and 283.81 (equals 28.8638/0.1017). The long-run elasticity reported in the text is calculated using this estimate for Υ_1.

TABLE 3.1 Alternative Short-run Farm Output Supply Functions Assuming Production Lags[a]

Supply Functions	Market Conditions
	Price Uncertainty Only
(1) $Q_{t+1} = S(\Theta_P)$	(a) Farmers are risk neutral.
	(b) Farmers are risk averse, but the cost of liquidity is low.[b]
(2) $Q_{t+1} = S[F_t(P_{t+1})]$	Farmers are risk averse but hedge in futures markets.
(3) $Q_{t+1} = S(\Theta_P, \sigma_P)$	Farmers are risk averse, the cost of liquidity is not low, and there is no hedging.[b]
	Price and Production Uncertainty
(4) $\Theta_Q = S(\Theta_G)$	(a) Farmers are risk neutral.
	(b) Farmers are risk averse, but the cost of liquidity is low.[b]
	(c) Farmers are risk averse, but hedge in futures market, assuming that $\Theta_P = F_t(P_{t+1})$ and $\lvert \rho(G_{t+1}, P_{t+1}) \rvert = 1$.
(5) $\Theta_Q = S(\Theta_G, \sigma_G)$	Farmers are risk averse, the cost of liquidity is not low, and there is no hedging.[b]
(6) $\Theta_Q = S[\Theta_G, \sigma_G, \rho(G_{t+1}, P_{t+1})]$	Farmers are risk averse, assuming that $\Theta_P = F_t(P_{t+1})$ and $0 < \lvert \rho(G_{t+1}, P_{t+1}) \rvert < 1$.

[a] See the text for notation and more detailed explanations.

[b] Dissociating consumption and profit is inexpensive when the cost of liquidity is low.

may be appropriate in the analysis of real-world production decisions (even if farmers are risk averse). In other cases, risk can have significant effects on supply, as found in models (3), (5), and (6) in Table 3.1. In these situations, risk aversion plus uncertainty increases the cost of operation.

This chapter has analyzed various means for reducing the cost of risk. Farmers may use financial transactions to decouple consumption expenditures from profit, diversify their output mix, and hedge through futures market transactions or forward contracting. The strategies available to farmers for cutting the cost of risk are themselves costly, however. Farmers are thus confronted with trade-offs, and it is important to keep risk-reducing strategies and the resultant trade-offs in mind when trying to understand how farmers make decisions and how farm markets perform.

PROBLEMS

3.1. You are given the following information for a farmer:

$$u_t = 2 + 2\Theta_\kappa - \sigma_\kappa^2$$

$$\Theta = \Theta_P q_{t+1} - q_{t+1}^2$$

$$\sigma^2 = q_{t+1}^2 \sigma_P^2$$

$$C_B = 10\Psi$$

$$\Theta_P = 20 \quad \text{and} \quad \sigma_P^2 = 4$$

Recall that C_B equals the banking cost (the cost of liquidity), which we assume here goes up by 1 unit of money for every 0.1 increase in Ψ. Find the optimal values for q_{t+1} and u_{t+1} given that $\Psi = 0$, $\Psi = 0.5$, and $\Psi = 1.0$.

3.2. Consider the following game. Two fair coins are tossed simultaneously only once. If the first coin comes up heads, you receive \$1.00. If the second coin comes up heads, you receive \$2.00. Tails yield nothing.

a. Compute the expectation and the variance of the payoff.

b. Now a poltergeist joins the action and forces the second coin always to come up opposite to the first, which remains a fair coin. Again compute the expectation and the variance. (*Note:* Perfect negative correlation does not drive variance to zero when the payoffs are unequal.)

3.3. You are given the following information for a farmer who has the option of producing both corn and soybeans:

$$u_t = 2\Theta - 0.05\sigma^2$$

$$\Theta = \Theta_{Gc} a_c + \Theta_{Gs} a_s$$

$$\sigma^2 = a_c^2 \sigma_{Gc}^2 + a_s^2 \sigma_{Gs}^2 + 2a_c a_s \sigma_{Gc} \sigma_{Gs} \rho(G_c, G_s)$$

$$400 = a_c + a_s$$

(Since $\Theta_{Gc} > \Theta_{Gs}$ in what follows, the farmer will only grow soybeans to reduce risk. From the answers to parts a, b, and c, you will see that as ρ falls from +1 to –1 soybean production becomes a more effective means of reducing risk.)

a. Let $\Theta_{Gc} = 275$, $\Theta_{Gs} = 100$, $\sigma_{Gc}^2 = 25$, and $\sigma_{Gs}^2 = 9$. Find the optimum levels for a_c, a_s, Θ, σ^2, and u_t for $\rho(G_c, G_s)$ equal to +1, 0, and –1.

b. Repeat part a, except let $\sigma_{Gc}^2 = 9$ and $\sigma_{Gs}^2 = 25$.

c. Repeat part a, except let $\sigma_{Gc}^2 = 36$ and $\sigma_{Gs}^2 = 9$.

3.4. One thousand speculators all have the following utility function:

$$u_t = 2 + 4\Theta - 0.1\sigma^2$$

Let g_s (g_b) equal the quantity of commodity futures a speculator is willing to sell (buy).

a. Derive the supply function for a seller who believes that $F_t(P_{t+1}) > \Theta_P$.

b. Derive the demand function for a buyer who believes that $F_t(P_{t+1}) < \Theta_P$.

c. For 400 speculators, $\Theta_P = 4$. For the remaining speculators, $\Theta_P = 2$. All believe that $\sigma_P^2 = 1$. Find the equilibrium price of futures.

3.5. You are given the following information for a farmer:

$$\pi_{t+1} = P_{t+1}q_{t+1} - q_{t+1}^2 - TFC + [F_t(P_{t+1})g_s - P_{t+1}\, g_s]$$

$$u_t = 6\Theta - 0.1\sigma^2$$

a. Derive the farmer's supply functions for q_{t+1} and g_s assuming that $F_t\,(P_{t+1}) > \Theta_P$. Verify your answers using Eqs. (3-13a) and (3-13b).

b. Let $F_t\,(P_{t+1}) = 10$ and $\sigma_P^2 = 1$. Find q_{t+1} and g_s for $\Theta_P = 5, 10,$ and 15. *(Hint:* A negative value for g_s may be interpreted as buying commodity futures.)

c. Let $F_t\,(P_{t+1}) = 10$ and $\sigma_P^2 = 60$. Again find q_{t+1} and g_s for $\Theta_P = 5, 10,$ and 15. *(Note:* A large value for σ_P^2 may be interpreted to mean that the farmer has little confidence in his or her ability to predict the future price P_{t+1}. In this situation the farmer is unlikely to both produce a crop and buy commodity futures as happened in part b.)

REFERENCES

Anderson, Jock R., John L. Dillon, and Brian Hardaker, *Agricultural Decision Analysis*. Ames, Iowa: Iowa State University Press, 1976.

Chavas, Jean-Paul, and Matthew T. Holt, "Acreage Decisions under Risk: The Case of Corn and Soybeans," *American Journal of Agricultural Economics*, 72, no. 3 (1990), 529–538.

Feder, Gershon, Richard E. Just, and Andrew Schmitz, "Futures Markets and the Theory of the Firm under Price Uncertainty," *Quarterly Journal of Economics*, 95, no. 2 (March 1980), 317–328.

Hieronymous, Thomas A., *Economics of Futures Trading for Commercial and Personal Profit*. New York: Commodity Research Bureau Inc., 1971.

Holthausen, Duncan M., "Hedging and the Competitive Firm under Price Uncertainty," *American Economic Review,* 69, no. 5 (1979), 989–995.

Leuthold, Raymond M., Joan C. Junkus, and Jean E. Cordier, *The Theory and Practice of Futures Markets*. Lexington, Mass.: Lexington Books, 1989.

Newbery, David M. G., and Joseph E. Stiglitz, *The Theory of Commodity Price Stabilization: A Study in the Economics of Risk*. New York: Oxford University Press, 1981.

Wright, Bruce, Joy L. Harwood, Linwood A. Hoffman, and Richard G. Heifner, *Forward Contracting in the Corn Belt and Spring Wheat Areas, 1988,* U.S. Department of Agriculture, Economic Research Service, December 1988.

4

Farm Output and Input Pricing

Chapter 2 centered on the determination and pricing of output in a farm market characterized by uncertainty. Little was said regarding farm inputs used in production and how the quantities and prices of these inputs are determined. This chapter centers on these issues. As will be seen shortly, however, the pricing and application of inputs is not independent of the pricing and production of output. Understanding the interdependence between output and input markets is of fundamental importance in this book and in the study of economics generally. Section 4.1 introduces the Cobb–Douglas production function, a concept that is particularly useful in developing algebraic models of input markets. Section 4.2 centers on the derivation of the market demands for farm inputs, using land input as an illustrative example. The approach taken involves both mathematical and graphic analysis, and some of the behavioral relationships that we derive will play important roles in later chapters. Section 4.3 sets forth and analyzes a simultaneous equations model of the farm sector that provides an explicit treatment of the interdependency among farm output and input markets. This section assumes that farmers are risk neutral or, alternatively, that farmers are able through hedging and other means to reduce risk to an insignificant level. Changes in the model required by the relaxation of this assumption are considered briefly in Section 4.4. A final section takes up the pricing of assets, which is one of the more difficult phenomena considered in economic theory. The pricing of land is modeled as an illustrative example.

4.1 COBB–DOUGLAS PRODUCTION FUNCTION

It is often convenient and useful to postulate specific functional forms in economic analysis. Linear equations, for example, often allow the derivation of hypotheses that would otherwise be extremely difficult to obtain. This explains why in this and succeeding chapters we will often make use of the Cobb–Douglas production function.

A production function is said to have the Cobb–Douglas form if

$$q_{t+1} = \alpha_0 a_t^{\alpha_1} k_t^{\alpha_2} h_t^{\alpha_3}(v_{t+1}) \tag{4-1}$$

where $\alpha_1 + \alpha_2 + \alpha_3 = 1, 0 < \alpha_i < 1, i = 1, 2, 3$, and a_t, k_t, and h_t, respectively, denote land, producer goods (capital), and labor input used at time t. The assumption $(\alpha_1 + \alpha_2 + \alpha_3) = 1$ corresponds to a technology exhibiting *constant returns to scale*, where a proportional increase in all inputs generates the same proportional increase in expected output. In this case, the production function is said to be linearly homogeneous. We will normally assume fixed family labor such that $h_t = h_{t0}$. We let v_{t+1} equal a random variable that measures deviations due to weather or other events, with expected value $\Theta_v = 1$ and standard deviation $\sigma_v > 0$. Expected output Θ_q, with the expectation formed in period t, is given by

$$\Theta_q = \alpha_0 a_t^{\alpha 1} k_t^{\alpha 2} h_{t0}^{\alpha 3}$$

A Cobb–Douglas (CD) production function for expected output has several interesting properties. The CD isoquant map is shown in Fig. 4.1. Smooth and convex to the origin, the isoquants do not touch either axis. Also, the expansion path is linear for constant prices of a and k, as shown by *XP*. The assumption of a CD production function is restrictive; analytical results might not hold for other production functions. We assume that aggregative models based on CD functions provide plausible approximations of the real world to simplify analysis. (For a more general statement, see Appendix C.)

To show how the farmer chooses optimal levels of inputs, drawing on the analysis given in Chapter 2, we let expected return per unit of expected (planned) output be given by Θ_G, where

$$\Theta_G = E_t(P_{t+1} v_{t+1}) \tag{4-2}$$

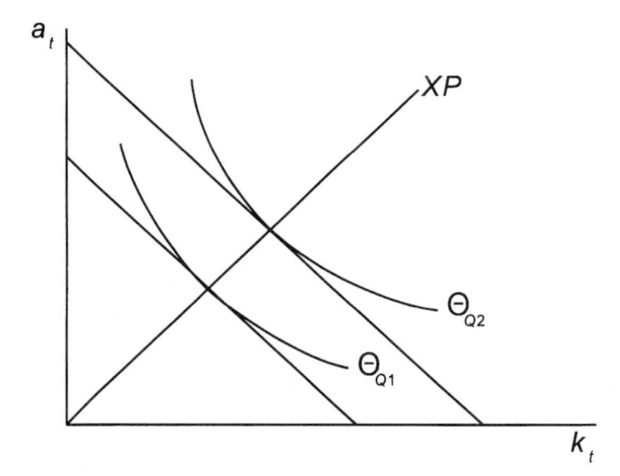

Figure 4.1 Cobb–Douglas isoquant map with isocost lines and expansion path.

Treating family labor h_t as fixed, expected profit may then be expressed as follows:

$$\begin{aligned}\Theta &= \Theta_G \alpha_0 a_t^{\alpha_1} k_t^{\alpha_2} h_{t0}^{\alpha_3} - R_t a_t - J_t k_t - TFC \\ &= \Theta_G \Theta_q - R_t a_t - J_t k_t - TFC\end{aligned} \tag{4-3}$$

where J_t and k_t equal, respectively, the price and quantity of a variable input, such as fertilizer; R_t and a_t equal land rent per acre and acreage planted; and the other variables are defined as in the previous chapters. The student should note that if v_{t+1} always equaled 1, that is, if production was not subject to the random influence of the weather, then the expected return per unit of expected output would equal the expected price. Life would be simpler. The present formulation allows, however, for negative covariance between price and needed rainfall. That is, yield and price may be negatively correlated. The rational farmer is aware, for example, that years of favorable weather are associated with relatively high yields and relatively low prices.

Assuming risk neutrality, we maximize Eq. (4-3) with respect to land a_t and producer goods k_t, which yields

$$\begin{aligned}R_t &= \Theta_G(\alpha_1 \alpha_0 a_t^{\alpha_1 - 1} k_t^{\alpha_2} h_{t0}^{\alpha_3}) = \Theta_G MPP_a \\ J_t &= \Theta_G(\alpha_2 \alpha_0 a_t^{\alpha_1} k_t^{\alpha_2 - 1} h_{t0}^{\alpha_3}) = \Theta_G MPP_k\end{aligned} \tag{4-4}$$

where MPP equals expected marginal physical product. Maximization of expected profit implies that the level of each variable input will be chosen such that its price equals its expected value of marginal product.

For a numerical example, let $\alpha_0 = 1.0$, $\alpha_1 = 0.4$, $\alpha_2 = 0.2$, and $h_{t0} = 1.0$. Then $\Theta_q = a^{0.4}k^{0.2}$. To simplify notation, we neglect for the moment the t subscript. Taking the partial derivative of Θ_q with respect to input a yields the expected marginal physical product of a: $MPP_a = 0.4a^{-0.6}k^{0.2}$. Similarly, we find that $MPP_k = 0.2a^{0.4}k^{-0.8}$. These expressions for MPP_a and MPP_k can be inserted in Eqs. (4-4). Using logarithms, we have

$$\log R = \log(0.4) - 0.6\log(a) + 0.2\log(k) + \log(\Theta_G)$$

$$\log J = \log(0.2) + 0.4\log(a) - 0.8\log(k) + \log(\Theta_G)$$

Conversion to logarithms yields a linear, two-equation system. (Recall that R, Θ_G, and J are viewed as constants by the farmer.) Solving for $\log(a)$ and $\log(k)$ and taking the antilogarithms yields

$$a = 0.0716R^{-2}J^{-0.5}\Theta_G^{2.5}$$

$$k = 0.0358R^{-1}J^{-1.5}\Theta_G^{2.5}$$

Assigning values to R, J, and Θ_G allows calculating the optimal quantities of a and k. These values determine the optimal level of planned output.

4.2 FARM INPUT DEMAND

The aggregate demand for an input may be defined as a function that shows the quantity of input an industry would be willing to buy at alternative input prices, allowing for the optimal variation of all other variable inputs and taking into account the changes in output price that are associated with changes in the price of the input. The objective in what follows is to show under conditions of uncertainty how such a relationship can be derived for a farm industry. Industry demand for an input is of basic importance, of course, in the study of farm input pricing. We begin by considering the input choices of an individual farmer.

Short-run Demand of an Individual Farmer

As in Chapter 2, actual output and expected output for an individual farmer are functions of inputs as follows:

$$q_{t+1} = q(a_t, k_t, h_t)v_{t+1} \tag{4-5a}$$

$$\Theta_q = q(a_t, k_t, h_t) \tag{4-5b}$$

We are not assuming a Cobb–Douglas production function for the moment, but we will do so later when we take up a numerical example. Profit and expected profit are given by

$$\pi_{t+1} = P_{t+1}q_{t+1} - J_t k_t - R_t a_t - TFC \tag{4-6a}$$

$$\Theta = \Theta_G \Theta_q - J_t k_t - R_t a_t - TFC \tag{4-6b}$$

where, as before, $\Theta_G = E_t[P_{t+1}v_{t+1}]$ is the expected gross return per unit of planned output. The product $\Theta_G\Theta_q$ equals the expected or planned total revenue. The fixed plant may be viewed as fixed family labor, h_{t0}, perhaps including a complement of farm machinery. Also, the expected marginal product of fertilizer is given by $\partial q/\partial k_t$, which is assumed to be a downward sloping function of k_t in light of the law of diminishing returns and similarly for land input a_t. (Think about the successive increments in expected output for successive increments in fertilizer at higher and higher levels of fertilizer use.)

Our intention is to derive the farmer's demand function for land, more precisely, for the use of land, and then to develop an aggregative model of the market for land use. For simplicity, we will assume that the farmer is risk neutral and therefore maximizes utility through simply maximizing expected profit Θ. This assumption likely provides useful approximations, particularly in markets where farmers have access to strategies that can be used to lessen the cost of risk, such as those analyzed in Chapter 3. We will proceed in a stepwise fashion that may seem at first blush to be awkward; the payoff will become clear, however, at a later point in the argument.

The maximization of expected profit Θ may be approached by noting that the optimal values of two variables, k_t and a_t, are to be chosen. What we propose to do is decompose the problem into two stages: first, choose k_t conditional on some value of a_t, and second choose a_t. In the first stage, we find the optimal value for k_t and the associated value of Θ for all possible values of a_t. The problem then comes down simply to choosing the optimal value for a_t, that is, the value of a_t yielding the highest Θ. Accordingly, to start with, let a_t equal some specific value; that is, let $a_t = a_{t0} > 0$. The expression for expected profit is then given by

$$\Theta_0 = \Theta_G q(a_{t0}, k_t) - J_t k_t - R_t a_{t0} - TFC \tag{4-7}$$

We write expected profit as Θ_0 to remind ourselves that $a_t = a_{t0}$. Also, h_t does not appear explicitly because we are assuming that family labor is fixed. To maximize Θ_0, we set the first derivative of Θ_0 with respect to k_t equal to zero thus:

$$\frac{\partial \Theta_0}{\partial k_t} = \Theta_G \frac{\partial q}{\partial k_t} - J_t = 0 \tag{4-8}$$

The first-order condition for a maximum states that the expected marginal revenue per unit of fertilizer, given by $\Theta_G(\partial q/\partial k_t)$, must equal the price of fertilizer per pound given by J_t. The expected marginal revenue is equal to the product of the expected gross return per unit of planned output, Θ_G, and the expected marginal product of k_t. The second-order condition for the maximization of expected profit is satisfied if the expected marginal revenue of fertilizer is a downward sloping function of k_t, which we assume is the case. It should also be noted that Eq. (4-8) gives the optimal value of k_t if expected profit cannot be increased by shutting down all operations and setting k_t equal to zero. More on this condition later.

Maximizing expected profit given $a_t = a_{t0}$ is equivalent to maximizing the difference between $\Theta_G \Theta_q$ and $J_t k_t$, that is, the difference between the expected gross returns and the total outlay on fertilizer. (The terms $R_t a_{t0}$ and, of course, TFC are to be viewed as fixed costs at this stage of the analysis.) Importantly, we now define expected *total revenue product* (TRP) as the maximized difference between the expected gross returns and the outlay on fertilizer, treating a_t as exogenous. In the preceding analysis we have found the value for TRP (i.e., TRP_0) for only one value of a_t (i.e., a_{t0}). This pair of values is plotted in Fig. 4.2.

Having found the expected total revenue product for one value of a_t, we may clearly repeat the process for as many values of a_t as we choose, always plotting the resulting points in Fig. 4.2. Connecting such points yields the curve labeled *TRP*. The result is a graphic derivation of the following important function:

$$TRP = f(a_t) \tag{4-9}$$

Notice that the *TRP* function shows the maximized value of *TRP* for all values of a_t, allowing the optimal choice of k_t for each value of a_t.

We are now in a position to move to the second stage of our expected profit maximization problem: choosing a_t. Taking advantage of Eq. (4-9), Eq. (4-6b) may be rewritten as follows:

$$\begin{aligned} \Theta &= TRP - R_t a_t - TFC \\ &= f(a_t) - R_t a_t - TFC \end{aligned} \tag{4-10}$$

Maximizing expected profit involves taking the derivative of Θ with respect to a_t according to Eq. (4-10) and setting it equal to zero, as follows:

$$\frac{\partial TRP}{\partial a_t} - R_t = 0 \tag{4-11}$$

The first-order condition for maximizing expected profit states that the expected marginal revenue product of a_t, given by $\partial TRP/\partial a_t$, equals land rent. The second-order condition for a maximum will be satisfied if $\partial TRP/\partial a_t$ is a downward sloping function of a_t. This condition holds if *TRP* increases at a decreasing rate in the relevant range, as in Fig. 4.2. If we let the linear curve *R* be the graphic counterpart to

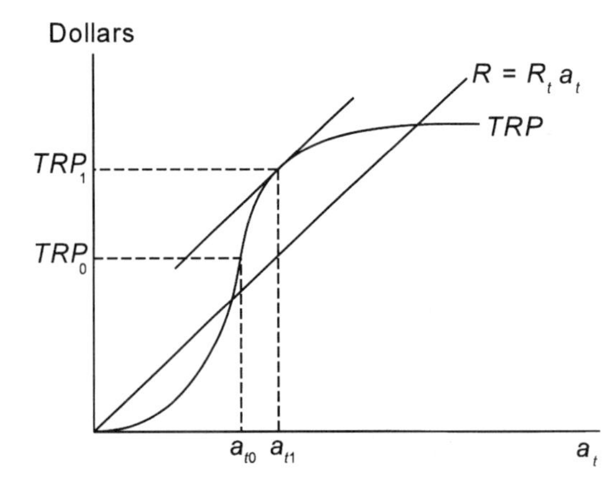

Figure 4.2 Total revenue product and total input cost curves for land input.

$R_t a_t$, then maximizing Θ comes down to finding the a_t associated with the maximum vertical difference between the *TRP* curve and the *R* curve. This occurs where the slopes of the two curves are equal, at $a_t = a_{t1}$. The maximization of Θ according to Eq. (4-10) can thus be accomplished either through calculus or graphics.

Recall from Chapter 3 that the derivation of the farmer's supply curve involves taking advantage of the farmer's marginal and average variable cost curves. We now develop the counterpart analysis of the farmer's demand curve for rented land a_t. We define the expected *average revenue product ARP* for a_t as *TRP* divided by a_t. Mathematically, we have

$$\begin{aligned} ARP &= \frac{TRP}{a_t} \\ &= \frac{f(a_t)}{a_t} \end{aligned} \tag{4-12}$$

Graphic representations of the expected average revenue and marginal revenue product functions are given by *ARP* and *MRP* curves in Fig. 4.3. (The student should keep in mind that *TRP*, *ARP*, and *MRP* are all expected values, with the expectations formed in period *t*.) We will not pause here to provide a rigorous analysis of the shapes of *ARP* and *MRP*. Suffice to say that these two curves are drawn in a manner consistent with the shape of *TRP* in Fig. 4.2. The crucial characteristics of these curves are first, for sufficiently large values of a_t, *MRP* is downward sloping and, second, *MRP* passes through *ARP* at the latter's maximum. The first condition must hold if Eq. (4-11) is to yield maximum expected profit; it is usually justified on the assumption of the law of diminishing returns to the fixed plant. The second condition can be shown to be a mathematical necessity.

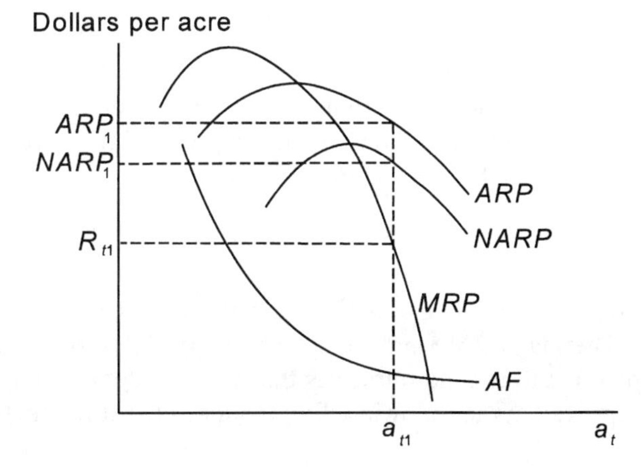

Figure 4.3 Short-run revenue product curves for land input.

It is also helpful for subsequent analysis to introduce two additional concepts. The average fixed cost per acre planted, TFC/a_t, is given as a function of a_t by the AF curve in Fig. 4.3. Subtracting AF vertically from ARP yields the expected *net average revenue product* curve for land input, labeled $NARP$. The student should note the similarity of the curves in Fig. 4.3 to the typical average cost curves used in deriving the firm's supply curve. The latter are expressed as U-shaped functions of output; the former are expressed as functions of an input, shaped like inverted U's.

Let rent per acre R_t equal the specific value R_{t1} as in Fig. 4.3. In this case the farmer maximizes expected profit by equating R_t and expected marginal revenue product MRP, setting a_t equal to a_{t1}. [See Eq. (4-11).] Because MRP is downward sloping, the second-order condition for profit maximization is met. With a_t equal to a_{t1}, $ARP = ARP_1$, which exceeds R_{t1}. What this means is that expected quasi-rent is positive. Recall that TRP is the maximized difference between expected gross returns and the outlay on fertilizer. As long as this difference exceeds the total rent paid for land, quasi-rent is positive.

We are now in a position to derive the farmer's demand curve for land input. The demand consists of that part of MRP that does not lie above ARP. This curve segment shows how much land the farmer will be willing to plant for alternative levels of land rent. If land rent exceeds the maximum of ARP, the farmer will shut down all operations, and MRP will not be equated to land rent. The student should recognize that movement along the farmer's demand curve for land does not, in this analysis, assume that fertilizer is held constant. On the contrary, fertilizer is always maintained at its optimal value and is allowed to adjust to changing market conditions.

Figure 4.3 also allows giving a graphic representation of expected profit. For $a_t = a_{t1}$, expected profit equals $(NARP_1 - R_{t1})a_{t1}$. To see this algebraically, we note that

$$\Theta_1 = \left[\frac{TRP_1}{a_{t1}} - \frac{TFC}{a_{t1}} - R_{t1}\right]a_{t1} \tag{4-13}$$

A numerical example is now considered that is important in understanding the derivation of input demand. We assume a Cobb–Douglas production function with $\alpha_0 = 1.0$, $\alpha_1 = 0.5$, $\alpha_2 = 0.4$, and $\alpha_3 = 0.1$. Letting $h_1 = 1.0$, we have

$$\Theta_q = a_t^{0.5} k_t^{0.4}$$

Furthermore, let $\Theta_G = 10$, $J_t = 4$, and $TFC = 2$. Various values for a_t may be considered, starting with $a_t = 4$. Then $\Theta_q = 2k_t^{0.4}$ and we have $\Theta_4 = 10(2k_t^{0.4}) - 4k_t - R_t 4 - 2$. (We give Θ the subscript 4 just to remind ourselves that we have set $a_t = 4$ arbitrarily.) This convention is repeated in the following. To maximize Θ_4, set the first derivative of Θ_4 with respect to k_t equal to zero, which implies that the optimal value of k_t is 3.1748. Since TRP is given by

$$TRP_4 = 20k_t^{0.4} - 4k_t$$

we have, for $a_t = 4$, $TRP_4 = 19.0488$. Also, $ARP_4 = 4.7622$. Since $NARP = ARP - (TFC/a_t)$, we have $NARP_4 = 4.2622$. All these points could be plotted in the appropriate diagrams, and we could repeat this process for many other values of a_t. The locus of such points would trace out the TRP, ARP, and $NARP$ curves. The MRP curve could be derived by estimating the slopes of TRP for various values of a_t.

An algebraic approach is not as cumbersome as this procedure. In the first stage of the analysis, we treat a_t as exogenous without assigning it a specific value. We have

$$\Theta = 10a_t^{0.5}k_t^{0.4} - 4k_t - R_t a_t - 2$$

We set the first partial derivative of Θ with respect to k_t equal to zero, which yields

$$k_t = a_t^{0.8333}$$

Notice that this result is a reduced form relationship. It shows the optimal value for a choice variable k_t for all values of an exogenous variable a_t. We use this result to get rid of k_t in the expression for Θ. We then have

$$\Theta = [10a_t^{0.5}a_t^{0.8333} - 4a_t^{0.8333}] - R_t a_t - 2$$

The bracketed term on the right-hand side of the equality is the algebraic expression for TRP. With further simplification, we have

$$TRP = 6a_t^{0.8333}$$

The MRP function is the derivative of TRP with respect to a_t. Therefore, $MRP = 4.9998a_t^{-0.1667}$. Also, $ARP = 6a_t^{-0.1667}$ and $NARP = 6a_t^{-0.1667} - 2a_t^{-1}$. The $NARP$ curve is shaped like an inverted U with a maximum at $a_t = 2.2968$. Also, ARP and MRP are downward sloping for all values of a_t. The MRP curve lies below that for ARP everywhere. In this specific example, the entire MRP curve is the farmer's demand for land input.

To continue the example, let $R_t = 4$. Since maximization of Θ implies equating R_t and MRP in the second stage of the analysis, where a_t is no longer treated as exogenous, we have $a_t = 3.8128$. This implies that $k_t = 3.0503$ and $\Theta = 1.0510$.

We now take up the question of the properties of the farmer's demand function for land input. First, the demand for land input must be downward sloping. Perhaps the easiest way to see this is to look at Fig. 4.2. If *TRP* were linear or if it increased at an increasing rate, there would be no level of a_t that yields more expected profit than any other. Profit maximization itself would become problematic. To have a unique level of land input that maximizes expected profit, the *TRP* curve must increase at a decreasing rate; this means that the farmer's demand for land must be downward sloping.

A second issue is how the demand for land input changes in response to an increase, say, in the expected gross returns per unit of expected output Θ_G. Here we must be particularly careful. Note that Θ_G might increase because the demand for food increases or because either technological change and/or a permanent change in the weather increases yield. Having said as much, all that we can safely say is that a ceteris paribus increase in Θ_G causes the *TRP* to shift upward as in Fig. 4.4, where an increase in Θ_G causes *TRP* to shift from TRP_0 to TRP_1 or to TRP_2. In most cases we might expect that, for any given level of a_t, the slope of *TRP* rises with increases in Θ_G, as happens if *TRP* shifts from TRP_0 to TRP_1. In this case, an increase in Θ_G causes the demand for a_t to increase, and we say by definition that land is a *normal input.*

It is theoretically possible, however, for *TRP* to rise with an increase in Θ_G and for its shape to change as well. The dashed curve TRP_2 in Fig. 4.4 suggests the following possibility. Although TRP_2 lies everywhere above TRP_0, the former's slope is less than the latter's for a wide range of relevant outputs. Here an increase in Θ_G causes the demand for land input to contract, and we would then refer to land, by definition, as an *inferior input.* In the remainder of this book, we will maintain the assumption that all farm inputs are normal unless explicitly noted otherwise. What this

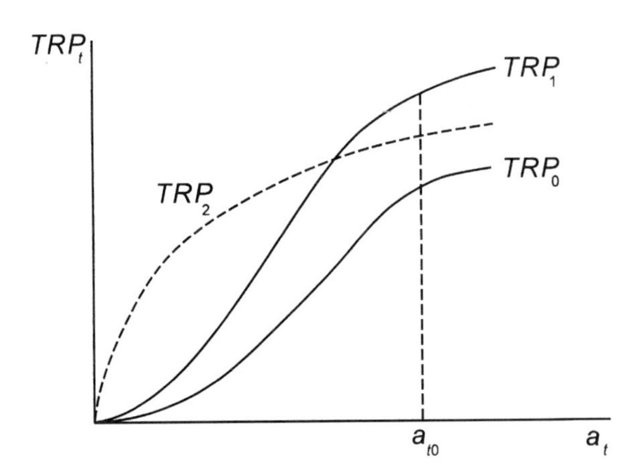

Figure 4.4 Total revenue product curves for land input under alternative technologies.

means is that we will usually assume that ceteris paribus increases in Θ_G elevate the demands for all inputs, including land input.

Finally, how does the farmer's demand for land vary with changes in the price of fertilizer? In a two-variable-input world, such as the one analyzed here, it can be shown under very general assumptions that an increase in the price of one input (fertilizer) increases the farmer's demand for the other (land) *if* the expected gross return per unit of planned output acre is held constant. The basic reason for this is that land and fertilizer are substitutes one for the other. In a typical isoquant diagram, each isoquant slopes downward at a decreasing rate. In an aggregative analysis, however, the expected gross return per unit of planned output will vary with changes in the prices of variable inputs. The analysis becomes both more complicated and interesting, as we will see later in this chapter.

The student should be aware, moreover, of the possibility of complementary inputs in a model that allows for more than two variable inputs. With many variable inputs, any two could be complements. Then a ceteris paribus increase in the price of one complement *decreases* the demand for the other. If the price of field sprayers rises, for example, the demand for insecticide falls.

Short-run Market Input Demand

Finally, it remains to be stressed that the properties of an individual farmer's demand for land input need not be the same as those of the demand of all farmers taken together. It will be shown in the next section, for example, that the effect of an increase in the price of fertilizer on the aggregate demand for land input depends in a crucial way on the elasticity of demand for output.

We now take up the problem of deriving the aggregate demand of all farmers for land input. First, hold constant for the moment the expected gross returns per acre at, say, Θ_{G0}. The expected marginal revenue product for land input for the individual farmer is given by MRP_0 in panel a, Fig. 4.5. Summing horizontally such curves for all N farmers yields the curve labeled $\Sigma\ MRP_0$ in panel b. Let land rent R_t equal R_{t0}. The individual farmer thus demands a_{t0}; together, farmers demand A_{t0}. We assume that for all farmers R_{t0} is not less than the maxima of their respective expected average revenue product curves. (Quasi-rent must not be negative.) We assume further that the planted acreage A_{t0} is consistent with expected gross returns per acre equal to Θ_{G0}. In other words, we start with a known point on the aggregate demand for land where $R_t = R_{t0}$ and $A_t = A_{t0}$. If rent falls from R_{t0} to R_{t1}, farmers in total will expand the quantity of land input demanded from A_{t0} to A_{t2} as long as Θ_G stays constant at Θ_{G0}. But will Θ_G remain constant? The answer is no, at least under free-market conditions. When rent falls, all farmers attempt to rent more land. As they succeed, expected production rises, but the expected product price and the expected gross return per unit of planned output decline. For normal inputs, declining expected gross returns per unit of planned output cause the MRP and the $\Sigma\ MRP$ curves to decline as well. This tends to choke off the increase in the quantity of land

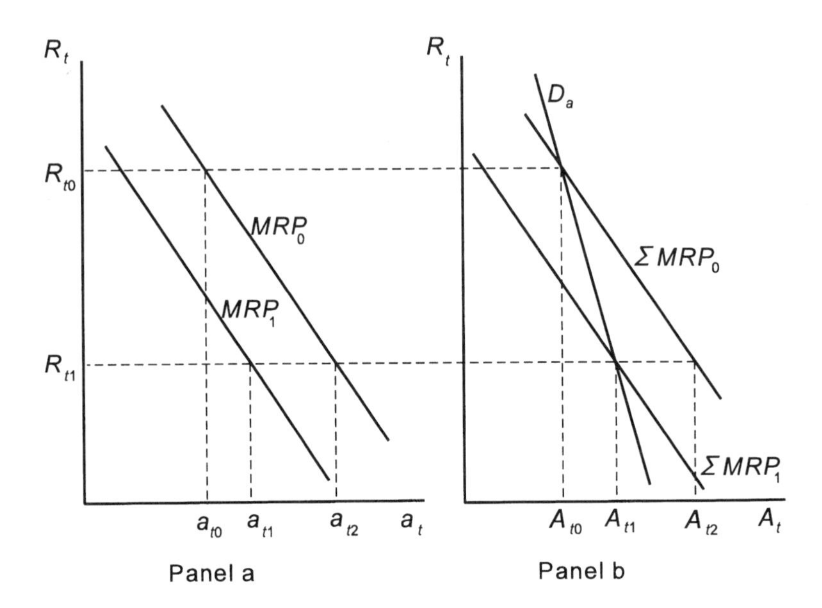

Figure 4.5 Individual and aggregate marginal revenue product curves and the aggregate short-run demand curve for land input.

input demanded. What we find is that, if rent equals R_{t1}, the equilibrium level of land input demanded equals a_{t1} for the individual farmer and A_{t1} for all farmers in the aggregate. The expected gross return Θ_G falls from Θ_{G0} to some new level Θ_{G1}, which is exactly consistent with farmers' plans to plant acreage equal to A_{t1}. In other words, we find a new point where a larger planted acreage A_{t1} (it is larger than A_{t0}, although less than A_{t2}) is optimal for and consistent with a lower level of Θ_G.

In this way we have derived two points on the aggregate demand curve for land input given by the curve labeled D_a in panel b, Fig. 4.5. Of course, many such points could be derived, which allows plotting D_a with as much precision as desired. Importantly, this demand is in the nature of a market equilibrium relationship. Movements along it hold neither the level of fertilizer application nor the expected gross return constant. The price of fertilizer is constant, but the application of it varies optimally as R_t changes and as the levels of acreage planted change. Expected gross return per unit of planned output is also allowed to vary as movements along D_a occur. To qualify as a point on D_a, the specific levels of R_t and A_t must be consistent with the corresponding level of Θ_G. The demand for land input, or for any input for that matter, must be downward sloping, unlike the consumer demand for output. (Recall from price theory that the consumer demand is upward sloping for the Giffin good, a good that is rarely encountered in the real world, its main use being to confuse freshmen.) As to the shape of the demand for an input, it can be shown in more advanced work that its elasticity (in absolute value) varies positively with (1) the elasticity of the demand for output (in absolute value), (2) the input's share of the total cost of production, and (3) the elasticity of the supply for the competing input. The shape of the demand for an input also depends on the existence of other inputs that may be more or less close substitutes (or complements) for land. (Appendix C provides a rigorous treatment of some of these issues.)

Equilibrium in the land input market is given by the intersection of the demand and supply for land input, D_a and S_a, in Fig. 4.6. Equilibrium rent and land planted equal R_{t0} and A_{t0}, respectively. Notice that equilibrium in the land market can only exist if we have simultaneous equilibria in the markets for food and fertilizer. This will be brought out more clearly in the next section.

Although we will not pause here to consider at any length the comparative static analysis of the land input market, the student should contemplate the following questions. What happens to the performance of the land input market if the supply of land expands or contracts? What happens if the demand for food expands or contracts? (The student is not advised to speculate on what will happen in the land market if the price of fertilizer rises or falls. This rather complex issue is taken up later.)

Up to this point we have been concerned with deriving a short-run demand for land input. Although we have chosen to center the analysis on land, we could just as well have derived the demand for fertilizer; the analysis is the same. Also, although we have centered on a case involving two variable inputs, the analysis can be generalized in a straightforward manner to a large number of inputs. An important point to bear in mind is that a market equilibrium demand for a variable input allows both for induced changes in expected gross return per unit of planned output and the optimal induced changes in the levels of all other variable inputs.

Long-run Input Demand

The long-run competitive demand for an input is now considered. We maintain the assumption discussed in Chapter 1 that, when a household enters farming, family labor be-

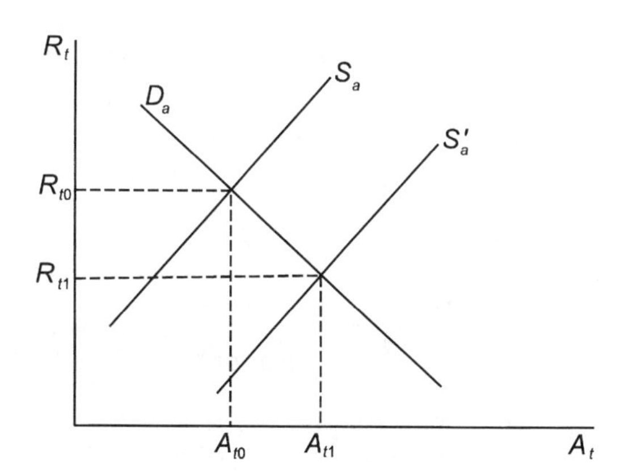

Figure 4.6 Market demand and supply curves for land input.

comes fixed in the farm sector. (Transfer earnings, defined as the minimal return to labor necessary to keep farmers in business, must be viewed as a fixed cost for any extant farmer, although the number of farmers is variable.) Let the short-run demand for land be given by D_{a0} in panel b, Fig. 4.7. The intersection of D_{a0} and the supply for land S_{a0} gives the short-run equilibrium rent R_{t0} and land planted A_{t0}. We next assume this point is also a long-run competitive equilibrium; the marginal farm earns no excess profit. This implies that the maximum of the marginal farmer's $NARP$ curve, given by $NARP_0$, in panel a, equals R_{t0}. At $R_t = R_{t0}$, there is neither an incentive for any farmer to exit the industry nor for any potential entrant to enter. In addition, the sizes of all extant farmers are optimal in that, with R_t equal to R_{t0}, they have no incentive to alter their fixed plants.

Now suppose that the supply of land input increases from S_{a0} to S_{a1} in year t (see Fig. 4.7). In the short run, R_t falls to R_{t2}, and total acreage planted expands to A_{t2}. The expansion of acreage planted will be even greater in the long run after all adjustments in fixed plants and the number of farmers have been completed. In Fig. 4.7, we suppose that in the long run acreage planted shifts from A_{t2} to, say, A'; rent rises from R_{t2} to R'. (The t subscript is dropped from long-run values because the length of time required to achieve long-run equilibrium is left open.) In the new long-run equilibrium, the maximum of the net average revenue product curve of the new marginal farmer equals R'. We now have two points on the long-run supply for land, given by the curve LRD_a in Fig. 4.7.

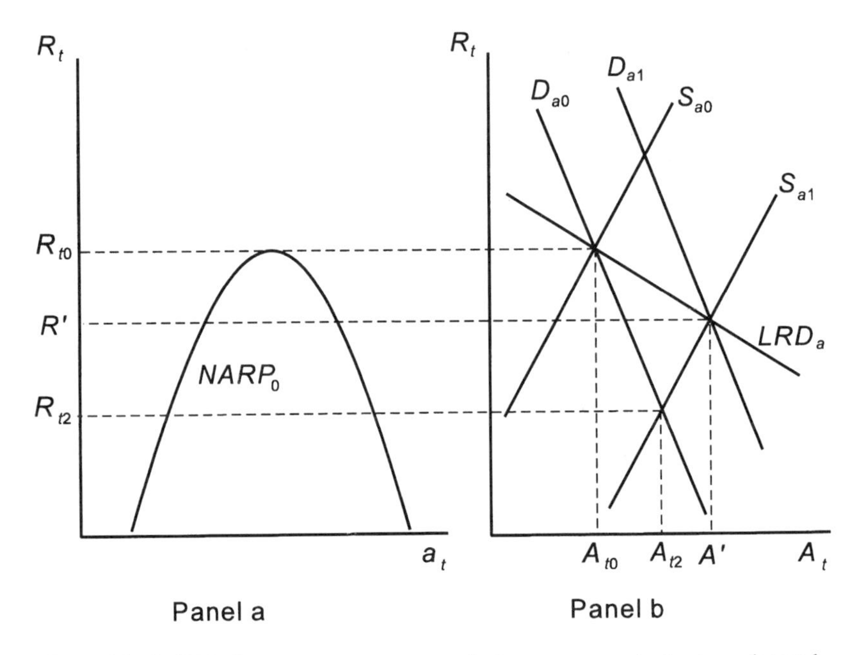

Figure 4.7 Individual net average revenue product curve, aggregate short-run demand and supply curves, and the aggregate long-run demand curve for land input.

This brings us to two important results regarding long-run analysis of input pricing. First, the competitive long-run demand for an input shows the maximum price that a competitive industry can and will pay for an input for various alternative levels of input, allowing for the induced variation in the expected gross return per unit of planned output and for the optimal variation of all other inputs, including the number of farms. Second, the demand for an input is more elastic in the long run than it is in the short run. The full justification for this assertion can be more fully appreciated after completing the next section, but the intuition behind the assertion may be set forth as follows: The expansion of the land supply encourages increased production together with the substitution of land for other inputs. Both effects tend to expand acreage planted, but in the short run both effects are constrained by a fixed number of farmers and fixed input (family labor). In the long run, the opportunities for renting more land to increase production and to displace inputs that have become relatively more costly are increased relative to the short-run opportunities. Take an example. Suppose that the demand for food is perfectly inelastic. An increase in the land supply allows substituting land for variable inputs in the short run, but not for fixed inputs such as family labor. In the long run, relatively cheap land can be substituted for all inputs, including family labor.

4.3 MODEL OF OUTPUT AND INPUT PRICING

We now analyze a model that takes more explicit account of the interdependence among output and input markets. The model will then be used to derive important hypotheses about the impacts of exogenous shocks on the farm sector. Many interesting questions are considered. For example, does technological change benefit or hurt farm labor? Would an increase in the price of fertilizer increase or decrease land rent? Would an increase in the nonfarm wage rate cause the number of farmers to decline? It is questions of this kind that the following analysis is designed to consider and in many cases answer.

Model of an Individual Farm

Let the profit function for the ith farmer be given by

$$\pi_{t+1} = P_{t+1}q_{t+1} - R_t a_t - J_t k_t - TFC \tag{4-14}$$

where, as before, π equals profit; a equals acreage farmed; k equals producer goods, such as diesel fuel, fertilizer, and hours of tractor use, purchased from the nonfarm sector; and P, R, and J equal, respectively, the price of output, land, and producer goods. We let TFC equal the transfer earnings for fixed labor supplied by the farm family, as explained in Chapter 1. The model could easily be expanded to include less aggregative input categories, as will be seen shortly.

We assume that the farmer's production function is Cobb–Douglas, as follows:

$$q_{t+1} = \alpha_0 a_t^{\alpha_1} k_t^{\alpha_2} h_{t0}^{\alpha_3}(v_{t+1}) \tag{4-15}$$

Assuming that the farmer seeks to maximize expected profit, we have

$$\begin{aligned} R_t &= \Theta_G(\alpha_1 \alpha_0 a_t^{\alpha_1 - 1} k_t^{\alpha_2} h_{t0}^{\alpha_3}) = \Theta_G MPP_a \\ J_t &= \Theta_G(\alpha_2 \alpha_0 a_t^{\alpha_1} k_t^{\alpha_2 - 1} h_{t0}^{\alpha_3}) = \Theta_G MPP_k \end{aligned} \tag{4-16}$$

as shown in Section 4.1. The term MRP_a is the marginal physical product of land input and similarly for input k.

In a moment we will make considerable use of what we will call the *imputed return to family labor* W, as follows:

$$W_{t+1} = \frac{P_{t+1}q_{t+1} - R_t a_t - J_t k_t}{h_{t0}} \tag{4-17}$$

The term W is nothing more than quasi-rent per unit of family labor. As an alternative to maximizing expected profit Θ, since TFC is a constant, we could just as well have chosen to maximize the expected quasi-rent or the expected imputed return to family labor Θ_w defined as follows:

$$\Theta_w = \frac{\Theta_G \Theta_q - R_t a_t - J_t k_t}{h_{t0}} \tag{4-18}$$

An Aggregate Model

All farmers are now assumed to have the same stochastic production function and face the same prices and price parameters. We can form the following identities:

$$Q_{t+1} = N_t q_{t+1}, \quad A_t = N_t a_t, \quad K_t = N_t k_t, \quad \text{and} \quad L_t = N_t h_{t0} \tag{4-19}$$

where N equals the number of farm operators or families, and where Q, A, K, and L equal, respectively, the aggregate quantities of output supplied, acreage, producer goods, and family labor. In the long run, the number of farms is variable. *This means that, although family labor per farm is fixed, the aggregate level of family labor is variable.*

We next derive aggregative relationships that are the counterparts to Eqs. (4-15) and (4-16). Using Eq. (4-19), Eq. (4-15) can be rewritten as

$$\frac{Q_{t+1}}{N_t} = \alpha_0 A_t^{\alpha_1} K_t^{\alpha_2} L_t^{\alpha_3}(v_{t+1})\left(\frac{1}{N_t}\right)^{\alpha_1 + \alpha_2 + \alpha_3} \tag{4-20}$$

But under constant returns to scale, the alphas sum to 1, and we can multiply both sides of Eq. (4-20) by N_t, yielding the aggregate production function:

$$Q_{t+1} = \alpha_0 A_t^{\alpha_1} K_t^{\alpha_2} L_t^{\alpha_3}(v_{t+1}) \tag{4-21}$$

The aggregative relationships corresponding to Eqs. (4-16) can be derived in similar fashion, as the reader may verify. We postpone writing out Eqs. (4-16) in aggregative form until a later point in the analysis.

It is now convenient to show that in long-run competitive equilibrium, the expected value of the marginal product of family labor equals its expected imputed return Θ_W. Under constant returns to scale, the production function is linearly homogeneous. According to Euler's theorem,[1]

$$\begin{aligned}\Theta_Q &= \frac{\partial \Theta_Q}{\partial A_t} A_t + \frac{\partial \Theta_Q}{\partial K_t} K_t + \frac{\partial \Theta_Q}{\partial L_t} L_t \\ &= \frac{R_t A_t}{\Theta_G} + \frac{J_t K_t}{\Theta_G} + \frac{\partial \Theta_Q}{\partial L_t} L_t\end{aligned} \tag{4-22}$$

(The partial derivatives of Θ_Q with respect to the inputs are the expected marginal physical products.) The second expression follows because, at the optimum, the ratio of an input's price to expected return per unit of output equals expected marginal product. Rearranging Eq. (4-22) using Eq. (4-18) yields

$$\begin{aligned}\frac{\partial \Theta_Q}{\partial L_t} \Theta_G L_t &= \Theta_G \Theta_Q - R_t A_t - J_t K_t \\ &= N_t h_{t0} \Theta_w\end{aligned} \tag{4-23}$$

Dividing both sides of Eq. (4-23) by L_t, we have

$$\Theta_w = \Theta_G \alpha_3 \alpha_0 A_t^{\alpha_1} K_t^{\alpha_2} L_t^{\alpha_3 - 1} = \Theta_G \frac{\partial \Theta_Q}{\partial L_t} \tag{4-24}$$

[1]Let $Y = f(X, Z)$. The function is said to be linearly homogeneous if $aY = f(aX, aZ)$, where a is a positive number. If, for example, X and Z are doubled, then Y would also double. Euler's theorem states that if $f(X, Z)$ is linearly homogeneous then

$$Y = \frac{\partial Y}{\partial X} X + \frac{\partial Y}{\partial Z} Z$$

For a proof, see Silberberg (1978).

What this result shows is that, even though family input is fixed to each farm, the number of farms will adjust in the long run such that the implicit expected return to family labor exactly equals its expected value of marginal product.

We proceed with the construction of our model by assuming that land is in perfectly inelastic supply; aggregate land input A_t is thus exogenous. Producer goods (capital), on the other hand, are assumed to be in perfectly elastic supply; the price of producer goods J, as opposed to their level, is exogenous. The supply function for family input is upward sloping, based on arguments advanced in Chapter 1. All farmers have rational price expectations. Expected return per unit of planned output is expressed as a function of A_t, K_t, and L_t on the basis of rational expectations. This latter proposition requires further explanation.

To simplify analysis, without sacrificing major results, we assume that demand for output is stochastic and linear:

$$P_{t+1} = \beta_0 - \beta_1 Q_{t+1} + \beta_2 Z_{t+1} + e_{t+1} \tag{4-25}$$

where Z is an exogenous shifter such as per capita income and e is random with expectation equal to zero. On the assumption of rational expectations, we know that Θ_G is given by

$$E_t\,(P_{t+1}v_{t+1}) = E_t(\beta_0\, v_{t+1} - \beta_1 Q_{t+1}v_{t+1} + \beta_2 Z_{t+1}v_{t+1} + e_{t+1}v_{t+1}) \tag{4-26}$$

Assuming that e_{t+1} and v_{t+1} are independently distributed, as are Z_{t+1} and v_{t+1}, and taking advantage of the aggregate production function, we have

$$E_t\,(P_{t+1}\, v_{t+1}) = \beta_0 + \beta_2\Theta_z - \beta_1(1 + \sigma_v^2)\,\Theta_Q \tag{4-27}$$

where σ_v^2 equals the variance of the weather variable. Clearly, the per unit return expected to prevail in period $t + 1$ depends on the levels of input applied in the production process in period t. Equation (4-27) will be referred to as the expected product return function. It is *not* the same as expected demand. More particularly, the expected product return falls more rapidly with increases in expected output than does the expected price.

Our aggregative model may now be given as follows:

$\Theta_Q = \alpha_0 A_t^{\alpha_1} K_t^{\alpha_2} L_t^{\alpha_3}$	expected aggregate production function
$R_t = \Theta_G \alpha_1 \alpha_0 A_t^{\alpha_1 - 1} K_t^{\alpha_2} L_t^{\alpha_3}$	rent equals land's expected value marginal product
$J_t = \Theta_G \alpha_2 \alpha_0 A_t^{\alpha_1} K_t^{\alpha_2 - 1} L_t^{\alpha_3}$	price of producer goods equals its expected value of marginal product

$$\Theta_W = \Theta_G \alpha_3 \alpha_0 A_t^{\alpha_1} K_t^{\alpha_2} L_t^{\alpha_3 - 1}$$ expected return to family labor equals its expected value of marginal product

$$\Theta_w = f(L_t, Y_t)$$ family labor supply function

--- (4-28)

$$\Theta_G = E_t(P_{t+1} v_{t+1})$$ rational expectation hypothesis

$$E_t\,(P_{t+1} v_{t+1}) = \beta_0 + \beta_2 \Theta_z - \beta_1(1 + \sigma_v^2)\Theta_Q$$ expected product return function

$$Q_{t+1} = \alpha_0 A_t^{\alpha_1} K_t^{\alpha_2} L_t^{\alpha_3}(v_{t+1})$$ aggregate production function

$$P_{t+1} = \beta_0 - \beta_1 Q_{t+1} + \beta_2 Z_{t+1} + e_{t+1}$$ aggregate farm level demand

The model given by Eqs. (4-28) offers a tentative explanation of how the performance of the farm industry is determined. The first equation is an expected aggregate production function showing how the expected output varies with land, producer goods, and farm family labor. The second and third equations indicate that in market equilibrium, and on the assumption that farmers seek to maximize expected profit, land and producer goods are employed up to the point when the expected values of their marginal products equal their prices. The fourth and fifth equations are crucial in understanding how the aggregate level of family labor is determined. Although we have assumed that family labor per farm is fixed, in a long-run formulation the number of farms N_t is variable. At issue, then, is how the number of farms gets determined. The fourth equation shows that the expected imputed return to family labor Θ_w is equated, through entry and exit of farmers, to the expected value of marginal product of family labor. The fifth equation is the family labor supply function, which we assume, based on arguments given in Chapter 1, is upward sloping. Mathematically, $\partial\Theta_w/\partial L_t > 0$. The variable Y_t is a shifter of the labor supply function such as the nonfarm wage rate. As explained in Chapter 1, to induce households to organize additional farms, bringing their labor into the farm sector, it may be necessary to assure them of higher expected returns to their labor. That is, in choosing to enter (or exit) the farm sector, households look at the expected returns to their labor inputs. Expected returns must rise in order to induce the entry of additional family labor given an initial market equilibrium. For simplicity, we are assuming that both entry and exit of farmers occurs readily, without the lags discussed in Chapter 2. The reason for this simplification is that our interest is centered here on the interdependence among output and input markets in long-run equilibrium. The time paths of adjustment are of secondary importance.

Analyzing the Aggregate Model

The block of equations given above the dashed line of Eqs. (4-28) consists of five equations and six endogenous variables: Θ_Q, K, L, R, Θ_G, and Θ_W. The solutions or reduced forms for five of these variables can be found, provided we treat Θ_G as exogenous. The task is doable, but not recommended; the algebra becomes quite tedious. It is instructive, nevertheless, to consider the reduced form for Θ_Q, still centering on the block above the dashed line, using general notation as follows:

$$\Theta_Q = S_\Theta\,(\Theta_G, A_t, J_t, Y_t) \tag{4-29}$$

Equation (4-29) may be referred to as the expected supply function, with A_t, J_t, and Y_t viewed as shifters. A graphic representation of Eq. (4-29) is given by the curve S_Θ in Fig. 4.8, holding the exogenous shifters constant. Higher expected product returns Θ_G prompt increases in expected output Θ_Q. An increase in A_t causes expected supply to increase; increases in J_t, on the other hand, tend to contract expected supply. An increase in Y_t due, say, to a fall in the nonfarm wage rate increases the supply of farm operators; the effect is to expand expected supply for output. Recall that Θ_G equals the expected per unit return to expected farm output. A graphic representation of Eq. (4-27), the expected product return function, is given by the f_1 curve in Fig. 4.8. The intersection of S_Θ and f_1 gives the competitive equilibrium levels Θ_{Qc} and Θ_{Gc}. The student should consider how changes in A_t, J_t, Y_t, Θ_z, and σ_v^2 affect the competitive equilibrium values for Θ_G and Θ_Q.

Given planned output, we note that the weather determines actual output. Suppose that actual output exceeds expected output, as in Fig. 4.8. The intersection of actual demand D and actual supply S in period $t + 1$ determines market price; $P_{t+1} = P_{t+1,c}$. (Demand could, of course, lie to the left of f_1.) Again, comparative sta-

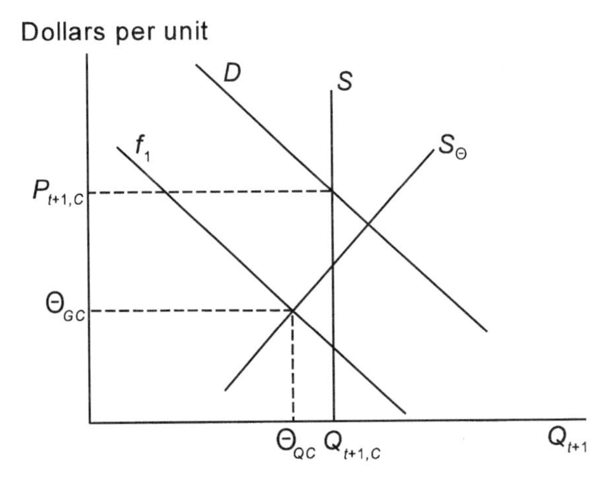

Figure 4.8 Aggregate expected gross return per unit, expected supply curve, and aggregate actual demand and supply curves.

tistics can be used to derive numerous hypotheses as to the price and quantity effects of changes in exogenous variables. More on this later.

The graphic analysis of output determination and pricing based on Fig. 4.8 displaces the usual output demand and supply analysis because of our assumptions with regard to uncertainty and production lags. The various functional relationships given in Fig. 4.8 are all either implicit or explicit in Eqs. (4-28). We develop the model given by Eqs. (4-28) and then manipulate it in a certain way to see how equilibrium in the product market is obtained. Importantly, we can do the same for any one of the input markets. To see this, we once again examine the market for land.

For this purpose, we simply set aside and ignore the last two equations of Eqs. (4-28). To simplify notation, we use the fifth equation to get rid of $E_t(P_{t+1}v_{t+1})$ in the sixth equation. The remaining block consists of six equations and the six endogenous variables Θ_Q, K_t, L_t, R_t, Θ_G, and Θ_W. Let the reduced form for R_t be given by

$$R_t = D_a[A_t, J_t, Z_t, \overline{\alpha}, E_t(Z_{t+1}), \sigma_v^2] \tag{4-30}$$

where $\overline{\alpha}$ is a vector given by α_i, $i = 0, 1, 2, 3$ (the parameters of the expected production function) and where $E_t(Z_{t+1}) = \Theta_z$. The demand for land input given by LRD_a in panel b, Fig. 4.7, is nothing more than a graphic representation of Eq. (4-30) with the proviso that all exogenous variables be held constant. If the supply function for land were upward sloping, we would then need to add it to Eqs. (4-28); A_t then could be treated as an endogenous variable. In the present version of the model, however, the supply for land is perfectly inelastic.

Models such as that given by Eqs. (4-28) are useful for various purposes, but we are interested mainly in the development of hypotheses about the effects of exogenous changes on the performance of the farm industry, recognizing the interdependence that exists among output and farm input markets. How is this to be done? The typical way to proceed in analyses of this sort is to solve algebraically the eight-equation system to find solutions for the eight endogenous variables. We could then partially differentiate the reduced form system and strive to deduce the signs of the partial derivatives using restrictions on the structural parameters. (An example of a restriction is that the sum of α_1, α_2, and α_3 equals 1.) This is the high road to take in the derivation of hypotheses, but it is a hard road to follow. Solving an eight-equation system simultaneously is not easy. Fortunately, thanks to the assumption of an expected Cobb–Douglas production function, there is a much simpler but still powerful method of analyzing the model. The method can be easily applied, moreover, to large equation systems for which the reduced form models could simply not be found.

We show first that the exponents of the inputs in the aggregate expected production function are production elasticities. Consider, for example, α_1:

$$\frac{\partial \Theta_Q}{\partial A_t}\frac{A_t}{\Theta_Q} = \frac{\alpha_1 \alpha_0 A_t^{\alpha_1 - 1} K_t^{\alpha_2} L_t^{\alpha_3}}{\alpha_0 A_t^{\alpha_1} K_t^{\alpha_2} L_t^{\alpha_3}} A_t = \alpha_1 \tag{4-31}$$

In words, Eq. (4-31) states that the percentage of change in output divided by the percentage of change in land equals α_1. We also know, however, from the equality between land rent and land's expected value marginal product, that $\partial\Theta_Q/\partial A_t = R_t/\Theta_G$. Therefore,

$$\alpha_1 = \frac{R_t A_t}{\Theta_G \Theta_Q} \tag{4-32}$$

In words, under a Cobb–Douglas technology, the production elasticity for land equals its share of expected total receipts. Similarly, it can be shown that

$$\alpha_2 = \frac{J_t K_t}{\Theta_G \Theta_Q} \quad \text{and} \quad \alpha_3 = \frac{\Theta_W L_t}{\Theta_G \Theta_Q} \tag{4-33}$$

The usefulness of Eqs. (4-32) and (4-33) will become apparent in the discussion that follows.

For the purpose of numerical example, we analyze a short-run version of the model given by Eqs. (4-28). We assume that the units of measurement are such that $A_t = L_t = \alpha_0 = 1$ and that $\Theta_Q = K_t^{0.5}$. Also, $Q_{t+1} = K^{0.5}(v_{t+1})$. On these assumptions, $d\Theta_Q/dK_t = 0.5K_t^{-0.5}$. Therefore, $J_t = 0.5K_t^{-0.5}\Theta_G$. Since $K_t = \Theta_Q^2$, we have $\Theta_Q = 0.5\Theta_G J_t^{-1}$. If $J_t = 10$, then $\Theta_Q = 0.05\ \Theta_G$, which is the industry's short-run expected supply function. (How would an increase in J_t affect short-run supply?) Suppose further that demand is given by $P_{t+1} = 10 - Q_{t+1}$, being nonstochastic, and that $v_{t+1} = 1.5$ with probability equal to 0.5 and $v_{t+1} = 0.5$ with probability equal to 0.5. The variance of v_{t+1} equals 0.25. Therefore, $\Theta_G = 10 - 1.25\ \Theta_Q$. Since we also know that $\Theta_Q = 0.05\ \Theta_G$ from above, it follows that $\Theta_G = 9.4118$ and $\Theta_Q = 0.4706$. Therefore, $K_t = 0.2215$. Half of the time $Q = 0.7059$ and $P = 9.2941$. Half of the time $Q = 0.2353$ and $P = 9.7647$. Notice that the expected price, 9.5294, is more than Θ_G. (Why is that?) We next notice that the product of J_t and K_t, which is 2.2145, equals approximately 0.5 times the product of Θ_G and Θ_Q. Expected quasi-rent equals $\Theta_G\Theta_Q$ minus the outlay on K, which amounts to 2.21. In years of high yield, actual quasi-rent equals 4.35; in years of low yield, actual quasi-rent equals 0.08. It may be noted that, in the relevant range, demand is highly elastic.

Returning to Eqs. (4-28), we now consider in some detail the effects of changes in exogenous variables on all endogenous variables. A useful first step in such analysis is to inquire how the exogenous shock in question affects the expected supply function and the expected product returns function (see the f_1 and S_Θ curves in Fig. 4.8). It should also be remembered that an input demand function is always downward sloping and that the share of expected total product returns going to an input equals that input's production elasticity. We will consider several illustrative cases.

First to be considered are the effects of an increase in J_t, the exogenous price of producer goods. An increase in the price of an input is unfavorable to production, since it drives production costs upward. If J_t rises, the S_Θ curve in Fig. 4.8 contracts to the left, but the expected product returns curve f_1 is unaffected. We can therefore be certain that the expected product return Θ_G rises and expected output Θ_Q falls. Also, since the demand for K_t, implicit in Eqs. (4-28), must be downward sloping, an increase in J_t causes K_t to decline.

The question of the effect of an increase in J_t on R_t, L_t, and Θ_W is more challenging. Here the factor share equations come into play. We know that $R_tA_t = \alpha_1(\Theta_G\Theta_Q)$, and we have seen that Θ_G rises and Θ_Q falls. The product Θ_G times Θ_Q rises, falls, or stays the same depending on whether the expected product returns curve, the f_1 curve, is inelastic, elastic, or unitarily elastic in the relevant range of output (see Fig. 4.8). Here we see clearly how the value of a structural parameter can condition in a very important way the effects of exogenous shocks. If the f_1 curve is inelastic in the relevant range, a ceteris paribus increase in J_t causes the product $\Theta_G\Theta_Q$ to rise. Since α_1 is constant, R_tA_t must also rise. This means that, with a given supply curve for land assumed to be perfectly inelastic in the present model, the market equilibrium demand for land must rise. An increase in the exogenous price of one input causes the demand for another substitutive input to increase if demand is inelastic. If the f_1 curve is elastic, on the other hand, an increase in J_t causes $\Theta_G\Theta_Q$ to fall. It follows that R_tA_t falls as well. Here an increase in the exogenous price of an input decreases the demand for another input.

Our analysis applies to the market for family labor as well. We note that $\Theta_WL_t = \alpha_3(\Theta_G\Theta_Q)$. If the f_1 curve is inelastic, an increase in J_t causes $\Theta_G\Theta_Q$ to rise; Θ_WL_t must rise as well. An increase in J_t with an inelastic f_1 curve causes the demand for farm family labor to increase. An increase in the price of fertilizer (or pesticides) might increase both the expected returns to and the quantity of farm family labor. If, on the other hand, the f_1 curve is elastic in the relevant range, $\Theta_G\Theta_Q$ falls, as does the product Θ_WL_t; the demand for family labor declines. This completes our analysis of a change (increase) in J_t. A summary of the results will be given later.

We now turn to an exogenous shock to the system that enters through the market for farm family labor. More particularly, suppose that the nonfarm wage rate increases. In Fig. 4.9 the supply curve for family labor contracts from S_L to S'_L. Why is this? An increase in the nonfarm wage rate makes farming a relatively less attractive career choice. To induce households to offer the same quantity of family labor to farming as before requires a higher expected return Θ_W. We see immediately that L_t falls and Θ_W rises.

Decreasing the supply function for any farm input causes the S_Θ curve in Fig. 4.8 to contract as well. Hence an increase in the nonfarm wage rate causes Θ_G to rise and Θ_Q to fall. The effects on K_t and R_t can be analyzed using the factor share equations, as in the case of our analysis of changes in J_t.

As a further illustration of the usefulness of the model given by Eqs. (4-28), we next take up the important question of technological change. Few topics have elicited as much analysis, speculation, and controversy as has the effects of technological change on agriculture. The reason for this is that technological change has been a driving force in agriculture worldwide.

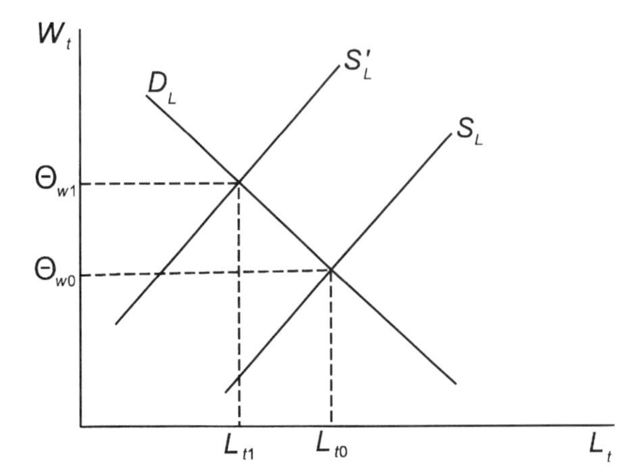

Figure 4.9 Aggregate demand and supply curves for farm family labor.

By definition, a technological change alters the production function in such a manner as to lower the minima of a farmer's average cost curves. Two types of technological change may be distinguished. The first, neutral change, can be explained with the help of panel a, Fig. 4.10. Neutral technological change shifts a representative isoquant inward from the Θ_{Q0} curve to the Θ'_{Q0} curve, where $\Theta_{Q0} = \Theta'_{Q0}$. The same level of planned output can be produced with smaller input bundles after the change and similarly for all other isoquants. With constant input prices, the new ex-

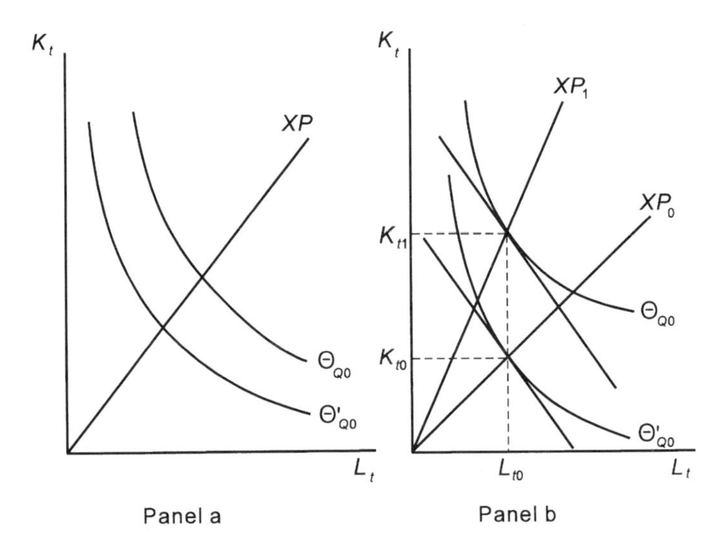

Figure 4.10 Isoquant and expansion path curves under neutral technological change and under technological change biased against labor.

pansion path coincides with the old one. This means that the ratio of the quantity of one variable input to the quantity of another will not change. The definition generalizes to the case of more than two inputs, and the crucial element is that, with constant input prices, *a neutral technological change leaves all input ratios unaffected.* Equivalently, we may say that a neutral technological change leaves the ratio of the marginal physical productivities of any two inputs the same as before.

We define technological change to be a biased change if the isoquants shift inward in such a manner that, with constant input prices, the new and the old expansion paths do not coincide. *A biased technological change alters the ratio of at least two variable inputs.* In panel b, Fig. 4.10, a technological change shifts a representative isoquant inward from the Θ_{Q0} curve to the Θ'_{Q0} curve where $\Theta_{Q0} = \Theta'_{Q0}$. Here the expansion path shifts from XP_1 to XP_0. Movements along XP_0, with constant input prices, are associated with the input ratio of K_t to L_t equal, always, to K_{t0}/L_{t0}. Movements along XP_1 are associated with the ratio of K_t to L_t equal to K_{t1}/L_{t0}. Since $(K_{t1}/L_{t0}) > (K_{t0}/L_{t0})$, we say that the technological change is *biased in favor of farm family labor and against producer goods.* A biased technological change alters the ratio of the marginal productivities of at least one pair of inputs.

Returning to our farm sector model, we note that an increase in the parameter α_0 is an example of a neutral technological change. The reason for this is that the ratio of marginal productivities of any two inputs is unaffected by changes in α_0. An approximate real-world example of a neutral technological change might be the introduction of hybrid seed, a change that increases yield with small effects on the organization of crop production activities such as plowing, seeding, and spraying.

Neutral technological change shifts the expected supply (the S_Θ curve in Fig. 4.8) to the right, lowering Θ_G and increasing Θ_Q. Since $\Theta_w L_t = \alpha_3 (\Theta_G \Theta_Q)$, the impact on expected returns to family input depends on the elasticity of the f_1 curve. An inelastic f_1 curve means that the product $\Theta_G \Theta_Q$ falls and both L_t and Θ_w fall as well. This doubtless explains in part why farmers have become increasingly sensitive to the introduction of new technologies. Of course, if the f_1 curve is elastic, then the product $\Theta_G \Theta_Q$ rises and Θ_w and L_t rise as well.

Assessing the effects of biased technological change is relatively difficult. Given a Cobb–Douglas production function, biased technological change occurs if the production elasticities (α_1, α_2, α_3) change. There are many examples of biased technological change in agriculture, including the introduction of tractors, combines, and milking machines. Since we have assumed constant returns to scale ($\alpha_1 + \alpha_2 + \alpha_3 = 1$), an increase in α_1, say, must be associated with a decrease in α_2 or α_3 or both. For the purpose of analysis, we assume a change that is biased against family labor in favor of producer goods. In other words, assume that α_3 falls by the same amount that α_2 rises. By definition, technological change shifts the expected supply to the right, causing Θ_G to fall and Θ_Q to increase. We note once again that $\Theta_w L_t = \alpha_3 (\Theta_G \Theta_Q)$.

If the f_1 curve is inelastic, $\Theta_G \Theta_Q$ falls as α_3 falls and as α_2 rises; clearly, the product $\Theta_w L_t$ falls as well. If the f_1 curve is elastic, on the other hand, $\Theta_G \Theta_Q$ rises as α_3 falls, and it is not possible to predict how the product $\Theta_w L_t$ will respond. It may rise, fall, or stay the same. The effect on producer goods is very different from that for family labor. We know that $J_t K_t = \alpha_2(\Theta_G \Theta_Q)$. An inelastic f_1 curve implies that $\Theta_G \Theta_Q$ falls as α_2 rises; it is not possible to predict safely whether K_t will rise, fall, or stay the same. Only if the f_1 curve is not inelastic can we predict that a technological change biased in favor of producer goods will actually increase the input of producer goods.

What about changes in the probability distribution for weather? Since such changes occur slowly over long periods of time (think of such things as the ice age), we will not give them much attention here. Of greater importance is the possibility of new crop varieties or production practices that, holding yield constant, increase crop resistance to variation in climate. A new variety that is drought resistant is an example. Note that as σ_v^2 tends toward zero, the present model tends toward a more traditional comparative static model. With regard to the expected product returns function, Eq. 4-27, we note that as the variance of v_{t+1}, σ_v^2, tends toward zero the product returns function tends toward the expected demand function.

Table 4.1 summarizes the main hypotheses that can be derived from the structural model given by Eqs. (4-28). The endogenous farm sector variables are given in a row along the top of the table. Exogenous shocks are given in a column along the

TABLE 4.1 Qualitative Effects of Exogenous Shocks on Endogenous Farm Sector Variables

	Effects on Endogenous Variables					
Exogenous Shock	Θ_G	Θ_Q	Θ_w	L_t	K_t	R_t
Expected demand expands[a]	↑	↑	↑	↑	↑	↑
Nonfarm wage rate rises[b]	↑	↓	↑	↓		
Inelastic f_1 function[c]					↑	↑
Elastic f_1 function					↓	↓
Producer goods price falls	↓	↑			↑	
Inelastic f_1 function			↓	↓		↓
Elastic f_1 function			↑	↑		↑
Neutral technological change[d]	↓	↑				
Inelastic f_1 function			↓	↓	↓	↓
Elastic f_1 function			↑	↑	↑	↑
Biased technological change[e]	↓	↑				
Inelastic f_1 function			↓	↓	?	↓
Elastic f_1 function			?	?	↑	↑

[a]Assume that Θ_z rises [see Eq. (4-25)].

[b]Assume that Y_t rises [see fifth equation, Eqs. (4-28)].

[c]The f_1 function gives the expected gross return per unit of planned output for alternative levels of planned output [see Eq. (4-27)].

[d]Assume that α_0 rises.

[e]Assume that α_3 falls and α_2 rises. The change is biased against labor.

left-hand margin. Arrows indicate whether a specified exogenous shock increases or decreases an endogenous variable. Question marks indicate uncertain results. The student will quickly observe that both upward and downward arrows appear in many columns. If exogenous shocks of many kinds buffet the farm sector more or less continually through time, we will have no way of knowing on the basis of theoretical analysis why systematic changes occur in farm sector variables. For example, we know that the input of family labor has been declining for many years. All Table 4.1 tells us is that many factors might explain this important phenomenon. What is required is empirical research that measures the quantitative magnitudes of the exogenous shocks together with their economic effects. Only in this way can we make meaningful statements about the likely causes of the decline in farm labor.

4.4 A NOTE ON INPUT PRICING IN THE PRESENCE OF RISK

The preceding analysis of the interdependent pricing of output and inputs was based on the assumption that risk-neutral farmers strived to maximize expected profit. The objective in what follows is to show briefly the implications of risk premiums for input demands under risk aversion. Assuming that the production function is given by Eq. (4-15), the cost of risk lessens the demands for inputs much as it contracts the supply for output.

$$\begin{aligned} \Theta &= \Theta_G \Theta_q - R_t a_t - J_t k_t - TFC \\ \sigma^2 &= \Theta_q^2 \sigma_G^2 \end{aligned} \tag{4-34}$$

where Θ_G and σ_G^2 are, respectively, the expected value and variance of the expression $(P_{t+1} v_{t+1})$ and Θ_q is expected output, given by a Cobb–Douglas production function under multiplicative production uncertainty. Although this specification of production risk is somewhat restrictive, it will be convenient for the analysis presented here. (More general forms of production uncertainty are considered in advanced work.) Following the development given in Chapter 2, let the utility function be

$$u_t = \phi_0 + \phi_1 \Theta - \phi_2 \sigma^2$$

Using the definition of risk premium given by

$$u(\Theta, \sigma) = u(\Theta - X, 0)$$

together with the above utility function, it can be shown that the risk premium X equals $[(\phi_2/\phi_1)\Theta_q^2 \sigma_G^2]$. It measures the implicit cost of private risk bearing for a risk-averse farmer ($\phi_2 > 0$).

The maximization of utility with respect to a_t and k_t yields

$$\Theta_G \frac{\partial \Theta_q}{\partial a_t} = R_t + \frac{2\alpha_1 X}{a_t} \tag{4-35}$$

$$\Theta_G \frac{\partial \Theta_q}{\partial k_t} = J_t + \frac{2\alpha_2 X}{k_t} \tag{4-36}$$

The far right term of Eq. (4-35) is the marginal risk premium associated with renting acreage ($2\alpha_1 X/a_t = \partial X/\partial a_t$) (this term is not the marginal risk premium of planned output) and likewise for producer goods ($2\alpha_2 X/k_t = \partial X/\partial k_t$). Importantly, the expected marginal revenue product of land is no longer equated to rent, but is instead equated to the sum of land rent and the marginal risk premium of acreage rented. A graphic representation of the expected marginal revenue product function, holding k_t and Θ_G constant and assuming a zero risk premium, is given by d' in Fig. 4.11. If rent equals R_{t2}, then the risk-neutral farmer rents a_{t2}. The risk-averse farmer rents a_{t1} (less than a_{t2}) such that the expected marginal revenue product exceeds rent by the marginal risk premium given by ($2\alpha_1 X/a_{t1}$). The risk-averse farmer's demand for land input given by d'' in Fig. 4.11, holding both k_t and Θ_G constant, lies to the left of the expected marginal value product curve. In long-run competitive equilibrium, the expected marginal revenue product of land equals the sum of land rent and the marginal risk premium of land. It must also be true that land rent equals the expected net average revenue product corrected for the average risk premium per acre, that is, the total risk premium divided by the equilibrium level of land. A similar analysis applies to producer goods.

As we saw in Chapter 2, risk aversion increases the cost of farming operations and reduces the level of expected output relative to the case of risk neutrality. The main conclusion of this section is that risk aversion can also contract the demands

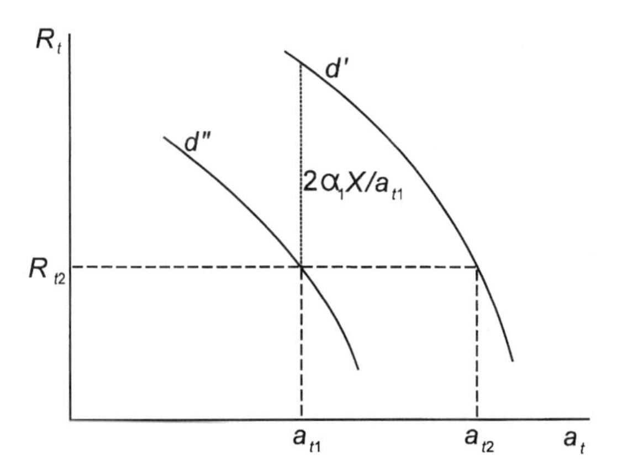

Figure 4.11 Individual demands for land input under risk neutrality and risk aversion.

for farm inputs. We have shown that the equilibrium quantities of farm inputs (or rent in the case of exogenous land) are less than in the case of risk neutrality. It should be noted, however, that a more general analysis would not assume the production function given by Eq. (4-15) and would allow for inputs that are peculiarly well suited for diminishing the undesirable effects of bad weather. Uncertain weather might actually increase the demand for such inputs, at the same time contracting the supply for output.

4.5 RESOURCE PRICING: THE CASE OF FARMLAND

The previous analysis has centered almost exclusively on the pricing and allocation of inputs and outputs at a particular time period. In this section we explore the theory of resource pricing using land as an important illustrative example. The first and perhaps most important stage of such an analysis has already been completed in that we have in hand a theory that purports to explain how land rent gets determined. We refer specifically to the aggregative model given by Eqs. (4-28) as analyzed in Section 4.3, where rent, expected output, farm inputs, and other variables were determined simultaneously. In principle, at least, the structural Eqs. (4-28) could be solved in order to obtain the reduced forms for all the endogenous variables, including that for land rent, as follows:

$$R_t = D_a[A_t, J_t, Z_t, \overline{\alpha}, E_t(Z_{t+1}), \sigma_v^2] \tag{4-37}$$

[This equation is the market equilibrium demand for land input given earlier by Eq. (4-30).] The intersection of the demand and supply curves for land input determines current rent. What we will find in the analysis that follows is that the expected land rent for future periods together with other variables determines the price of land itself. It is useful for what follows to note that under certainty $\sigma_v^2 = 0$, and the value for Z_{t+1} would be known in year t.

Investment under Certainty

Consider an investor with an initial holding of money or wealth, M_0, who is considering alternative investment options in a world of certainty. (Uncertainty is considered later.) The investor must make decisions at the end of year 0, and the investment returns are realized at the end of year 1. The objective is to maximize wealth at the end of year 1. The analysis that follows could be generalized to cover many years or periods but a 1-year analysis illustrates the basic principles. Also, we abstract from complications caused by consumption decisions. Perhaps the investor lives with a family that takes care of all of his or her consumption needs.

One option open to the investor is simply to buy a 1-year government bond (or deposit money in a savings account) where the annual interest, given by i, is guaranteed. If B_0 equals the amount of money thus invested, $B_0 \le M_0$, then at the end of the year the return on the investment equals iB_0.

Alternatively, the investor could buy farm land $\bar{a}_0$ at the price per acre P_{a0}, collecting rent R_1 for the year and selling the land at the price P_{a1} at year's end. Of course, $P_{a0}\bar{a}_0 \le M_0$. (We place a bar above a_0 to remind the reader that land is purchased, not rented.) The total return from this investment consists of $R_1\bar{a}_0$ (rent) plus a capital gain (or loss) equal to $(P_{a1} - P_{a0})\bar{a}_0$. We let $P_{a1} = P_{a0}$, thus ignoring capital gains (losses). The rate of return, that is, the return to the investment in land per dollar invested, is given by i_a, where

$$\begin{aligned} i_a &= \frac{R_1\bar{a}_0}{P_{a0}\bar{a}_0} \\ &= \frac{R_1}{P_{a0}} \end{aligned} \tag{4-38}$$

Notice that both R_1 and P_{a0} are stated in terms of money per acre of land. Hence the ratio R_1/P_{a0} is a pure number; it is the interest rate received from land investment, i_a.

We next assume the existence of a great number of investors like the one described, all surveying the rates of return from alternative options. Under conditions of certainty the land price will be in equilibrium only if $i = i_a$. Why is this? Consider some P_{a0} such that $i > i_a$. Then investors will strive to sell land, intending to buy governmental bonds or to put the proceeds in the bank. This, of course, drives the land price down, which makes land investment more attractive. Alternatively, suppose that $i < i_a$. Then investors will rush to buy land, which causes the land price to rise. How an individual investor allocates her or his money between the bank and land ownership is a matter of complete indifference in equilibrium. The fact is that, under certainty, all investments yield exactly the same rate of return if only those rates are properly measured. Since $i = i_a$ in equilibrium, we have[2]

[2]Capital gains and losses can be easily incorporated in the analysis. If we treat P_{a1} as an exogenous variable, then

$$i_a = \frac{R_1\bar{a}_0 + (P_{a1} - P_{a0})\bar{a}_0}{P_{a0}\bar{a}_0}$$

and $i_a = i$ implies that

$$P_{a0} = \frac{1}{1+i}(R_1 + P_{a1})$$

$$P_{a0} = \frac{R_1}{i} \tag{4-39}$$

Taking advantage of Eq. (4-37) appropriately modified for the case at hand (certainty), we have

$$P_{a0} = \frac{1}{i} D_a\left(A_0, J_0, Z_0, \overline{\alpha}, Z_1\right) \tag{4-40}$$

This is the demand function for land to own. The price of land varies directly with land rent. Thus the analysis summarized in Table 4.1 applies immediately to the analysis of land prices. For example, P_{a0} falls with increases in A_0. If product demand is inelastic, land rent falls with neutral technological change, as does the price of land. Note also that P_{a0} falls as the interest rate i rises.

Investment under Uncertainty

We now take up the case of uncertainty. As before, B_0 and $P_{a0}\overline{a}$ equal investment in bonds and farm land ownership, respectively. We assume that

$$M_0 = B_0 + P_{a0}\overline{a}_0 \tag{4-41}$$

The return from the investment in bonds is iB_0. The return from land investment is

$$\begin{aligned} H_1' &= \left[R_1 + \left(P_{a1} - P_{a0}\right)\right]\overline{a}_0 \\ &= H_1\overline{a}_0 \end{aligned} \tag{4-42}$$

where $H_1 = R_1 + P_{a1} - P_{a0}$. We now interpret H_1 as the random return per acre of land purchased. It is random because the price of land 1 year hence cannot be known with certainty. Although R_1 is known at the end of year 0 (or at the beginning of year 1), future rents are random variables, which explains why P_{a1} is random as well.

It is assumed that both the expectation of H_1, $E_0(H_1) = \Theta_H$, and the variance of H_1, $V_0(H_1) = \sigma_H^2$, are determined exogenously. There may be as many subjective estimates of these two variables as there are investors, and the student may recall our earlier discussion of the many and varied ways economic agents might proceed to estimate the expected values for exogenous variables. It must be admitted that this is not a wholly satisfactory theory, but given the present state of development of the literature, it is not clear how else the modeler can proceed. Although the present analysis leaves the determination of Θ_H and σ_H up in the air, we do assume that the estimates of investors respond in sensible ways to changes in the interest rate and in the exogenous variables that affect rent according to Eq. (4-37). For example, we assume that investor estimates of Θ_H rise with all exogenous changes in the farm sector that elevate land rent.

Let M_1 equal the investor's wealth at the end of year 1. Then

$$M_1 = (1 + i)B_0 + H_1\bar{a}_0 + P_{a0}\bar{a}_0 \tag{4-43}$$

Eliminating B_0 using the constraint given by Eq. (4-41), we have

$$M_1 = (1 + i)M_0 + (H_1 - iP_{a0})\bar{a}_0 \tag{4-44}$$

where $\bar{a}_0$ is a choice variable. Letting $E_0(M_1) = \Theta_M$ and $V_0(M_1) = \sigma_M^2$, we have

$$\Theta_M = (1 + i)M_0 + (\Theta_H - iP_{a0})\bar{a}_0 \tag{4-45}$$

$$\sigma_M^2 = \bar{a}_0^2\sigma_H^2 \tag{4-46}$$

where $\Theta_H = E_0(H_1)$ and $\sigma_H^2 = V_0(H_1)$. Substituting σ_M/σ_H for $\bar{a}_0$ in Eq. (4-45) yields

$$\Theta_M = (1+i)M_0 + \frac{(\Theta_H - iP_{a0})\sigma_M}{\sigma_H} \tag{4-47}$$

Expected wealth Θ_M is seen to be a linear function of the standard deviation of wealth σ_M. The function is upward sloping if Θ_H exceeds iP_{a0}. Otherwise, no land would be purchased, as will soon become apparent. Let $u_0 = u(\Theta_M, \sigma_M)$ denote the utility function of the risk-averse decision maker, where $u'_\Theta = \partial u/\partial\Theta_M > 0$ and $u'_\sigma = \partial u/\sigma_M < 0$. A graphic representation of Eq. (4-47) is given by the curve labeled $f(\sigma_M)$ in Fig. 4.12 along with indifference curves corresponding to the utility function

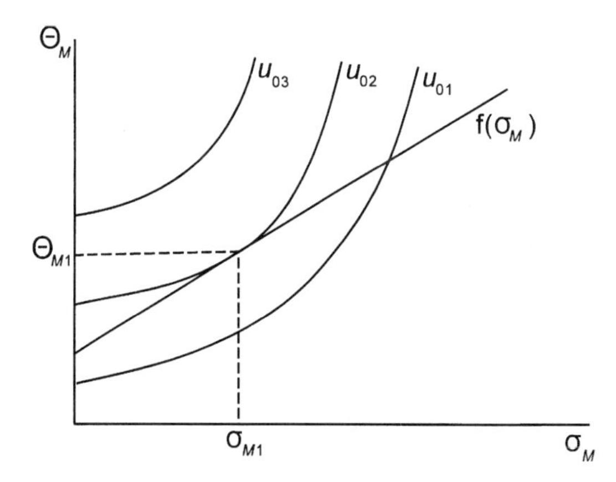

Figure 4.12 Indifference curve map and a linear relationship between expected wealth and the standard deviation of wealth.

$u(\Theta_M, \sigma_M)$. The investor maximizes $u(\Theta_M, \sigma_M)$, where Θ_M and σ_M are given, respectively, in Eqs. (4-45) and (4-46). It is clear from the figure that to maximize utility the investor locates a point $(\Theta_{M1}, \sigma_{M1})$ where $f(\sigma_M)$ is tangent to the indifference curve labeled u_{02}. This gives the first-order condition for utility maximization, as follows:

$$\Theta_H = iP_{a0} - \frac{\sigma_H u'_\sigma}{u'_\Theta} \tag{4-48}$$

where $-u'_\sigma/u'_\Theta$ is the ratio of marginal utilities (the marginal rate of substitution of σ_M for Θ_M), which equals the slope of the indifference curve. It is obvious from Fig. 4.12 that the second-order condition for maximum utility is satisfied at the tangency position. The term on the left, Θ_H, is the expected return per acre of land purchased, as noted. The first term on the right, iP_{a0}, is the opportunity cost of buying an acre of land in that it measures the wealth that could have been accumulated through the purchase of bonds. The last term to the right, equaling the product of σ_H and the ratio of marginal utilities, is the marginal risk premium. The sum of the two terms on the right measures the marginal cost of buying an acre of land. Thus Eq. (4-48) states that in order to optimize the investor equates the expected return per acre to the marginal cost of acquiring acres.

Another way of looking at the first-order condition for utility maximization is to rewrite Eq. (4-48) by moving the risk premium to the left of the equality sign. We would then say that the investor equates the net marginal return per acre of land, Θ_H minus the marginal risk premium, to the opportunity cost per acre.

Although the expected net return per acre of land and the opportunity cost of land acquisition are independent of the amount of land acquired, the same cannot be said for the marginal risk premium. The latter is expressed as an upward sloping function of $\bar{a}_0$ given by the MX curve in Fig. 4.13. Summing vertically the curves for the opportunity cost of land purchase, iP_{a0}, and the marginal risk premium, MX,

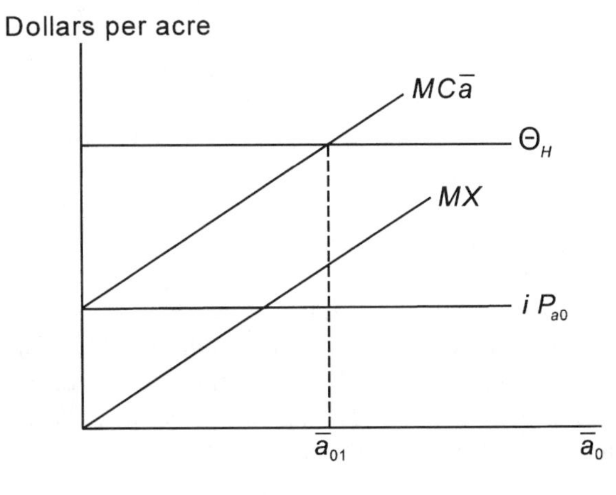

Figure 4.13 Marginal risk premium, opportunity cost of land acquisition, and the marginal cost of land acquisition curves for a risk-averse land investor.

yields the $MC\bar{a}$ curve that shows the marginal cost of land acquisition for various levels of land. Equating Θ_H to $MC\bar{a}$ yields the optimum land purchase $\bar{a}_{01}$. It is possible, of course, returning to Fig. 4.12, that utility maximization might entail setting $\bar{a}_0$ equal to zero; the optimal point then lies on the vertical axis, and no tangency position exists. It is also possible that the choice of $\bar{a}_0$ is constrained by the amount of money that the investor has for investment, recognizing that $\bar{a}_0 \leq (M_0 - B_0)/P_{a0}$. Aside from these two special cases, Fig. 4.12 shows that utility maximization involves choosing a finite value for σ_M. We can be sure that this will occur, however, if the MX curve in Fig. 4.13 slopes upward at a constant or increasing rate. On the assumption that utility maximization involves the purchase of some land, but less land than the maximum, the optimal land purchase may be expressed as follows:

$$\bar{a}_0 = a(P_{a0}, \Theta_H, \sigma_H, i) \tag{4-49}$$

Equation (4-49) is the investor's demand function for land at the end of period 0. From Fig. 4.13, we see that an increase in P_{a0} elevates the $MC\bar{a}$ curve, thus lowering the purchase of land; the demand for land is downward sloping. It is possible, and indeed likely, that the demand for land intersects the price axis. If price is too high, the investor avoids land purchase entirely. It is also possible that the demand for land might have a perfectly inelastic segment for low land prices because of the constraint imposed by the available funds.

Shifting attention to other arguments in the demand for land, we note that a ceteris paribus increase in the expected returns to land, Θ_H, will increase the quantity of land demanded. An increase in σ_H causes the slope of the MX curve in Fig. 4.13 to increase, thus cutting back on the extent of land purchase. A ceteris paribus increase in the interest rate elevates the opportunity cost of land acquisition. The curve labeled iP_{a0} in Fig. 4.13 shifts upward, which shifts $MC\bar{a}$ upward as well. The purchase of land declines.

We now take up an aggregative or market model of land pricing. In Fig. 4.14 the curve labeled D_0' shows how much land investors would be willing to buy (or own) in period 0 for alternative land prices, P_{a0}, holding the demand shifters Θ_H, σ_H, and i constant. [By implication, the shifters of the demand for land input in Eq. (4-37) are also held constant.] We will refer to the D_0' curve as a ceteris paribus demand for land ownership in period 0. It is also assumed that all shifters are fixed at levels consistent with the land price P_{a0}', given by the intersection of the ceteris paribus demand for land D_0' and the supply for land in period 1 given by S_0'. In short, we start with an equilibrium price of land. What would the price of land in period 0 have been, however, if the land supply had been given by S_0'' instead of S_0'? As noted previously, investor estimates of Θ_H and σ_H are determined exogenously in the present formulation. It seems reasonable to suppose, however, that investor estimates of Θ_H fall with increases in the supply of land, the reason being that land rents fall. The effect of an increase in the land supply on investor estimates of σ_H is problematic. There is no particular reason for supposing that the effect, whether positive or negative, will be large,

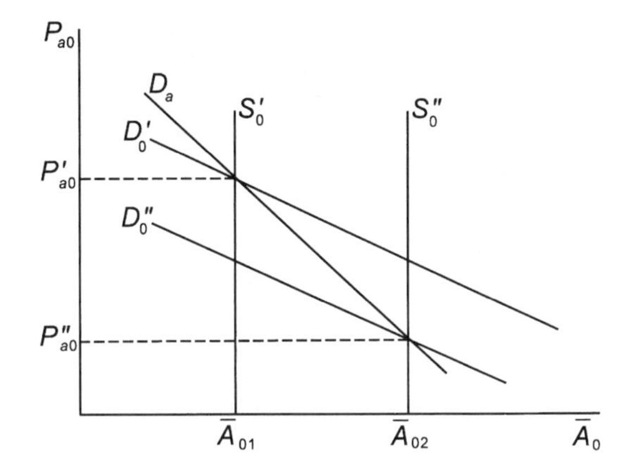

Figure 4.14 Aggregate ceteris paribus demand curves and market equilibrium demand curve for land ownership.

and as a first step we may suppose that an increase in the land supply from S_0' to S_0'' in Fig. 4.14 lowers investor estimates of Θ_H, but leaves estimates of σ_H unaffected. The ceteris paribus demand for land falls from D_0' to D_0''. In the new equilibrium, the price of land falls to P_{a0}''. If estimates of σ_H fall as a result of the increase in land supply, the ceteris paribus demand would likely fall, but not so far as in Fig. 4.14. Investment in land would be less risky. If estimates of σ_H rise as a result of the increase in land supply, on the other hand, the ceteris paribus demand would drop even farther than in Fig. 4.14. For simplicity, however, we will suppose that estimates of σ_H remain the same and that the ceteris paribus demand falls from D_0' to D_0''.

In this way, we derive two points on the *market equilibrium* demand for land, $(\bar{A}_{01}, P_{a0}')$ and $(\bar{A}_{02}, P_{a0}'')$. Many such points could be generated by assigning different values to the supply of land. The locus of such points given by the curve labeled D_a in Fig. 4.14 is the market equilibrium demand for land. The market demand shows the alternative prices for land that investors would be willing to pay as a function of the amount of land available, allowing for the effects of changes in available land on both the expected value and variance of the return to land in year 1. The intersection of the market demand curve D_a and a perfectly inelastic supply curve determines the equilibrium price of land.

Our analysis allows for the immediate entry and exit of investors in the land market. The number of investors is not held constant. The resulting demand should therefore be interpreted as a long-run demand that shows how much investors will pay for land as a function of how much is available, recognizing that the greater the quantity of land available, the lower the level of rent farmers will pay for its use. As to demand shifters, the student should again consider the exogenous variables appearing in Eq. (4-37) and the role of the interest rate as analyzed with the aid of Fig. 4.13.

A more general theory and a more complete explanation of farmland pricing would relax many of the above assumptions. Such a theory would recognize that the grand total of the wealth of the entire nation (not to mention the wealth of foreign nations!) must be held in one form or another—in common stocks, corporate and government bonds, commercial and residential real estate, precious metals, oil and mineral deposits, and cash. The general theory of asset pricing is a complex subject in which the interest rate itself is an endogenous variable and in which the demand and supply for money and the monetary policies of the central bank (the Federal Reserve System, for example) play a crucial role. The task of setting forth such a theory is formidable and is well beyond the scope of this book. The analysis given here may be thought of as a microeconomic foundation for a more complete macroeconomic approach.

PROBLEMS

4.1. A risk-neutral farmer's production function is given by

$$q_{t+1} = a_t^{0.3} k_t^{0.5} (v_{t+1})$$

Let $R_t = 2, J_t = 5$, and $TFC = 4$. Also, $E_t(v_{t+1}) = 1$.

a. Find the equation for the expansion path for planned output. (*Hint:* Use the first-order conditions for minimizing the cost of producing output.)

b. What is the least-cost method of producing planned output equal to 100?

c. Suppose that the expected gross return per unit of planned output, Θ_G, equals 20. Find the levels of inputs and planned output that maximize expected profit.

4.2. A risk-neutral farmer's production function is given by

$$q_{t+1} = a_t^{0.5} k_t^{0.2} (v_{t+1})$$

where $E_t(v_{t+1}) = 1$. Let $J_t = 2$ and $TFC = 0.5$.

a. Derive the farmer's *TRP* function and demand function for a_t.

b. Derive the *ARP* and *NARP* functions.

c. Suppose that the government fixes the price at 8 and let $R_t = 4$. Find the optimal levels for a_t, k_t, Θ_q, and Θ. Calculate *TRP*, *MRP*, *ARP*, and *NARP* for the optimal level of a_t.

4.3. The model given by Eqs. (4-28) assumes that land input is in perfectly inelastic supply. Assume instead that $A_t = S(R_t, B_t)$, where B_t is an exogenous variable and $\partial A_t / \partial R_t > 0$ and $\partial A_t / \partial B_t < 0$.

a. Add the appropriate column for A_t to Table 4.1 and, where possible, find the signs of the effects of the specified exogenous shocks on A_t.

b. Suppose that the supply of land shifts to the left (B_t increases). How will this exogenous shock affect the endogenous variables?

4.4. The production function for a farmer is given by

$$q_{t+1} = a_t^{0.5} k_t^{0.5} v_{t+1}$$

where $E_t(v_{t+1}) = 1$. In a short-run model, $a_t = 16$. The farmer's utility function is given by $u_{t+1} = \phi_0 + \phi_1\Theta - \phi_2\sigma^2$. Derive the farmer's short-run demand for k_t, treating Θ_G and σ_G^2 as constants. [Notice that $(\partial k_t/\partial\phi_2) < 0$ and $(\partial k_t/\partial\sigma_G) < 0$.]

4.5 Food demand in a world of certainty is given by $Q = P^{-0.6}$. The aggregate short-run production function is $Q = A^{0.5}$. Investors assume that the future price of land will always equal the current price. Letting the interest rate be given by i, derive the investor demand for land ownership.

REFERENCES

Barry, Peter J., John A. Hopkins, and C. B. Baker, *Financial Management in Agriculture,* 4th ed. Danville, Ill.: Interstate Printers and Publishers, Inc., 1988.

Chavas, Jean-Paul, "The Ricardian Rent and the Allocation of Land under Uncertainty," *European Review of Agricultural Economics*, 20 (1993), 451–469.

Gardner, B. L., "Determinants of Supply Elasticity in Interdependent Markets," *American Journal of Agricultural Economics,* 61, no. 3 (April 1979), 463–475.

Helmberger, Peter G., *Economic Analysis of Farm Programs*. New York: McGraw-Hill Book Co., 1991.

Layard, R. R. G., and A. A. Walters, *Microeconomic Theory*. New York: McGraw-Hill Book Co., 1978.

Silberberg, Eugene, *The Structure of Economics: A Mathematical Analysis*. New York: McGraw-Hill Book Co., 1978.

PART II

Marketing Farm Output and Farm-level Demand

5

Farm-level Demand

Farm-level demand for output plays a central role in agricultural pricing, and in this and the next chapter we examine in some detail the nature of this relationship. Our examination will necessarily shift attention from farm production to the myriad markets that make up the marketing channel for farm output, a channel that stretches from the farmer at one end to the ultimate consumer at the other. Although agricultural marketing is an important subject in its own right, we will soon see that changes in marketing can and do have profound impacts on agricultural pricing.

The farm-level demand for output may usefully be viewed as an aggregate, representing the sum total of the demand for domestic consumption, the demand for exports, and the demand for storage. The theoretical underpinnings of these three demands are vastly different, however, and in what follows we shall examine the underpinnings of each in some detail. To simplify analysis, production lags in the conversion of farm outputs into retail outputs suitable for consumer purchase are ignored in this chapter. In other words, little attention is given to uncertainty in marketing channels. The attention given to uncertainty at the farm level is also limited in order to highlight marketing phenomena. The objective is to identify and analyze the demand shocks that play an important role in determining farm prices, shocks that may spring from very different sources. Section 5.1 centers on the domestic demand for consumption and sets the stage for the second, which centers on competitive pricing of heterogeneous products. Sections 5.3 and 5.4 take up the demand for farm exports, with some attention given to farm imports as well. Section 5.5, which is something of a digression, applies the theory of export demand to interregional competition. The demand for storage is the subject of Chapter 6.

5.1 FARM-LEVEL DEMAND FOR DOMESTIC CONSUMPTION

With minor exceptions (roadside stands, for example), farmers produce raw materials, not finished consumer food products. We therefore envisage a marketing firm that buys raw material at the farm gate and converts the raw material into finished consumer goods sold at retail to ultimate consumers. Our imaginary firm is rather like a retail store that is integrated backward to the farm gate, performing all the economic activities required to convert raw farm products into finished consumer goods. The student should realize at the outset that this is a considerable simplification. On the way to the final consumer, farm output ordinarily passes through several markets representing different sets of buyers and sellers as the raw material moves through various processes. Grain elevators, for example, facilitate the collection and assembly of grain from many relatively small farmers. Inspectors from the Department of Agriculture grade crops and animals. Breweries, meat packers, and bakers convert farm raw materials into beer, wieners, and bread. Trucks, barges, and railroad cars move outputs along from the country to the city. Retail store employees shelve final products and work cash registers.

To see the economic importance of marketing in the food business, consider the following estimates reported by the U.S. Department of Agriculture: In 1993, consumer expenditures on domestically produced food equaled $491.3 billion. The marketing bill equaled $382.1 billion, accounting for 78 percent of food expenditures; the value of farm raw materials accounted for the remaining 22 percent. As a component of food expenditures, the marketing bill equaled only 59 percent in 1950, a percentage that has increased steadily over the years. The growing importance of the marketing bill reflects many trends, including the increasing importance of food marketing services that substitute for food preparation in the kitchen. The commercial baking of bread is an example. The expenditures on the labor required to market domestically produced food equaled $177.5 billion in 1993, accounting for 46 percent of the marketing bill and exceeding the farm value. Other components of the food marketing bill, with percentage shares of the total, are packaging materials (10 percent), intercity rail and truck transportation (6 percent), fuels and electricity (4 percent), corporate profits before taxes (4 percent), and a miscellaneous category including depreciation, rent, advertising, interest, and the like (30 percent).

Short-run Analysis

To develop a theoretical explanation of the impacts of changes in the marketing sector on agricultural pricing, we begin by examining the behavior of an idealized marketing firm in a short-run model. The profit function for this firm is

$$\pi = P_r q_r - \Sigma\, V_i x_i - Pq - TFC \qquad (5\text{-}1)$$

where P_r and P equal, respectively, price at retail and price at the farm level; q_r and q equal output sold at retail and output purchased from farmers; and V_i and x_i equal, respectively, price and quantity of the ith input, $i = 1, 2, \ldots, k$. Notice that we have dropped the t subscripts that appeared in Chapters 2 and 3. We do this to simplify analysis and to center attention on the issues of interest in this chapter. From one point of view, farm output may be viewed by the marketing firm as just another input, and the derivation of both the short- and long-run aggregate demands for an input given in Chapter 4 applies with little modification.

As an alternative to the usual analysis, however, it is useful to recognize that, for many farm products, farm output q is roughly proportional to retail output q_r, particularly if technology is held constant. A hundredweight of milk, for example, yields so many pounds of cheese; as a practical matter, labor and power do not substitute for raw milk. A thousand broilers produced at the farm level yield a thousand broilers at retail. For many commodities, it is instructive to assume that $q_r = \delta q$, where δ is the (constant) transformation rate of the farm commodity into the corresponding retail commodity, $0 < \delta \leq 1$. As assumptions go, at least in economics, this one often appears plausible. Considerable complexity caused by complements and substitutes for farm output in the marketing process can be sidestepped, with a handsome payoff in terms of hypotheses derived. Note that the assumption of proportionality does not preclude the possibility of substitution among the inputs $x_1, \ldots, x_k$. Strictly for convenience, we begin by further assuming that $\delta = 1$. The student should be mindful that in this analysis we treat technology as exogenous. In the real world, a large permanent price increase for a commodity might induce new technologies designed to conserve the commodity in processing and marketing operations. An increase in the price of broilers, for example, might encourage the design of new equipment to reduce bruising or to improve refrigeration.

Armed with our new assumption regarding processes that move commodities from the barnyard to the table, the marketing firm's profit may be rewritten thus:

$$\pi = (P_r - P)q - \Sigma\, V_i x_i - TFC \qquad (5\text{-}2)$$

What we suppose is that food at retail embodies a set of marketing services such as collection, cleaning, processing, transportation, and retailing. In a very real sense, the marketing sector does not produce food; it produces food services that are essential to a modern society. The quantity of services can be measured by the quantity of food moving along a grand conveyer belt. The price received by the marketing firm for services rendered equals $(P_r - P)$, which is often referred to as the *marketing margin*. The marketing margin may be viewed as the price received for the output of marketing services.

Assuming perfect competition in marketing, we note that all firms must be efficient; they must strive to minimize the cost of providing whatever level of services they elect to provide. For simplicity, we further assume that all firms are identical. The total variable cost (TVC), marginal cost (MC), and average variable cost (AVC) functions for a representative marketing firm are given by

$$\begin{aligned} TVC &= aq + bq^2 \\ MC &= a + 2bq \qquad (5\text{-}3) \\ AVC &= a + bq \end{aligned}$$

The profit function is then given by

$$\pi = (P_r - P)q - (aq + bq^2) - TFC \tag{5-4}$$

Profit maximization implies that $(P_r - P)$ equals marginal cost MC. In the short run it is reasonable to assume that the marginal cost curve for each marketing firm is upward sloping in the relevant region because of diminishing returns to fixed plants. With fixed plants, large levels of output may entail paying overtime, running extra shifts, and paying higher prices for packaging materials that may be in short supply. Linearity is assumed to simplify the algebra.

Let m equal the number of marketing firms such that aggregate quantity is $Q = mq$. Since $MC = a + 2bq$ is the representative firm's supply function, the aggregate supply function is given by

$$MC' = a + 2b'Q \tag{5-5}$$

where b' equals b/m. A prime on MC is used to remind the reader that Eq. (5-5) is an aggregate marginal cost function. A graphic representation of the aggregate short-run supply function for marketing services is given by S_m (equals MC') in panel a, Fig. 5.1. The S_m curve is the lateral summation of the marginal cost curves of all m marketing firms. The average variable cost function for the industry is

$$AVC' = a + b'Q$$

The graph of this function is given by AVC' in panel a, the lateral summation of the AVC curves of the m marketing firms. The AFC' and ATC' curves are the average fixed cost and average total cost curves for the industry, where $AFC' = (m)(TFC)/Q$. The average total cost (ATC') curve is the vertical summation of AFC' and AVC'. Although S_m (equals MC') and AVC' are linear, ATC' has the traditional U-shape because of the average fixed cost curve. Importantly, these aggregative cost curves must be interpreted as cost curves for marketing services measured by Q or Q_r. Going back to Eq. (5-4), we note that the product (Pq) is not treated as a component of total variable cost. The curves in panel a are not, therefore, to be interpreted as cost curves for producing retail food. (The cost curves for producing retail food could, of course, be derived as well, but we have no need for them here.)

The domestic retail demand for food (i.e., consumer demand) is given by D_r in both panels of Fig. 5.1. In panel *a* we subtract vertically the short-run supply for marketing services S_m from the demand for retail food given by D_r. The result is plotted in panel b as curve D, which is the short-run input demand for farm food.

We may now consider market equilibrium, ignoring both the demand for export and the demand for stocks. The quantity of farm output is predetermined. As explained in Chapters 2 and 3, output depends on expected price or product return and, possibly, on random elements such as the weather or the incidence of plant or

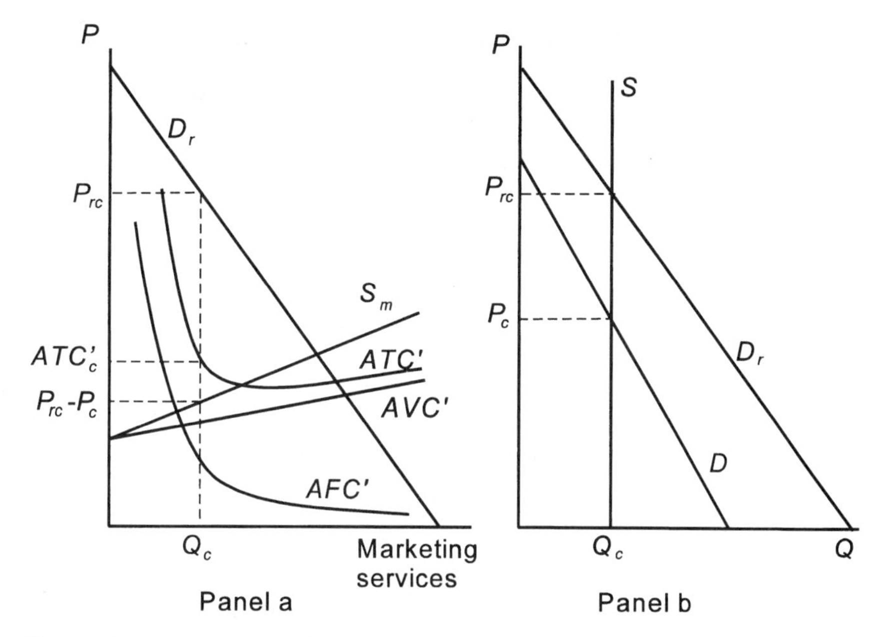

Figure 5.1 Aggregate cost relationships for the marketing industry and the retail and derived farm-level demand curves.

animal diseases. Thus, in panel b we let actual farm output in competitive equilibrium equal Q_c. The intersection of S and D determines the competitive farm-level price given by P_c and the competitive retail price given by P_{rc}. The difference between the two, $(P_{rc} - P_c)$, which is the competitive marketing margin, is equated to the marginal cost of providing marketing services (see panel a). The price spread or margin in short-run analysis gives a measure of the marginal cost of providing marketing services.

Panel a shows that the retail price P_{rc} is not sufficiently high to allow marketing firms to pay farmers P_c and still cover the average total cost of marketing given by ATC'_c. To put it somewhat differently, the price received for providing a level of marketing services needed to move Q_c from farmers to consumers, $(P_{rc} - P_c)$, is less than the average total cost of providing those services (ATC'_c). In the short run, marketing firms need not and often do not make profit. It must be stressed, however, that if farm output were larger by a sufficient amount than Q_c in panel b, then marketing firms would enjoy excess profit. Under competitive conditions marketing firms enjoy the greatest profits in years of large supply, ceteris paribus, when the demand for the services they provide is relatively strong. At this juncture, the student should reflect on the factors that might shift the various curves in Fig. 5.1 and on the effects these changes would have on farm and retail prices. For example, an increase in the wage rate in the marketing sector elevates S_m (in panel a), causing D (in panel

b) to fall. This lowers the farm-level price and, if time is allowed for farm output to respond accordingly, increases the price at retail.

We now take up a model in which δ is less than 1, but larger than zero, using algebra instead of graphics to both specify and solve the model. Assume that a part of each unit $(1 - \delta)$ of farm output is wastage. Profit for the representative marketing firm is written as

$$\begin{aligned} \pi &= P_r q_r - (aq + bq^2) - Pq + TFC \\ &= (\delta P_r - P)q - (aq + bq^2) + TFC \end{aligned} \tag{5-6}$$

where $q_r = \delta q$ and $TVC = aq + bq^2$. (The reader may verify that the results of what follows would be the same if TVC were expressed as a function of q_r.) Profit maximization now implies that $(\delta P_r - P)$ equals marginal cost MC. Again letting $Q = mq$, we have

$$\delta P_r - P = a + 2b'Q \tag{5-7}$$

Equation (5-7) is an example of an arbitrage condition that links prices of a farm raw material at two stages as its form is converted through marketing activities. It states that the difference between the retail and farm-level prices, corrected for wastage, trimming, cooking, or whatever, equals the marginal cost of providing the necessary marketing services. The marketing margin MM is now defined as the difference $(\delta P_r - P)$. Profit maximization implies that the marketing margin $(MM = \delta P_r - P)$ is equated to the marginal cost of marketing.

In this and succeeding chapters, it will be seen that arbitrage conditions not only link prices of farm commodities in different forms (wheat and flour, for example), but they also link as well prices formed at different geographic locations and at different points in time. These conditions are basic to understanding price determination.

A mathematical specification of the preceding model of agricultural marketing, with $\delta < 1$, follows:

$$\begin{aligned} Q_r &= \delta Q && \text{retail output as a proportion of farm output} \\ MM &= \delta P_r - P && \text{definition of marketing margin} \\ MM &= a + 2b'Q && \text{supply for marketing services} \\ P_r &= d - eQ_r && \text{retail demand for food} \end{aligned} \tag{5-8}$$

The four endogenous variables of this four-equation system are Q_r, MM, P_r, and P. (Farm output Q is predetermined.) The arbitrage condition is implicit in the second and third equations taken together. The reduced form functions for P and P_r are as follows:

$$P = \delta d - a - (2b' + e\delta^2)Q$$
$$P_r = d - e\delta Q \quad (5\text{-}9)$$

The first of these two equations is the farm-level demand for output, written in the price-dependent form. It is the algebraic counterpart to D in panel b, Fig. 5.1. The stage is set for comparative static analysis. Suppose that labor unions jack up wage rates in the marketing sector, causing the parameter a to increase. Farm price falls by the amount of the increase in the marketing margin. With farm output treated as predetermined, the retail price is unaffected. An increase in retail demand, on the other hand, elevates both the farm and retail price. Suppose that a technological change occurs that reduces wastage, such that δ moves closer to 1. To analyze the effects of this change, differentiate P with respect to δ as follows:

$$\frac{\partial P}{\partial \delta} = d - 2e\delta Q \quad (5\text{-}10)$$

The right-hand side of Eq. (5-10) is the marginal revenue function associated with retail demand. An increase in δ causes P to rise or fall according to whether the retail demand in the relevant region is elastic or inelastic. Clearly, P_r falls with an increase in δ.

The importance of these comparative static results can easily be exaggerated. It should be understood that the analysis takes farm output as predetermined, centering on exogenous change that occurs after farm production plans have come to fruition and without taking into account the effects of exogenous shocks on price expectations and future levels of production. It is a very short run analysis.

Before leaving this analysis, we show that for any given level of farm output the farm-level demand is more inelastic than is the retail demand for food. Taking advantage of the first and last equations of Eqs. (5-8) together with Eqs. (5-9), we find that

$$E_r = 1 - \frac{d}{e\delta Q} < 0$$
$$E_Q = 1 - \frac{\delta d - a}{\left(2b + e\delta^2\right)Q} < 0 \quad (5\text{-}11)$$

where $E_r = (\partial Q/\partial P_r)(P_r/Q)$ and $E_Q = (\partial Q/\partial P)(P/Q)$ are, respectively, the elasticities of retail and farm-level demands. Some algebra reveals that $|E_Q| < |E_r|$ or, equivalently, that $E_Q > E_r$. For any plausible value for farm output Q, demand at the farm level is less elastic than is the demand at retail.

The student should be aware of the possibility, however, that the assumption of proportionality might not always apply; good substitutes for some farm raw material might exist. In such cases, the farm-level demand could be more elastic than retail demand. Take cotton as an example. Because synthetic fibers substitute for cotton, the farm-level demand for cotton might be more elastic than is the retail demand for clothing.

It is sometimes argued that the farm-level demand for output in the relevant range is highly inelastic. This together with predetermined supply might lead one to believe, as it did the famous agricultural economist Willard Cochrane, that farm prices are highly volatile, verging on being downright chaotic. Small changes in yield could cause farm-level prices to skyrocket or fall to the basement. Importantly, however, the analysis to this point overlooks the strategic role of the demand for stocks as a stabilizing element in farm price determination. More on this in Chapter 6.

The analysis to this point has centered on short-run models with a fixed number of marketing firms, each with a fixed plant. It has been argued that, in light of the law of diminishing returns, the supply for marketing services, the S_m curve in panel a, Fig. 5.1, is upward sloping. In the long run, of course, every firm can adjust the size of the plant, and the number of firms is variable. The long-run supply curve could still be upward sloping if the prices of the inputs used in marketing tend to rise in the long run as quantities purchased go up. In other words, the long-run supply curve for marketing services will be upward sloping if at least some of the input supply functions to the marketing sector are upward sloping. Are they? Some are and some are not, it seems. It is possible that transportation rates might rise with increased demand for transportation services (barge, railroad, and trucking) in the long run. It seems likely, however, that the supply of labor to the agricultural marketing sector has a great deal of elasticity.

Long-run Analysis

Long-run analysis requires modification of the marketing models considered previously. Let the long-run total cost function for the representative marketing firm exhibit constant returns to scale such that $TC = kq$, where k is a parameter that equals the per unit marketing cost for a firm. We assume the marketing industry is an increasing cost industry, however, and that $k = c + zQ$. Industry total cost is then given by $(cQ + zQ^2)$. As total output expands, the linear total cost curves of the marketing firms shift upward. In the long run, competition drives profit to zero; industry total cost $(cQ + zQ^2)$ equals industry total revenue $(\delta P_r - P)Q$. Therefore, in long-run equilibrium, the marketing margin MM, which by definition equals $(\delta P_r - P)$, must also equal $(c + zQ)$. After incorporating a farm-level supply function in the analysis, the long-run model of marketing is as follows:

$$\begin{aligned} Q_r &= \delta Q && \text{retail output as a proportion of farm output} \\ MM &= \delta P_r - P && \text{definition of marketing margin} \\ MM &= c + zQ && \text{supply for marketing services} \\ P_r &= d - eQ_n && \text{retail demand for food} \\ P &= w + sQ && \text{farm-level supply} \end{aligned} \tag{5-12}$$

In regard to the farm-level supply in this model, we are ignoring price expectations and production lags in order to facilitate comparative-static analysis. The farm-level demand implicit in the model is

$$P = \delta d - c - (z + e\delta^2)Q \tag{5-13}$$

The five-equation system can be solved for the reduced forms. When this is done, it can be shown that $(\partial Q/\partial c) < 0$, $(\partial P/\partial c) < 0$, and $(\partial P_r/\partial c) > 0$. (The mathematical derivations are left to the student.) What these results show is that lowering the marketing margin increases farm output and the farm price, but lowers the retail price. A useful exercise for the student is to compile a list of factors that could change the value of c. It can also be shown that $(\partial P_r/\partial \delta) < 0$, but that both $(\partial P/\partial \delta)$ and $(\partial Q/\partial \delta)$ could be larger than, equal to, or less than zero. Increasing δ lowers the retail price and increases retail consumption, but the greater consumption could come from the increased productivity of the marketing firms, rather than from increased farm production. Presumably, the more elastic the retail demand response is, the greater the likelihood that farm output and price will increase.

Domestic Demand for Food

The shape and dynamic behavior of the domestic consumer demand for food are of critical importance in understanding the performance of the farm sector. Consider the following estimated U.S. domestic demand for food:[1]

$$\begin{aligned} X_1 = 60.16 &- \underset{(0.1028)}{0.5969X_2} + \underset{(0.04149)}{0.2992X_3} + \underset{(0.0779)}{0.1066X_4} + \underset{(0.3202)}{1.0933X_5} \end{aligned}$$

[1]The relationship was estimated using multiple regression and annual time series data for 1966 through 1989 as reported by the Council of Economic Advisors (1990) and the U.S. Department of Agriculture (1991). The numbers in parentheses are the standard errors of the estimated structural coefficients. The coefficient of multiple determination (R^2) equaled 0.943.

where X_1 = per capita food consumption in the United States
X_2 = consumer price index for all food, prices at retail
X_3 = per capita disposable income (after income taxes) in current hundred dollars
X_4 = consumer price index for all items except food
X_5 = percentage of females who participate in the labor force

As expected, an increase in the price of food (X_2) lowers per capita consumption. The elasticity of demand with respect to price (evaluated at the means of the variables) equals approximately –0.44. A 10 percent increase in the price of food lowers per capita consumption by 4.4 percent. The retail demand for food is price inelastic.

Food appears to be a normal good in that an increase in disposable income (X_3) increases per capita food consumption. The elasticity of demand with respect to income equals +0.22, which is not very large. A 10 percent increase in income causes per capita food consumption to rise by a mere 2.2 percent. In this connection, it is important to realize that food at retail (or in the restaurant) contains a considerable amount of food services. Research suggests that as incomes rise consumers opt to buy the food services that they find embodied in such products as TV dinners, cake mixes, and frozen pizzas. The increase in food intake per se, as measured in nutrients, for example, is very small. Notice how these findings tie in with the effect of increased female participation in the labor force as measured by the variable X_5. (The percentage of females participating in the labor force increased from 40.3 in 1966 to 57.4 in 1989.) A 10 percent increase in X_5 causes per capita food consumption to rise by 5.3 percent. Working men and women haven't the time to peel potatoes; they reach for frozen French fries instead. As noted, the food services manufactured in food processing plants substitute for services produced in the home kitchen.

Introspection suggests that food has few good substitutes, a view supported by the preceding equation. An increase in the consumer price index for nonfood items tends to increase food consumption, but the effect tends to be insignificant. The cross-elasticity of the demand for food with respect to the price of items other than food is probably less than +0.08, meaning that a 10 percent increase in nonfood prices likely causes food consumption to rise by less than 0.8 percent.

The nature of the domestic demand for food has several implications for understanding the performance of the U.S. food production and marketing sector. Three merit emphasis. First, the domestic demand for food contributes little elasticity to the farm-level demand. We know from Chapter 4 that the elasticity of farm-level demand conditions in a crucial way the impacts of exogenous shocks on output and input levels and prices.

Second, the domestic demand for food is not very sensitive to changes in per capita income, particularly after allowances are made for the demand for food processing services. What this means is that as the U.S. economy has become more affluent, consumers have tended to spend their additional incomes on nonfood items; the nonfood sector has grown relative to the farm sector. As Schultz (1957) pointed out many years ago, this explains in part why sluggish growth of the farm sector relative to the nonfarm sector has been a natural consequence of economic growth (of rising per capita real incomes) in the United States, as well as in the other advanced nations of the world.

Third, as a special case of the marketing models considered previously, let us suppose that the farm-level demand is perfectly inelastic. In this extreme case, where consumers indirectly demand a specific amount of "raw" food, food shorn of food services, lowering the marketing margin decreases the retail price, but has no effect on the farm price or on the level of farm output. Increases in δ, on the other hand, actually lower both the farm price and farm output. More generally, increases in marketing efficiency will not elevate the farm price.

Domestic Demand for Meat

According to the theory of consumer behavior, the quantity of a good demanded by a consumer depends on the price of the good, the prices of related goods (substitutes and complements), and money income. The importance of substitution of one consumer good for another is not particularly apparent for an aggregate of goods such as food because there are no good substitutes for food. The same cannot be said, however, for a less aggregative food category such as poultry, beef, or pork. Table 5.1 gives estimates of own-price and cross-price elasticities for these three important foods. Estimated income elasticities are also provided. The own-price elasticities for poultry, beef, and pork are, respectively, −0.761, −0.916, and −0.734. Turning to substitution, we note that the cross-price elasticities for pork consumption with respect to the price of poultry and beef equal 0.296 and 0.225, respectively. A *ceteris paribus* 10 percent increase in the price of pork, for example, would lower pork consumption by 7.34 percent and increase poultry and beef consumption by 2.96 and 2.25 percent, respectively.

TABLE 5.1 Elasticities of Quantity Demanded (Poultry, Beef, and Pork) with Respect to Meat Prices and Consumer Income

	With Respect to			
Consumption of	Poultry Price	Beef Price	Pork Price	Income
Poultry	−0.761	0.296	0.296	0.170
Beef	0.085	−0.916	0.225	0.606
Pork	0.082	0.217	−0.734	0.435

Source: Jean-Paul Chavas,"Structural Change in the Demand for Meat," *American Journal of Agricultural Economics,* 65, no. 1 (February 1983), 148–153.

The estimates given in Table 5.1 support an important conclusion as regards agricultural price analysis. Realistic models of agricultural pricing will often involve many more equations than those given by Eqs. (5-12) because of the need to take account of commodities that are related both in production and consumption. Although the development and estimation of such models are the business of researchers, students should be on guard against hastily jumping to hypotheses on the basis of single-product models.

To take an example, according to Table 5.1, the income elasticity for poultry is positive, which means poultry is a normal good. This might lead one to suppose that an increase in incomes would increase poultry consumption, a condition that might be true. It is possible to develop multiple-product models, however, in which rising incomes shift farm production inputs from poultry to beef, because beef's income elasticity is relatively large, thus actually lowering the equilibrium consumption of poultry. The effectiveness of economic theory in the derivation of hypotheses is greatly diminished as the analyst moves from single-product to multiple-product models in the absence of empirical estimates of crucial parameters. In this regard, the student might think back to the single-output, multiple-input farm production model analyzed in Chapter 4.

5.2 PRODUCT HETEROGENEITY AND COMPETITIVE PRICING

A basic assumption of competitive price theory is that firms produce a homogeneous product. *A product is homogeneous if every unit of it is exactly the same as any other.* On this strict interpretation, product homogeneity is a theoretical term, useful in developing theory, but without a real-world counterpart. Is this assumption strictly necessary in the analysis of competitive pricing? In this section we show that the answer is no and that, indeed, important phenomena encountered in the study of agricultural pricing can only be explained on the assumption of product heterogeneity.

Allowing for product heterogeneity opens the door to the question of product quality, grading, and differentiation. The following analysis mainly centers on product quality (heterogeneity) at the farm level and seeks to explain how a competitive system determines the levels of outputs of varying quality or grades, together with the associated structure of prices. An important numerical example is given showing that the farmer must choose both the optimal quality and quantity of output to maximize profit. For simplicity, we assume that marketing firms play a rather passive role in the process, acting as a clearinghouse of information on the preferences of consumers for various grades and on the farm costs of providing various grades. We will return in Chapter 8, however, to the question of product heterogeneity and the active role marketing firms play in the determination of product qualities and prices at the retail level.

To begin, we assume that at harvest time each farmer's harvested crop q can be broken up into component parts $q_o = \delta_o q$ and $q_m = \delta_m q$, where $\delta_o + \delta_m \leq 1$. We may think of q_o and q_m as the quantities of oil and meal per bushel of soybeans or the fat and nonfat solids per hundredweight of milk. Alternatively, q_o and q_m might equal the quantities of small and large apples in a bushel of orchard-run apples. Indeed, the component quantities might reflect the results of a wide array of sorting processes that could be chemical or mechanical in nature or that might involve mainly hand labor and human judgment. The characteristics that could be used to break up output into its component parts could involve chemical composition, product size, color, taste, blemishes, or whatever. Letting $(\delta_o + \delta_m) < 1$ allows for waste or unwanted by-products that, we assume, can be disposed of free of charge. Clearly, we are proposing to measure quantity of output by the variable q and the quality of output by the two variables δ_o and δ_m. For the time being, we will assume that δ_o and δ_m are given, an assumption that is relaxed later.

Farm Output of a Single Quality

Suppose for the moment that δ_o and δ_m are the same for all farmers. The number of small and large apples per bushel of apples, for example, might be the same for all producers even though the product is not strictly homogeneous according to our previous definition. Turning directly to a long-run formulation, we let the total cost function for each marketing firm be given by $TC = kq$. If k were independent of the aggregate level of output, then the long-run industry supply for marketing services would be perfectly elastic; the marketing margin would equal k. Following previous analysis, however, we suppose that $k = c + zQ$. Then total industry cost equals $(cQ + zQ^2)$.

On the demand side, assume that consumers have information about the components q_o and q_m and that each component is perceived differently. As a result, there are consumer demands for both components with associated market prices P_o and P_m. Perfect competition in marketing implies that in long run equilibrium industry total cost equals industry total revenue. The arbitrage condition for marketing services becomes

$$P_o Q_o + P_m Q_m - PQ - (cQ + zQ^2) = 0 \tag{5-14}$$

where P_o and P_m are, as noted, the component prices at the retail level and P is the farm price. The marketing margin is now defined as

$$MM = (\delta_o P_o + \delta_m P_m) - P \tag{5-15}$$

Assuming that Q is exogenous for the moment, a model of agricultural pricing may be set forth as follows:

$$\begin{aligned}
Q_o &= \delta_o Q && \text{component relationship} \\
Q_m &= \delta_m Q && \text{component relationship} \\
MM &= (\delta_o P_o + \delta_m P_m) - P && \text{definition of marketing margin} \\
MM &= c + zQ && \text{supply for marketing services} \\
P_o &= a - bQ_o && \text{component demand} \\
P_m &= e - gQ_m && \text{component demand}
\end{aligned} \tag{5-16}$$

The six endogenous variables of this six-equation system are Q_o, Q_m, MM, P, P_o, and P_m.[2] Solving this system of equations for the farm-level price P gives the reduced form relationship:

$$\mathrm{P} = \delta_o a + \delta_m e - c - (z + b\delta_o^2 + g\delta_m^2)Q \tag{5-17}$$

Equation (5-17) may be interpreted as a price-dependent demand for farm output. It shows clearly that the price received by the farmer reflects the component yields of farm output together with the demands and prices for the components. The variable Q may now be viewed as an endogenous variable. The model is completed through postulating a farm-level supply function under certainty. The supply function together with Eq. (5-17) can be used to determine Q and P, the market equilibrium levels for farm quantity and price.

Alternatively, with production lags at the farm level, that is, under conditions of uncertainty, it becomes necessary to attach $t + 1$ subscripts to price P and output Q in Eq. (5-17) and to add the random variable e_{t+1}. We may then invoke whatever farm-level supply function is thought to be the most appropriate, drawing on the several supply models set forth in previous chapters. In a nutshell, the stochastic farm-level demands appearing in previous market models may now be interpreted as being equivalent to Eq. (5-17), appropriately modified to allow for production lags at the farm level.

Getting back to the question of grading, we note that farm output could be sold in an unprocessed or ungraded form or that components other than those con-

[2]The components Q_o and Q_m that appear in Eq. (5-16) are unrelated goods; the quantity demanded of one good is not influenced by the price of the other. Examples might be butter and nonfat dry milk or concentrated lemon juice and lemon oil. Alternatively, the quantity demanded of each component might be expressed as a function of both prices. The latter formulation might be appropriate when Q_o and Q_m are oranges of different sizes and we can be certain that the one component good is an excellent substitute for the other. The algebra would become more complex in this alternative formulation, but the principles would remain much the same.

sidered might be relevant to some buyers. Figure 5.2 illustrates a competitive equilibrium in which none of the commodity is marketed in an upgraded form. The D_g curve is the demand for farm output on the part of marketing firms that break up output into its component parts according to Eq. (5-17). The D_n curve, on the other hand, is the farm-level demand for mixed or ungraded output, that is, output that is not broken up into its component parts. The total demand D is given by the horizontal sum of D_g and D_n. Take a simple example. Some marketing firms (retail stores) might offer two kinds of apples, large and small, that sell at different prices. Other marketing firms avoid the cost of sizing by offering to sell at a single price apples that have not been separated into two classes or grades. In the competitive equilibrium illustrated in Fig. 5.2, D_n is low relative to D_g, so low, in fact, that marketing firms that do not grade output cannot compete with those that do. The equilibrium farm price is given by P_c, and all the output is broken up into component parts. In this illustrative case, the returns to sized apples are high enough to cover the cost of sizing. But this need not have been the case. Other demand–supply configurations are clearly possible. If grading is costly or even if it is not very costly, but buyers are indifferent between large and small apples, for example, then the output might be sold entirely in its mixed form. It is also possible that output is sold in both forms.

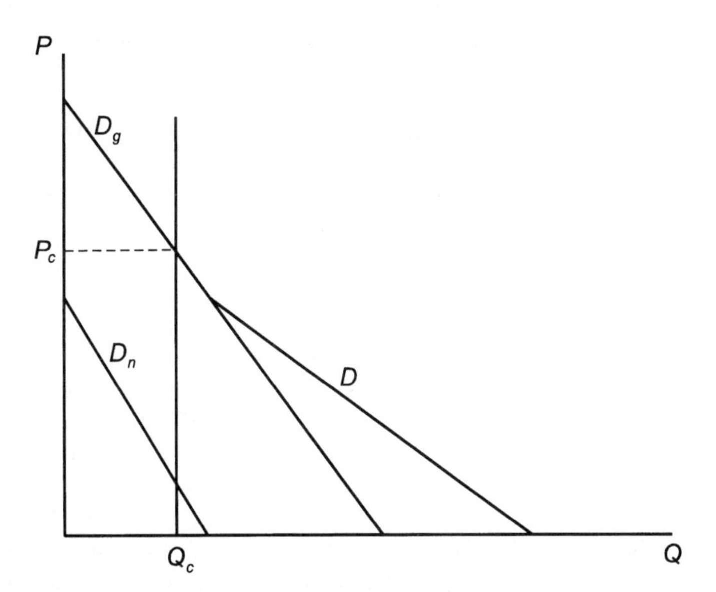

Figure 5.2 Farm-level demands for marketing firms that grade output and for firms that do not grade and total demand.

Farm Output of Multiple Qualities

We now relax the assumption of output homogeneity in a more substantial way. In particular, assume that δ_o and δ_m vary among farmers, but that, at least for the present, all δ's are known. Aggregate output is taken as given. We again assume that in the long run the total cost function for the representative marketing firm is linear, that $TC = kq$, where q equals the aggregate output purchased from n farmers by one marketing firm and k is again a parameter measuring the firm's per unit cost of marketing. Algebraically, $q = q_1 + q_2 + \cdots + q_n$ is the total farm quantity purchased by the marketing firm. For simplicity, we assume that the cost of marketing q is independent of the composition of q_i, $i = 1, \ldots, n$. The total cost of converting milk into butter and nonfat dried milk depends on the quantity of raw milk processed, but not on the fat and nonfat solids content of the milk. (The model could be generalized somewhat by assuming that the price of butter is a net price, net of such allocable marketing costs as packaging materials and similarly for the price of nonfat dried milk.) Letting $k = c + zQ$, as before, the average industry cost or margin required to market output is given by $MM = c + zQ$. If competition reduces profit to zero in the marketing sector in the long run, then the following aggregative arbitrage condition must hold:

$$P_o \sum_{i=1}^{N} \delta_{oi} q_i + P_m \sum_{i=1}^{N} \delta_{mi} q_i - MM \sum_{i=1}^{N} q_i - \sum_{i=1}^{N} P_i q_i = 0 \tag{5-18}$$

where P_i equals the price received by the ith farmer and where it is now necessary to sum over all N farmers in the sector, not just the n farmers delivering to a single marketing firm ($N > n$). Equation (5-18) can be rewritten thus:

$$\sum_{i=1}^{N} \left(\delta_{oi} P_o + \delta_{mi} P_m - MM - P_i\right) q_i = 0 \tag{5-19}$$

A moment's reflection will show, however, that the market will only be in equilibrium if

$$\delta_{oi} P_o + \delta_{mi} P_m - P_i = MM, \qquad i = 1, 2, \ldots, N \tag{5-20}$$

If the left-hand side of Eq. (5-20) is greater than MM for any one farmer, the relevant marketing firm could increase profit by not buying from that farmer. The marketing firm in question is not maximizing profit. If the left-hand side cannot be larger than the marketing margin for any one farmer, then it cannot be less for any farmer; otherwise, marketing firms would incur a loss. Keen competition assures that marketing firms cannot "rob" one farmer to "subsidize" another. That being the case, Eq. (5-19) holds only if Eq. (5-20) holds.

Before proceeding with the analysis, we pause to consider an interesting question: Why is the ith farmer paid a price per unit of output P_i instead of two prices, one price per unit of component q_{oi} and another price per unit of component q_{mi}? The basic reason for this is the existence of joint marketing costs that cannot be allocated, except in arbitrary ways, between the two final products sold by marketing firms (between Q_o and Q_m). The easiest way to see how product composition influences farm-level pricing is to consider a whimsical example. In a rather primitive society, perhaps in an era gone by, marketing firms separate the raw milk purchased from farmers into cream and skim milk and then sell these two products directly to consumers. Suppose that all farmers have black and white cows that produce exactly the same kind of milk. For reasons unknown, Farmer Jay sells his cows and buys a herd of brown and white cows that produce milk with a higher value for δ_o (contains more fat) and a lower value for δ_m (contains less protein and water). The cost of separating Farmer Jay's milk per hundredweight is the same as for any other farmer's milk. How will the marketing industry respond to this extraordinary development?

Prior to the purchase of the new herd, Farmer Jay received the price P_j per hundredweight according to the following:

$$P_j = \delta_{oj} P_o + \delta_{mj} P_m - MM$$

where, in our example, $\delta_{oj} + \delta_{mj} = 1$. (Under our assumptions, all farmers initially receive the same price.) Let δ_{oj} rise and δ_{mj} fall because of the switch in herds. This gives rise to a price differential as follows:[3]

$$dP_j = (P_o - P_m)\, d\delta_{oj}$$

After the change, Farmer Jay receives $(P_j + dP_j)$, the base price P_j plus *a price differential* dP_j reflecting the increased fat test of his or her milk. (If $P_o > P_m$, for example, the differential dP_j will be positive.) Perfect competition in the marketing industry forces this result because dP_j measures the change per hundredweight in the value of the milk to the marketing firm. If the actual price differential differs from dP_j, then the marketing firm that purchases Farmer Jay's milk will either incur a loss or enjoy a profit. Neither outcome is compatible with perfectly competitive long-run equilibrium.

Before leaving this example, it should be clearly understood that nothing has been said in regard to whether the purchase of a new breed of dairy cows by Farmer Jay is profitable. Much depends on the magnitude of dP_j relative to the cost per unit of producing high-fat test milk. This takes us back to the mainstream of the analysis.

Turning to the level and quality of aggregate farm output Q, we note that all farmers might use the same production techniques, with variation in the deltas being

[3]We have $P_j = \delta_{oj} P_o + (1 - \delta_{oj}) P_m - MM$. Taking the differential of P_j with respect to δ_{oj} yields the result given in the text.

attributable to stochastic elements. A more likely and interesting possibility, however, is that farmers may be able to affect the composition of their outputs through altering production techniques. Hogs can be fat or lean. Corn can be wet or dry. Although the farmer may alter the δ's through choice of farm inputs, we maintain for simplicity the assumption that marketing inputs are not substitutes for farm output. The farmer may vary the fat content of his or her milk, but a pound of butter at retail still requires a fixed amount of milkfat.

To explore such possibilities further, we let the profit function for the ith farmer be given by

$$\pi_i = (\delta_{oi} P_o + \delta_{mi} P_m - MM) q_i - C(q_i, \delta_{oi}, \delta_{mi}) \tag{5-21}$$

where the price received is $P_i = \delta_{oi} P_o + \delta_{mi} P_m - MM$ from Eq. (5-20), and total cost C is a function of q_i together with the deltas. We abstract from uncertainty to simplify analysis. (Alternatively, the student might think of the prices P_o, P_m, and P_i as being expected prices.) It is assumed that the partial derivative of total cost C with respect to each of its arguments is positive. Transfer earnings are ignored. We assume the δ's sum to less than 1, as might be the case when some proportion of farm output is discarded as waste. The law of one price, often emphasized in competitive theory, must be jettisoned. Farmers receive various prices depending on the grades they produce. Maximizing profit involves choosing the optimal values for q_i, δ_{oi}, and δ_{mi}. Both the quantity and the quality of output are optimized. The first-order conditions are

$$\begin{aligned} \delta_{oi} P_o + \delta_{mi} P_m - MM - MC_q &= 0 \\ P_o q_i - MC_o &= 0 \\ P_m q_i - MC_m &= 0 \end{aligned} \tag{5-22}$$

where MC_q, MC_o, and MC_m equal the marginal costs given by the partial derivatives of $C(\cdot)$ with respect to q_i, δ_{oi}, and δ_{mi}. The first equation tells us that in order to maximize profit the level of output must be chosen such that price received P_i equals the marginal cost of production MC_q, holding product quality constant. Since P_i is positively related to P_o and P_m, it is clear that output q_i rises with increases in P_o or P_m. It is equally clear that q_i falls with increases in the marketing margin.

Using the last two equations, we see that to optimize quality the farmer must equate the ratio of the component prices P_m/P_o to the ratio of their marginal costs MC_m/MC_o. In Fig. 5.3, the *Output Grade Transformation* curve RPT' shows the maximum values for δ_{oi} and δ_{mi} holding output constant. (For the moment, ignore the RPT'' curve.) It can be shown that the first derivative of RPT', which is the marginal

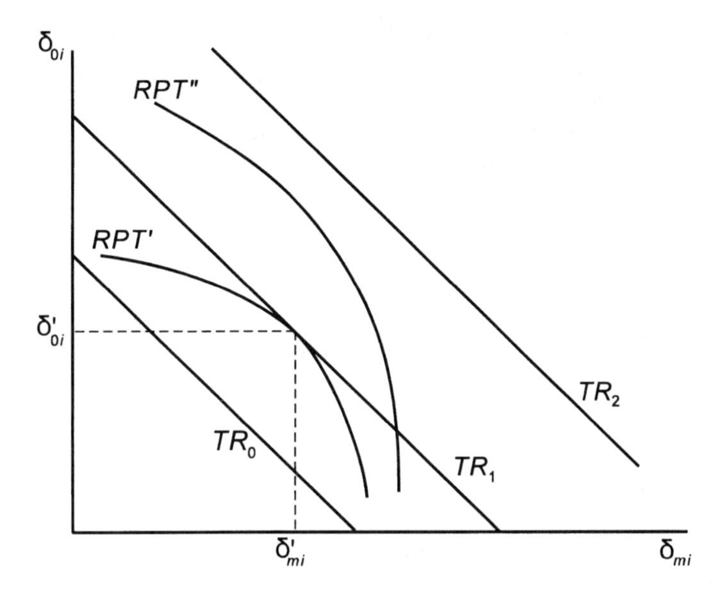

Figure 5.3 Output grade transformation curves for different technologies and isorevenue lines.

rate of grade transformation $\partial\delta_{oi}/\partial\delta_{mi}$, equals the ratio $-MC_m/MC_o$.[4] Isorevenue lines are given by the curves labeled *TR* for total revenue. Each *TR* curve shows the various grades or combinations of δ_{oi} and δ_{mi} that yield the same level of revenue, holding q_i fixed. It is left to the student to show that the slope of each *TR* curve equals $-P_m/P_o$, holding constant both total revenue and output. (*Hint:* Write the expression for total revenue in terms of component prices and δ_{oi} and δ_{mi} and then express δ_{oi} as a linear function of δ_{mi}.) Optimal values for δ_{mi} and δ_{oi}, holding output constant, can be found by locating a point on the highest isorevenue line obtainable, which is TR_1. At this point, where $\delta_{oi} = \delta'_{oi}$ and $\delta_{mi} = \delta'_{mi}$, the slope of the RPT' curve

[4]Let C and q_i be held constant such that $C = C'$ and $q_i = q'_i$. Then

$$C' = C(q'_i, \delta_{oi}, \delta_{mi})$$

Using the implicit function rule,

$$\frac{d\delta_{oi}}{d\delta_{mi}} = -\frac{\dfrac{\partial C}{\partial \delta_{mi}}}{\dfrac{\partial C}{\partial \delta_{oi}}} = -\frac{MC_m}{MC_o}$$

equals the slope of TR_1, and the ratio P_m/P_o equals MC_m/MC_o. An increase in P_o, holding output constant, flattens the TR lines. This has the effect of increasing the optimal value of δ_{oi} and decreasing the optimal value of δ_{mi}. Of course, a ceteris paribus increase in P_m has the opposite effect.

We now consider a numerical example. As a special case of $C(\delta_o, \delta_m, q)$, suppose that

$$C = (\delta_o + \delta_m^2)q^2$$

where $\delta_o = 0.75 - \delta_m^2; 0 \leq \delta_o \leq 0.75$; and $0 \leq \delta_m \leq 0.86$. The slope of the Output Grade Transformation curve (see Fig. 5.3) equals $-MC_m/MC_o$. Since profit maximization implies $MC_m/MC_o = P_m/P_o$, we have

$$\delta_m = \frac{P_m}{2P_o}; \quad \delta_o = 0.75 - \left(\frac{P_m}{2P_o}\right)^2$$

If we let $P_o = 4$ and $P_m = 2$, then $\delta_m = 0.25$; and $\delta_o = 0.6875$. Using Eq. (5-21), letting $MM = -1$, we have

$$\pi = 2.25q - 0.75q^2$$

Maximizing π implies that $q = 1.5$. Profit then equals 1.69. If P_m rises from 2 to 6, with $P_o = 4$, it can be shown that $\delta_m = 0.75$ and $\delta_o = 0.1875$. Optimal output rises to 2.8333 and profit increases to 6.10. A *ceteris paribus* increase in P_m increases output and shifts quality in favor of the component $q_m = \delta_m q$.

Returning to the theoretical analysis, we now consider a technological change that increases δ_{oi} for given levels of δ_{mi}, q_i, and all farm inputs. What this means is that the value of δ_{oi} rises for each and every value of δ_{mi} in Fig. 5.3, holding output constant at the level associated with RPT'. The RPT curve shifts from RPT' to RPT''. For any δ_{mi}, the steepness of RPT'' is less than the steepness of RPT'; the ratio MC_m/MC_o becomes smaller. As a consequence, it is clearly the case that the optimal value for δ_{oi} increases. Through experimenting with various diagrams, the student should demonstrate that, ceteris paribus, the effect of the contemplated technological change on the optimal value of δ_{mi} is uncertain, that a decrease in MC_o relative to MC_m could cause the optimal value of δ_{mi} to rise, fall, or stay the same.

Solving the three-equation system given by Eq. (5-22) for the choice variables for all N farmers yields

$$\begin{aligned} q_i &= f_{qi}(P_o, P_m, MM) \\ \delta_{oi} &= f_{oi}(P_o, P_m, MM) \qquad i = 1, \ldots, N \\ \delta_{mi} &= f_{mi}(P_o, P_m, MM) \end{aligned} \tag{5-23}$$

Recalling that $q_{oi} = \delta_{oi}q_i$ and $q_{mi} = \delta_{mi}q_i$ and making substitutions using Eqs. (5-23), we have

$$\frac{\partial q_{oi}}{\partial P_o} = \frac{\partial \delta_{oi}}{\partial P_o} q_i + \frac{\partial q_i}{\partial P_o} \delta_{oi} > 0$$
$$\frac{\partial q_{oi}}{\partial P_m} = \frac{\partial \delta_{oi}}{\partial P_m} q_i + \frac{\partial q_i}{\partial P_m} \delta_{oi} \qquad (5\text{-}24)$$

Drawing on our analysis, we see that an increase in P_o causes q_{oi} to rise. The effect of an increase in P_m on q_{oi} is uncertain. (As an important exercise, the student should develop the expressions for $\partial q_{mi}/\partial P_m$ and $\partial q_{mi}/\partial P_o$.)

The various components of our aggregative model of competitive pricing with product heterogeneity may now be brought together as follows:

$$\left.\begin{aligned} q_i &= f_{qi}(P_o, P_m, MM) \\ \delta_{oi} &= f_{oi}(P_o, P_m, MM) \\ \delta_{mi} &= f_{mi}(P_o, P_m, MM) \\ P_i &= \delta_{oi}P_o + \delta_{mi}P_m + MM \end{aligned}\right\} \quad \text{farmers' behavioral relationships; } i = 1, 2, 3, \ldots, N$$

$$Q = \sum_{i=1}^{N} q_i, \quad Q_o = \sum_{i=1}^{N} \delta_{oi} q_i, \quad \text{and} \quad Q_m = \sum_{i=1}^{N} \delta_{mi} q_i \quad \text{aggregation identities} \qquad (5\text{-}25)$$

$$MM = c + zQ \quad \text{supply for marketing services}$$

$$P_o = a - bQ_o \quad \text{component demand}$$

$$P_m = e - gQ_m \quad \text{component demand}$$

The complete model consists of $4N + 6$ equations and an equal number of endogenous variables. The model is large because, potentially at least, each farmer produces a unique quantity and quality of output and receives a unique price. [If the cost function $C(q_i, \delta_o, \delta_m)$ were the same for all farmers, the number of equations could be greatly reduced.]

The availability of outputs of various qualities reflects consumer preferences, for fat versus lean beef, for example, together with farm production costs. At this juncture the student should give some thought to how farm prices and the quantities and qualities of farm outputs vary with exogenous shocks, such as changes in (1) the demands for Q_o and Q_m, (2) prices of marketing inputs, (3) prices of farm inputs, and (4) technology at both the farm level and in the marketing channel for farm outputs. Suppose, for example, that Q_o and Q_m equal, respectively, milk fat and milk nonfat solids consumption and that the demand for fat increases. The price of fat P_o rises.

From Eqs. (5-24), we see this causes farmers to increase milk fat production Q_o. The production of milk nonfat solids Q_m and its price P_m may either increase or decrease depending on the nature of the underlying technology.

The importance of linking farm output quality and pricing can be appreciated by the further consideration of milk as an example. The Babcock test, introduced in 1890, was a simple and inexpensive method for measuring the fat content of milk. Prior to that time, some farmers added water to their milk; hence the expression "as suspicious as a trout in the milk." The widespread use of the Babcock test discouraged the watering and adulteration of milk and encouraged farmers to improve dairy herds. By the early 1940s, agricultural economists began stressing the need to take into consideration both the fat solids and nonfat solids (protein) content of milk in determining the pay prices to farmers. Such practices are now widespread and will doubtless have further impacts on dairy herds in light of the increasing importance of milk as a source of protein.

Although our analysis centers on only two valuable product components, the analysis can be generalized to include as many components as desired. It is entirely possible that different marketing firms will be interested in purchasing different packages of components. A market mechanism will ferret out those components that must be segregated if efficiency is to be achieved, consigning to limbo those components that are too costly to segregate in light of the demands for them.

The preceding formulation is based on the assumption that all deltas are known, but this assumption can easily be relaxed. Collecting information is a kind of production. The per unit marketing cost k may be interpreted as including the cost of determining the values of the deltas for each farmer. On this assumption, the previous analysis is valid as it stands. Assuming that the deltas are unknown does raise many interesting questions, however, two of which are explored briefly here. First, what assurance will farmers have that the tests or procedures used by marketing firms are accurate and will be applied fairly? Second, what happens if it is prohibitively expensive to determine the deltas for each farmer prior to combining one farmer's output with that of others?

There are several ways a marketing firm could allay fears of incorrectly estimating the deltas for a farmer. One way is to establish a reputation for fair dealing. An alternative is to have the grading performed by a disinterested third party, such as the U.S. Department of Agriculture. In fact, the department develops grade standards for cattle, poultry, dairy products, grain, and numerous other farm products. Since in most cases the grading is performed on a voluntary basis, those who use the service pay for it. It is of obvious importance in the establishment of grades to identify and measure accurately the deltas that are relevant to buyers and sellers and to ignore those that are not. The Department of Agriculture has a well-defined procedure that invites the opinions of experts and the general public as to grade definitions. Even so, the definition of grades by the department can be the source of dissatisfaction and controversy, as you might expect.

As a final issue in this introduction to pricing with product heterogeneity, it must be recognized that estimating the deltas for each individual farmer might be difficult or prohibitively expensive. This situation can lead to a serious economic problem if farm production practices that minimize cost of output also lead to poor quality. If farmers all receive a flat price with no differentials for quality, new private or public mechanisms may need to be created to avoid inefficient and perhaps even dangerous production. Production practices that use certain chemicals might be prohibited outright. Marketing firms may integrate backward into farm production through ownership and control or through contracts. There can be little doubt, for example, but that the widespread use of processor–grower contracts in the production of fruits and vegetables for processing reflects, in part, the desire of processors to coordinate farm production and processing operations in order to secure the desired quality of finished product.

5.3 EXPORTS AND IMPORTS OF FARM COMMODITIES

Having explored the theoretical underpinnings of the domestic demand for farm output, we now take up the farm-level demand for exports. This assumes, of course, that the country of interest is an exporter. As shown next, however, the analysis is easily modified for the case when the country is an importer. Of central interest are the effects of exogenous shocks in the international market on the prices of a country's domestic farm outputs.

Demand for Exports

We ignore transportation cost and the use of different currencies by different countries in the graphic derivation of the demand for exports given here. These complications are taken up later in an algebraic treatment. The supply and demand for food for the rest of the world (ROW) are given by S_w and D_w in panel a, Fig. 5.4. With no international trade, equilibrium price and output in the ROW equal P_{w1} and Q_{w1}. We assume for the moment, however, that the ROW does not have a comparative advantage in food production, that the price P_{w1} is higher than the price of U.S. food available in the international market. The ROW thus imports food. To obtain the U.S. demand for exports (for imports from the ROW viewpoint), we subtract laterally S_w from D_w in panel a, plotting the result in panel b. The D_x curve is the demand for U.S. exports. One aspect of this curve that the student should understand is that its slope in absolute value is less than that for D_w. In general, for any $Q_w > Q_{w1}$, the D_w curve will be more inelastic than will be the D_x curve for the corresponding import quantity. A numerical example is instructive. The ROW demand and supply are given by $Q_{wd} = 10 - P_w$ and $Q_{ws} = 2P_w$. (In numerical applications of the model it is

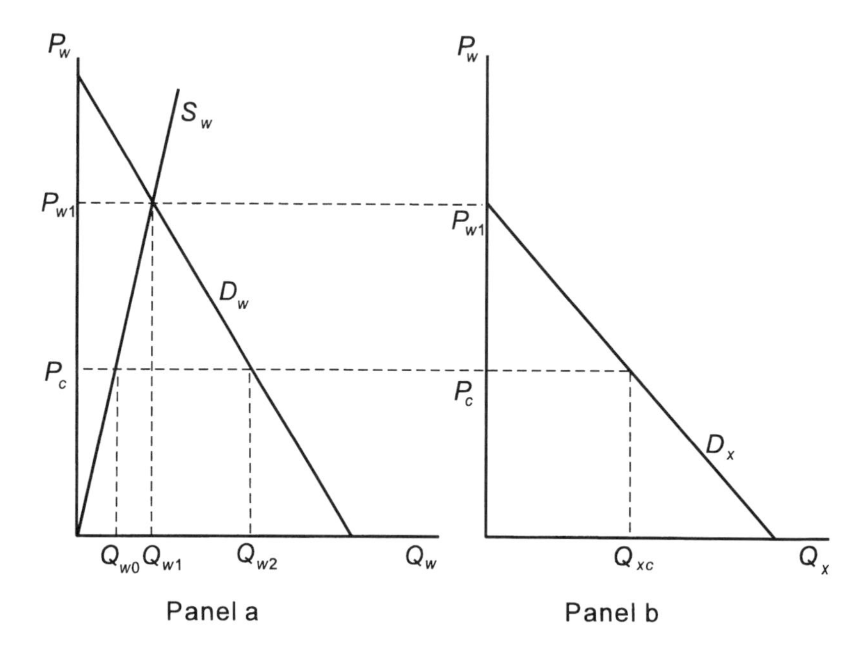

Figure 5.4 ROW domestic demand and supply for food and export demand.

convenient to use quantity-dependent functions.) The price corresponding to P_{w1} in Fig. 5.4 equals 3.33. The demand for U.S. exports is given by $Q_{xd} = Q_{wd} - Q_{ws}$, which, after the appropriate substitutions, yields $Q_{xd} = 10 - 3P_w$ for all $P_w \leq 3.33$. (This expression for Q_{xd} would become a supply for exports for $P_w > 3.33$, but more on this in a moment.) Consider a price $P_w = 2.0$. The corresponding quantities of Q_{wd} and Q_{dx} are 8.0 and 5.0. The corresponding elasticities of the ROW demand for food and the demand for U.S. exports are $-\frac{1}{4}$ and $-\frac{3}{2}$. The important point is this: If time is allowed for ROW output to respond to price changes, the demand for exports likely tends to be more elastic than the domestic demand. Under free-trade conditions, the demand for exports would likely add elasticity to the farm-level demand for U.S. farm output, and we have already seen the profound importance of demand elasticity in the determination of the effects of exogenous shocks on the farm sector.

To get on with the analysis, however, we next turn our attention to the United States. The U.S. supply and demand for food are given by S_u and D_u in panel b, Fig. 5.5. We note that for any $P_u > P_{u0}$ the United States produces more than it consumes. To derive the supply for U.S. exports, we subtract laterally U.S. demand from supply, plotting the result for all prices not less than P_{u0} in panel a. The demand for U.S. exports D_x from Fig. 5.4 is reproduced in Fig. 5.5.

Competitive equilibrium is given by the intersection of the demand and supply for U.S. exports, by the intersection of D_x and S_x in panel a, Fig. 5.5. Equilibrium

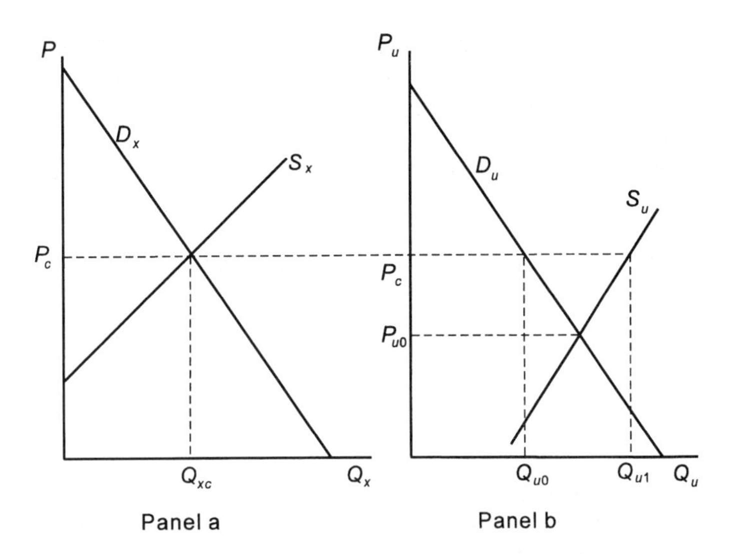

Figure 5.5 U.S. domestic demand and supply for food and export demand and supply.

price and quantity are given by P_c and Q_{xc}. By construction, $Q_{xc} = Q_{u1} - Q_{u0}$. Going back to Fig. 5.4, we see that Q_{xc} equals, also by construction, $(Q_{w2} - Q_{w0})$.

Simple though it may be, this analysis is powerful in explaining how exogenous factors affect U.S. agricultural prices. Two points merit emphasis. First, the analysis greatly expands the list of exogenous factors that can affect prices. All the exogenous variables that shift the farm product demand and supply functions in foreign countries are now seen to be potentially important in understanding U.S. price movements. For example, rising per capita incomes in the developing nations, as in Korea or Mexico, might expand significantly the demands for U.S. exports.

Second, because the demand for U.S. exports has in recent years grown in importance, relative to the domestic demand for consumption, the analysis of the manner in which farm input pricing is affected by exogenous shocks is greatly complicated. For one thing, some exogenous shocks affect both U.S. demand and supply, and this complicates comparative statics. For example, a worldwide technological change both decreases the demand for U.S. exports and increases the supply for U.S. output. Although it is easy to see that this exogenous shock lowers the U.S. price, the effect on U.S. output is uncertain.

In addition, we have seen in Chapter 4 that the elasticity of farm-level demand conditions in crucial ways the effects of exogenous shocks on farm input prices. Recall, for example, that an exogenous increase in the price of producer goods increases (decreases) farm labor and farm labor wages if demand is inelastic (elastic). Although we can be reasonably certain, on the basis of research, that the farm-level demand for domestic consumption is highly inelastic, the research on export demand elasticity is much less conclusive. It is probably fair to say that at present we do not know whether the long-run farm-level demand in the United States is elastic or inelastic. We will come back to a more detailed analysis of how export demand and domestic demand elasticities affect the elasticity of total farm-level demand, but first we show how our analysis can be generalized to allow for shipping costs and foreign exchange rates.

We let the long-run farm product demand and supply for the rest of the world (ROW) be given by

$$\begin{aligned} Q_{ws} &= a + bP_w \\ Q_{wd} &= c - dP_w \end{aligned} \tag{5-26}$$

where Q_{ws} equals quantity supplied, Q_{wd} equals quantity demanded, and P_w equals the price of wheat in German marks. Wheat serves as a proxy for farm output, and we use German marks for illustration. (Any foreign currency will do in a world with perfect foreign exchange markets.) With no trade, $P_{w1} = (c - a)/(b + d)$. Assuming that $P_w \le P_{w1}$, $(Q_{wd} - Q_{ws})$ yields what we define as the ROW's *excess demand:*

$$\begin{aligned} Q_x &= (c - a) - (b + d)P_w \\ &= i - jP_w \end{aligned} \tag{5-27}$$

The parameters i and j equal $(c - a)$ and $(b + d)$, respectively. This function is the counterpart of D_x in panel b, Fig. 5.4.

We now take advantage of arbitrage, defining an arbitrage condition as follows:

$$P_w = \alpha(P_u + T) \tag{5-28}$$

where the foreign exchange rate is given by α. More particularly, we let α equal the number of marks required to buy one dollar. We let P_u equal the U.S. price of wheat and T equal the per unit cost of transportation. Both P_u and T are stated in dollars. We assume that T is exogenous. The best way to understand arbitrage is to take an example. Let the U.S. price of wheat equal \$6.00 and T equal \$2.00. Suppose that 1 mark buys 2 dollars; 1 dollar buys 0.5 mark. Then $\alpha = 1/2$ and, according to Eq. (5-28), the ROW price of wheat equals 4 marks. If the ROW price were more than 4 marks, then profit could be made by spending marks on dollars and using dollars to import U.S. wheat. The effect of this would be to increase the U.S. price and decrease the ROW price. What happens if the ROW price were less than 4 marks is left to the student. Under perfectly competitive conditions and free trade, there will be neither pure profit nor losses from wheat transfers in long-run equilibrium. This implies that Eq. (5-28) must hold.

Armed with Eqs. (5-27) and (5-28), we find that the derivation of the demand for U.S. exports is easy. Use Eq. (5-28) to get rid of P_w in Eq. (5-27). This gives

$$Q_x = i - j\alpha(P_u + T) \tag{5-29}$$

which is the demand for U.S. exports. Because, in the U.S. market, commodities are purchased with dollars, the price is expressed in dollars.

The model for U.S. wheat is

$$\begin{aligned} &Q_d = m - nP_u && \text{U.S. domestic demand} \\ &Q_x = i - j\alpha(P_u + T) && \text{demand for U.S. exports} \\ &Q_s = w + vP_u && \text{U.S. supply} \\ &Q_s = Q_d + Q_x && \text{market clearing condition} \end{aligned} \tag{5-30}$$

We have four equations and four unknowns. The total farm-level demand for U.S. wheat can be found by adding Q_d and Q_x. Then solve for P_u:

$$P_u = \frac{1}{j\alpha + v + n}\left(m + i - w - j\alpha T\right) \tag{5-31}$$

If price P_u is to be positive, then $(m + i - w - j\alpha T) > 0$. Equation (5-31) is a reduced form function. It can be easily shown that $\partial P_u/\partial T < 0$ and $\partial P_u/\partial \alpha < 0$. (Note that $m + i - w$ must exceed zero.) An increase in T lowers export demand and decreases P_u. An increase in α, meaning that the number of marks needed to buy a dollar goes up, causes the export demand to decline; P_u falls. It is left to the student to show that both Q_s and Q_x decline as T or α increases. The reduced form for P_w is derived by substituting the right-hand side of Eq. (5-31) for P_u in Eq. (5-28). This yields

$$P_w = \frac{\alpha\left[m + i - w + (v + n)T\right]}{j\alpha + v + n} \tag{5-32}$$

We have

$$\begin{aligned} \frac{\partial P_w}{\partial T} &= \frac{\alpha(v + n)}{j\alpha + v + n} > 0 \\ \frac{\partial P_w}{\partial \alpha} &= \frac{(v + n)\left[m + i - w + (v + n)T\right]}{(j\alpha + v + n)^2} > 0 \end{aligned} \tag{5-33}$$

Although increases in T and α cause the U.S. price P_u to decline, they cause the ROW price P_w to rise. Wheat production in the ROW goes up; ROW consumption falls.

Demand for Imports

The preceding analysis may be easily modified to handle the case when the United States is an importer of food instead of an exporter. Go back to Eq. (5-26). Instead of supposing that $(Q_{wd} - Q_{ws})$ is positive in competitive equilibrium, we assume that $(Q_{ws} - Q_{wd})$ is positive. In other words, we assume that the ROW ships food to the United States. The arbitrage condition given by Eq. (5-28) must also be modified. Now P_w equals αP_u *minus* αT. Take an example. Suppose that P_w equals 2 marks, T equals 2 dollars, and $\alpha = \frac{1}{2}$. Then the U.S. price of wheat must equal 6 dollars. The ROW exporter buys wheat for 2 marks and ships the wheat to the United States, where it fetches 6 dollars that can be converted into 3 marks in the market for foreign exchange. From the 3 marks the exporter must deduct 1 mark (equals 2 dollars) to pay the cost of shipping. The exporter breaks even.

The model of the U.S. wheat market given by Eqs. (5-30) is now replaced by

$$
\begin{aligned}
Q_d &= m - nP_u && \text{U.S. domestic demand} \\
Q_m &= (c - a) + \alpha(b + d)(P_u - T) && \text{ROW export supply} \\
Q_s &= w + vP_u && \text{U.S. supply} \\
Q_s &= Q_d - Q_m && \text{market clearing condition}
\end{aligned}
\qquad (5\text{-}34)
$$

where Q_m equals the ROW exports to the United States (equals U.S. imports). The farm-level demand for U.S. production can now be found by subtracting the right-hand side of the equation for Q_m from that for Q_d. The U.S. demand for imports Q_m can be found using $Q_m = Q_d - Q_s$. Comparative static analysis of the effects of exogenous changes on the U.S. wheat market are left to the student.

5.4 ELASTICITIES OF EXPORT DEMAND AND TOTAL DEMAND

In what follows we first show that the elasticity of export demand is closely linked to the elasticities of the ROW demand and supply. We then show that the elasticity of total demand is a weighted average of the elasticities for the demand for domestic consumption and the demand for exports.

The ROW demand and supply are given by

$$Q_{wd} = D_w(P_w)$$
$$Q_{ws} = S_w(P_w) \tag{5-35}$$

Using the arbitrage condition given by Eq. (5-28), the demand for U.S. exports becomes

$$Q_x = D_w(\alpha P_u + \alpha T) - S_w(\alpha P_u + \alpha T) \tag{5-36}$$

We next differentiate Q_x with respect to P_u and convert the resulting expression into one involving elasticities. This gives

$$\frac{\partial Q_x}{\partial P_u}\frac{P_u}{Q_x} = \left(\frac{\partial D_w}{\partial P_w}\frac{P_w - \alpha T}{Q_{wd}}\right)\frac{Q_{wd}}{Q_x} - \left(\frac{\partial S_w}{\partial P_w}\frac{P_w - \alpha T}{Q_{ws}}\right)\frac{Q_{ws}}{Q_x} \tag{5-37}$$

The left-hand term is the elasticity of export demand. As T approaches zero, this elasticity approaches the elasticity of the ROW demand for consumption, weighted by Q_{wd}/Q_x, minus the elasticity of the ROW supply, weighted by Q_{ws}/Q_x. This result drives home the importance of distinguishing between short- and long-run export demand. Notice, further, that a large elasticity of ROW supply makes for an elastic demand for U.S. exports.

Turning to total U.S. farm-level demand, we have

$$Q_u = Q_d + Q_x$$
$$= D_d(P_u) + D_x(P_u) \tag{5-38}$$

We next differentiate Q_u with respect to P_u. Converting the resulting expression into one involving elasticities, we have

$$\frac{\partial Q_u}{\partial P_u}\frac{P_u}{Q_u} = \left(\frac{\partial D_d}{\partial P_u}\frac{P_u}{Q_d}\right)\frac{Q_d}{Q_u} + \left(\frac{\partial D_x}{\partial P_u}\frac{P_u}{Q_x}\right)\frac{Q_x}{Q_u} \tag{5-39}$$

The elasticity of total demand equals the elasticity of the demand for domestic consumption, weighted by the ratio of domestic consumption Q_d to domestic production Q_u, plus the elasticity of the demand for exports, weighted by the ratio of exports Q_x to domestic production Q_u. At present, there is widespread disagreement as to the magnitude of export demand elasticity, particularly in the long run. As a consequence, there is also much uncertainty as to the elasticity of total demand.

Thus far, we have abstracted from production lags and uncertainty in our discussion of the export demand for U.S. farm output. The insights gained from our analysis tend by and large to carry over to the more realistic cases that allow for uncertainty. For example, in an uncertain world, an increase in the ROW population tends to increase both the expected product returns to U.S. farmers and U.S. planned production.

The existence of production lags and uncertainty does, however, call attention to a dynamic issue that is important in understanding the pricing of farm outputs. Time is required for the response of U.S. and ROW outputs to changes in prices. If we hold constant planned production both here and abroad and ignore the demand for stocks, the stage is set for unstable markets and for wide price swings depending on the weather. Favorable weather both here and abroad increases U.S. output at the same time export demand contracts. Also, in the very short run, with fixed farm output abroad, the demand for U.S. farm exports is likely highly inelastic. The price in the United States, and in the ROW, too, could collapse because of high yields. It is just as easy, of course, to set the stage for sharp price hikes due to bad weather worldwide. The model implicit in these remarks is one of disarray and chaos. Before the student jumps from this model to the need for government intervention to stabilize prices and markets, however, it is of vital importance to recognize that we have yet to introduce the demand for stocks. This is the subject of Chapter 6, where it will be seen that speculation and storage in a competitive marketing system tend to stabilize prices and markets to a considerable degree, perhaps even to an optimal degree.

5.5 INTERREGIONAL COMPETITION

Sections 5.3 and 5.4 centered on international trade models for a single product. As we might expect, such models are similar to those for interregional trade except that, in the latter, we need not worry about foreign exchange. Much of what was said with regard to international trade applies to interregional trade as well. By the same token, much of what is said next with regard to interregional trade applies as well to the international scene. Although brief, our discussion of interregional trade models is intended to alert the student to an area of research of considerable importance in agricultural economics. As demand and supply conditions, transportation systems, and farm programs change, the farmers and marketing firms in some areas of the country find that they can no longer compete with their counterparts in other areas. At the time of this writing, for example, Midwest dairy farmers are concerned that they might not be able to compete very well with the large dairy operations in the Southwest. It would be a great saving to individuals and society as a whole if changes in the regional pattern of competitive production could be anticipated in order to forestall bad investments in farming and marketing facilities. As a consequence, agricultural economists are frequently called on to determine what regions have the comparative advantage in producing some particular product or set of products.

Our main objective in what follows is to generalize trade models to include more than two regions. We may start, however, by going back to the equations leading up to and culminating in Eq. (5-30). We assume two regions within a single country, with the subscript w indicating region W and the subscript u indicating region U. The parameter α is set equal to 1, and we let T equal the per unit cost of transporting output from one region to the other. Competitive pricing implies arbitrage such that the condition $(P_w - P_u) \leq T$ [or alternatively, that $(P_u - P_w) \leq T$] holds in equilibrium. The demand and supply curves from Fig. 5.5 are reproduced in Fig. 5.6, where we assume that region W is the high-cost region. The S_x curve in panel a is the excess supply curve for region U. The D_X curve is the excess demand curve for region W. Instead of letting $T = 0$, as in Fig. 5.5, we let T equal the length of the line segment T_0 in Fig. 5.6. In the absence of interregional trade, price in region U equals P_{u0}; price in region W equals P_{w1}. Since $(P_{w1} - P_{u0}) > T_o$, exports flow from region U to region W under competition. The equilibrium level of shipments equals Q_{xc}, which equals $(Q_{u2} - Q_{u1})$. With this level of shipments, $(P_{wc} - P_{uc}) = T_0$. Only with this set of prices will all participants be in competitive equilibrium. It is obvious that if T were large enough no trade would occur. It is also true, of course, that with different demand and supply conditions region W might ship to region U.

We now take up the case involving more than two regions. It is both fun and instructive to do so within the context of a numerical example. The demand and supply functions for regions 1, 2, and 3 are given by $D_1 = 20 - P_1$ and $S_1 = -4 + \frac{1}{2}P_1$;

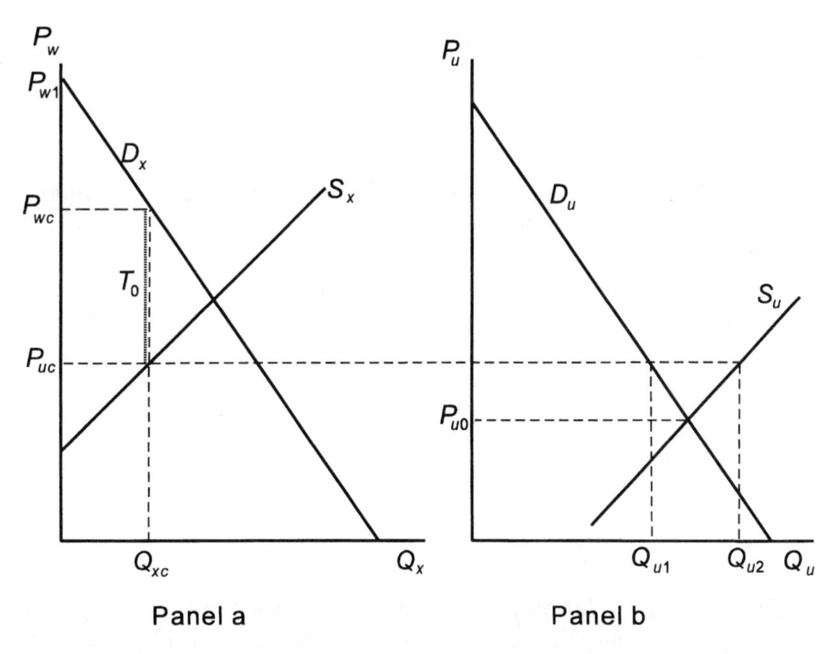

Figure 5.6 Demand and supply for shipments in high-cost region and local demand and supply in low-cost region.

$D_2 = 5 - \frac{1}{4}P_2$ and $S_2 = -4 + \frac{1}{4}P_2$; and $D_3 = 11 - P_3$ and $S_3 = -1 + P_3$. Letting T_{ij} equal the per unit cost of shipping from region i to region j, we assume that $T_{12} = T_{21} = 2$, $T_{13} = T_{31} = 4$, and $T_{23} = T_{32} = 3$. In the total absence of trade, we have $P_{10} = 16$, $P_{20} = 18$, and $P_{30} = 6$. Our problem is as follows: Given this information, how can we find competitive equilibrium allowing for interregional trade or shipments? To solve the puzzle, we propose using trial and error. Clearly, money can be made through shipping from region 3, the low-cost region, to either regions 1 or 2. Various possibilities present themselves. In competitive equilibrium, region 3 ships (1) to region 1 but not to region 2, (2) to region 2 but not to region 1, and (3) to both regions 1 and 2. Other possibilities come to mind, but these three seem to be the most likely. We check out the first alternative. Letting exports from region 3 equal X_3, the excess supply equation for region 3 is $X_3 = S_3 - D_3$ or $X_3 = -12 + 2P_3$. Letting M_1 equal region 1 imports, the excess demand equation for region 1 is $M_1 = D_1 - S_1$ or $M_1 = 24 - 1.5P_1$. Similarly, $M_2 = 9 - 0.5P_2$. Notice that the excess supply and demand equations do not appear to be affected by transportation costs, but appearances can sometimes be misleading, as we shall soon see.

Our first guess, not a very shrewd one as it turns out, is that region 3 ships to region 1, but not to region 2. According to this guess, the arbitrage conditions are $P_1 = P_3 + 4$ and $P_2 \leq P_3 + 3$. Also, $X_3 = M_1$, which implies that $[-12 + 2P_3 = 24 - 1.5(P_3 + 4)]$. The proposed competitive solution follows:

Region	Price	Consumption	Production	Exports	Imports
1	12.5714	7.4286	2.2857	0	5.1428
2	18	0.5	0.5	0	0
3	8.5714	2.4286	7.5714	5.1428	0

We quickly see that our guess is wrong, that this is not a competitive solution. Excess profit can be made by shipping from region 1 to region 2 and from region 3 to region 2. We must try again.

Our second guess is that region 3 ships to both regions 1 and 2. According to this guess, the arbitrage conditions are $P_1 = P_3 + 4$ and $P_2 = P_3 + 3$. Also, we have $X_3 = M_1 + M_2$, which implies that $[-12 + 2P_3 = 24 - 1.5(P_3 + 4) + 9 - \frac{1}{2}(P_3 + 3)]$. The proposed competitive solution follows:

Region	Price	Consumption	Production	Exports	Imports
1	13.375	6.625	2.6875	0	3.9375
2	12.375	1.9063	−0.9063	0	2.8125
3	9.375	1.625	8.375	6.75	0

This proposed solution looks plausible except for one glaring problem. Production in region 2 is negative. This example reminds us that prices, consumption, production, exports, and imports must be larger than or equal to zero; they cannot be negative. These nonnegativity constraints are obvious and trivial from the viewpoint of economic theory, but they are of basic importance in computational methods used to find solutions to numerical problems.

What we have discovered in our algebraic meandering is that very likely region 2 produces nothing in competitive equilibrium because the cost of production is too high relative to transportation cost and the cost of production elsewhere. This leads us to make still another guess. We propose that region 3 ships to regions 1 and 2 and that region 2 produces nothing. The excess supply equation for region 3 and the excess demand for region 1 are not affected by these new assumptions, but the excess demand for region 2 is now given by $M_2' = D_2 - S_2 = D_2$. In other words, the excess demand for region 2 is the demand. According to our current guess, the arbitrage conditions are again $P_1 = P_3 + 4$ and $P_2 = P_3 + 3$. Also, we have $X_3 = M_1 + M_2'$, which implies that $[-12 + 2P_3 = 24 - 1.5(P_3 + 4) + 5 - 0.25(P_3 + 3)]$. The proposed solution follows:

Region	Price	Consumption	Production	Exports	Imports
1	13.1333	6.8667	2.5667	0	4.3
2	12.1333	1.9667	0	0	1.9667
3	9.1333	1.8667	8.1333	6.2666	0

This proposed solution is the actual solution, as the student may easily verify. The student should recognize that excess demand and supply equations, arbitrage conditions, and nonnegativity constraints are the key concepts used in finding a solution that satisfies all the conditions of competitive equilibrium in the spatial dimension.

Drawing on these examples together with economic theory, we now consider a country divided into N regions. Let T_{ij} equal the cost of shipping from region i to j (or from j to i). In competitive equilibrium, arbitrage assures that $(P_i - P_j) \leq T_{ij}$, for all i and j. In each region, consumption equals either (1) production, (2) the sum of production plus imports from one or more regions, or (3) production minus exports to one or more regions. Obviously, negative prices and quantities are impossible.

Spatial patterns of prices and product shipments are inherent in these competitive conditions. Higher transportation costs tend to increase price differences and restrict product shipments. A worthwhile exercise is to consider the effects of shifts in demand or supply for one region or for all regions taken together. A technological change, for example, might lower supply functions in some regions more than in others, calling for a new spatial pattern of consumption, production, and product shipments.

As already noted, farmers and marketing firms are often concerned with regard to how changing economic conditions favor one part of the country over another. Agricultural economists are often called on to assess the future comparative advantage of some states or regions over others. An analyst with estimates of demand and supply functions for each region together with transportation costs could find the competitive solution through trial and error if a small number of regions is involved. Otherwise, computer software is available that allows finding solutions for trade models vastly more complicated than the three-region case considered here. The basis for such a general approach is presented in Appendix D.

PROBLEMS

5.1. Let $Q_r = 0.8Q$. There are 100 perfectly competitive marketing firms. The representative marketing firm's total variable cost function for marketing services is given by

$$TVC = 0.2q + 0.2q^2$$

where q measures marketing services produced. Total fixed cost per firm equals 20. The retail demand is

$$P_r = 1000 - 0.5Q_r$$

a. Derive the supply for marketing services both for a representative marketing firm and for the industry. Also, find the average variable cost function for the firm and the industry. [*Hint:* The marketing firm's profit may be written as

$$\pi = (0.8P_r - P)q - TVC - 20$$

where TVC equals the cost of marketing services and Pq equals the outlay on the raw material purchased from farmers.]

b. Derive the farm-level demand for farm output.

c. For $Q = 1500$, find the equilibrium values for P, P_r, and the marketing margin. Also, find the marketing firm's profit.

5.2. Farm-level demand and short-run supply are given by

$$P = 10\delta - 0.1 - (0.1 + 0.5\delta^2)Q$$

$$P = 1 + 0.1Q$$

A technological change increases δ by a small increment, thus "saving" or using a greater proportion of farm output.

a. Show that this technological change will increase P (and therefore farm income) if δ initially equals 0.5.

b. Show that this change will decrease P (and therefore lower farm income) if δ initially equals 0.9.

c. Find the level of δ such that $dP/d\delta$ equals zero.

5.3. Two grades of a farm commodity are produced. The price-dependent short-run supply functions for the two grades are

$$P' = 1 + 0.5Q'$$

$$P'' = 2 + 0.2Q''$$

For Q', $Q_o'/Q' = 0.6$ and $Q_m'/Q' = 0.2$. For Q'', $Q_o''/Q'' = 0.1$ and $Q_m''/Q'' = 0.7$. Let Q_w'' and Q_w' be waste components (perhaps water). The retail demands for the components are

$$P_o = 20 - 0.5Q_o$$

$$P_m = 15 - 0.2Q_m$$

The marketing cost per unit of Q' and Q'' is the same, equaling 0.3. Find the competitive equilibrium values for P', P'', Q', Q'', P_o, P_m, Q_o, and Q_m.

5.4. You are given the following information:

$$C = (\delta_o^2 + \delta_m^2)q^2$$

where $\delta_o^2 + \delta_m^2 = 0.5; 0 \le \delta_o \le 0.707;$ and $0 \le \delta_m \le 0.707$.

a. Treating δ_o as the dependent variable, find the equation for the Output Grade Transformation curve and demonstrate that

$$\delta_o = \left(\frac{0.5P_o^2}{P_o^2 + P_m^2} \right)^{0.5}$$

b. Find and compare the optimal values for δ_o,δ_m, and q for $P_o = 2$, $P_m = 4$, and $MM = 1$ and for $P_o = 4$, with P_m and MM unchanged.

5.5. Demand and supply for the rest of the world are given by

$$Q_{wd} = 25 - 0.5P_w$$

$$Q_{ws} = 2 + P_w$$

Domestic demand and supply for the United States are given by

$$Q_d = 18 - 0.5P_u$$

$$Q_s = 3.5P_u$$

Price P_w is measured in British pounds and P_u is measured in dollars. Let 1 pound buy \$1.55 in the market for foreign exchange. It costs 80 cents per unit to ship output from the United States to the rest of the world.

a. Derive the export demand for the United States.

b. Find the equilibrium levels for P_w, P_u, Q_d, Q_s, Q_{wd}, Q_{ws}, and the level of U.S. exports Q_x.

c. The exchange rate changes such that 1 pound buys \$1.60. Find the new export demand function and compare it with that found in part a. Does the export demand rise or fall?

REFERENCES

Bredahl, M., W. Meyers, and K. Collins, "The Elasticity of Foreign Demand for U.S. Agricultural Products," *American Journal of Agricultural Economics,* 61, no. 1 (1979), 58–63.

Cochrane, Willard W., *Farm Prices: Myth and Reality*. Minneapolis: University of Minnesota Press, 1958.

Council of Economic Advisors, *Economic Report of the President*. Washington, D.C.: U.S. Government Printing Office, 1990.

Ladd, George W. *Research on Product Characteristics: Models, Applications, and Measures,* Iowa Agriculture and Home Economics Experiment State Research Bulletin 584, University of Iowa, Ames, February 1978.

Rosen, Sherwin, "Hedonic Prices and Implicit Markets: Product Differentiation in Pure Competition," *Journal of Political Economy,* 82 (1974), 34–55.

Schultz, Theodore W., "The Farm Problem and Its Relation to Economic Growth and Development," *Policy for Commercial Agriculture: Its Relation to Economic Growth and Stability*, Joint Economic Committee, 85th Congress, 1st Session, 1957.

Takayama, Takashi, and George G. Judge, *Spatial and Temporal Price and Allocation Models*. Amsterdam: North-Holland Publishing Co., 1971.

U.S. Department of Agriculture, *Food Consumption, Prices, and Expenditures 1968–89,* Economic Research Service, Statistical Bulletin No. 825, 1991.

U.S. Department of Agriculture, *Food Costs: From Farm to Retail in 1993,* Economic Research Service, Agricultural Information Bulletin No. 698, April 1994.

6

The Theory of Commodity Storage and Pricing

In textbooks and courses in microeconomics, from freshman through Ph.D. levels, the theory of demand for current consumption is given a great deal of attention, while the theory of the demand for storage, which allows for consumption later, is hardly ever mentioned. This lack of balance is most unfortunate from the viewpoint of understanding how farm commodity prices get determined and how farm markets work in practice. In the period following crop harvest, for example, the demand for storage greatly exceeds the demand for current consumption. Storage also plays a key role in other commodity markets, including those for animal products. In addition, although farmers, farm leaders, and economists have long expressed concern about instability of farm prices and incomes over time, stability issues can only be analyzed properly, as shown later, if the demand for storage is assigned a central role. Why the demand for storage has received scant attention in the traditional literature is a question that need not detain us here. We merely note that the development of the theory of commodity pricing and storage was fostered considerably by three events that occurred since World War II: the development of the rational expectations hypothesis, the development of numerical methods for solving certain dynamic problems, and the invention of the high-speed computer.

The main objectives of this chapter are to use theory to elucidate the role of storage in the determination of market performance and in the stabilization of price and consumption through time. We show how this role is affected by changes in storage cost, the interest rate, and other exogenous variables. The theory of storage and pricing is complex because it must take account of both price expectations and dynamics. The following analysis starts with simple cases used as steppingstones to those that are more complex but also more realistic. The first case involves intrayear storage only and does not allow for the substitution of crop produced in the current year for production in the next. Ignoring intrayear storage, Section 6.2 centers on interyear storage, partly to set the stage for farm policy analysis in Chapter 9. Sections 6.3 and 6.4 center on a quarterly model of pricing, bringing together and integrating important ideas from this and earlier chapters.

6.1 INTRAYEAR STORAGE

Suppose that after an annually produced crop is harvested the output is stored for two periods, two months, say. After that the output spoils and must be discarded. As in the case of late summer potatoes or onions, intrayear storage is possible, but interyear storage is not.

Let consumption demand in the first, postharvest month be given by

$$Q_1 = \beta_0 - \beta_1 P_1 + e_1 \tag{6-1}$$

where Q_1 is the consumption quantity, P_1 is the price, and e_1 is a random or stochastic variable. The subscript 1 indicates period 1. Although known in the first period, the value of e_1 is not known in the initial period when production occurs. The expectation of e_1 formed in the initial period, $E_0(e_1)$, equals zero.

Clearing the market requires

$$Q_1 + I_1 = H_0 \tag{6-2}$$

where I_1 equals the aggregate quantity of output committed to storage in the first period for consumption in the second and H_0 equals total harvest in period 0, which is known at the beginning of the first period. We now have two equations, (6-1) and (6-2), and three endogenous variables, Q_1, P_1, and I_1.

To complete the model, we analyze the decisions of arbitrageurs who organize storage operations. As noted in Chapter 5, arbitrage involves purchasing a product in a low-price market and selling it in a high-price market. The two markets might be linked in product form through processing operations, through space by a transportation system, or, as in the case at hand, through time by commodity storage. Carried out on a competitive scale, arbitrage equates market prices subject to differentials that arise because of the marginal cost of processing, transfer, or storage.

Storage in the Absence of Hedging

Profit for an arbitrageur who buys the product at the beginning of period 1 and sells it at the beginning of period 2 is given by

$$\pi_2 = P_2 s_1 - P_1 s_1 - i(P_1 s_1) - C(s_1) - TFC \tag{6-3}$$

where π is profit, P is price, s_1 is the quantity of the commodity purchased in period 1 for sale in period 2, i is the interest rate per period, and $C(s_1)$ is the total variable cost function for the use of bin space. The subscripts 1 and 2 indicate periods 1 and 2. (In a departure from our usual convention, we let s_1 equal the level of stocks held by a single arbitrageur, with I_1 being the corresponding aggregate.) The total cost of carrying stocks from period 1 to period 2, often called the carrying cost or charge, consists of four components: $P_1 s_1$ is the cost of acquiring the commodity in the first period; the product $i(P_1 s_1)$ is the opportunity cost of the capital tied up in the holding of stocks, capital that could have earned interest in a savings account; $C(s_1)$ is the cost of bin space, which is often assumed in the literature to be a linear function of the level of bin space rented or operated; and TFC is total fixed cost. It is assumed that the cost of bin space is paid in period 2 when stocks are sold. Second-period demand is uncertain, which means that P_2 must be viewed as a random variable.

For analysis, it is convenient to rewrite the profit function as follows:

$$\pi_2 = [P_2 - (1 + i)P_1]s_1 - C(s_1) - TFC \tag{6-4}$$

where the bracketed expression $[P_2 - (1 + i)\, P_1]$ may be thought of as an uncertain net price. To keep the analysis simple, we assume that arbitrageurs are risk neutral and strive to maximize expected profit; we leave until later analysis that allows for hedging. Maximizing expected profit implies the first-order condition

$$\Theta_p - (1 + i)\, P_1 = C'(s_1) \tag{6-5}$$

where $\Theta_P = E_1(P_2)$ and $C'(s_1) = dC/ds_1$. Solving Eq. (6-5) for s_1 yields the arbitrageur's *supply function for storage*:

$$s_1 = s[\Theta_P - (1 + i)P_1] \tag{6-6}$$

This assumes that the optimal level of s_1 is positive, an assumption that is relaxed later.

With N arbitrageurs, aggregate supply is given by

$$\begin{aligned} I_1 &= Ns_1 \\ &= Ns[\Theta_P - (1 + i)P_1] \end{aligned} \tag{6-7}$$

which shows the level of storage in the first period as a function of the price expected in the second period Θ_p, the interest rate, and the first period price P_1. The aggregate supply curve is upward sloping in Θ_p^i and downward sloping in i and P_1. Also, an upward shift in the total variable cost function for bin space would contract the aggregate supply for storage.

This brief foray into the economics of storage decisions has as its payoff an aggregate supply for storage, Eq. (6-7), that can be added to Eqs. (6-1) and (6-2), thus yielding a three-equation system. If we treat the expected price $\Theta_p = E_1(P_2)$ as exogenous, this system could be used to solve for Q_1, P_1, and I_1. But, again, if our intention is to explain market performance, treating expected price as exogenous is unsatisfactory. We have added an equation, but we have also added another variable, $E_1(P_2)$, that must be treated as endogenous. In the modeling of farm commodity markets, one complication often leads to another. To make Θ_p endogenous, we again invoke Muth's hypothesis of rational expectations.

To simplify analysis and to bring out more clearly the basic principles of interest, we will assume in what follows a specific cost function for the use of bin space. In particular, suppose that $C(s_1) = \gamma s_1^2$. Then $dC/ds_1 = 2\gamma s_1$ and, since the maximization of expected profit implies that

$$\frac{dC}{ds_1} = 2\gamma s_1 = \Theta_p - (1+i)P_1$$

we have

$$\begin{aligned} I_1 &= Ns_1 \\ &= \frac{N}{2\gamma}\left[\Theta_p - (1+i)P_1\right] \end{aligned} \tag{6-8}$$

The aggregate supply function for storage is a linear function of the expected net price $[\Theta_p - (1 + i)P_1]$. Assume, for the moment, that the expected net price is nonnegative to assure that $I_1 \geq 0$. The carryout cannot be negative.

At this point in the argument, we consider the model of pricing in the second period. We have

$$\begin{aligned} P_2 &= \beta_0' - \beta_1' Q_2 + e_2 \qquad &\text{inverse demand} \\ Q_2 &= I_1 \qquad &\text{market clearing} \end{aligned} \tag{6-9}$$

The expectation of e_2 formed in period 1, $E_1(e_2)$, equals zero by assumption. Solving for P_2 and taking the expectation as of period 1, we have

$$E_1(P_2) = \beta_0' - \beta_1' I_1 \tag{6-10}$$

According to previous convention, Eq. (6-10) is called the expected price function. It relates the expected second-period price to the carryout I_1. Under rational expectations, arbitrageurs understand the pricing system. Although they do not know what e_2 will be, they do know what the parameters of Eq. (6-9) are. They realize fully that next period's price will depend on how much output is carried forward from period 1. They also know the probability distribution for e_2, which will be seen to be an important point later.

The structural model for period 1 may now be specified as follows:

$$Q_1 = \beta_0 - \beta_1 P_1 + e_1 \qquad \text{first-period demand} \qquad (6\text{-}11a)$$

$$Q_1 + I_1 = H_0 \qquad \text{market clearing} \qquad (6\text{-}11b)$$

$$I_1 = \frac{N}{2\gamma}\left[E_1(P_2) - (1+i)P_1\right] \qquad \text{supply for storage} \qquad (6\text{-}11c)$$

$$E_1(P_2) = \beta_0' - \beta_1' I_1 \qquad \text{expected price function} \qquad (6\text{-}11d)$$

This four-equation model has four endogenous variables: Q_1, P_1, I_1, and $E_1(P_2)$.[1] Using the last equation to eliminate $E_1(P_2)$ leaves the first-period demand, the market clearing condition, and a function that is called the *demand for storage*, given by

$$I_1 = \frac{N}{2\gamma + N\beta_1'}\left[\beta_0' - (1+i)P_1\right] \qquad (6\text{-}12)$$

The demand for storage shows how much arbitrageurs will commit to storage depending on the first-period price. An increase in β_0' expands expected second-period demand and increases the demand for storage in the first period. If carrying costs increase because the cost of bin space rises (γ increases), the demand for storage falls. An increase in the interest rate has a similar effect. *The demand for storage is an important concept in understanding the performance of farm markets, and the student should know that it is derived by using the expected price function to eliminate the expected price from the aggregate supply for storage.*[2]

With the elimination of expected price, the market model consists of Eqs. (6-11a), (6-11b), and (6-12). This three-equation system can be solved for the solutions or reduced form equations for the three endogenous variables Q_1, I_1, and P_1. For example, the reduced form for P_1 is given by

$$P_1 = \frac{(2\gamma + N\beta_1')(\beta_0 - H_0 + e_1) + N\beta_0'}{\beta_1(2\gamma + N\beta_1') + N(1+i)} \qquad (6\text{-}13)$$

[1]If arbitrageurs were risk averse, the aggregate quantity of storage I_1 would become a function of both $E_1(P_2)$ and $V_1(P_2)$. A variance of price function would then need to be added to Eqs. (6.11a) through (6.11d) to complete the model. In this simple linear model, $V_1(P_2) = V_1(e_2)$.

[2]It is sometimes asserted in agricultural economics textbooks that the quantity and price of stocks are determined by the supply and demand for storage, where the price is defined as the difference between the expected price (or futures price) and the current price. This assertion rests on a flawed theory that does not provide an adequate explanation of expected prices. As we have seen, the level of stocks is determined by competition among buyers of a commodity for consumption (exports, processing, seed, etc.) *and* for storage, with the supply of the commodity being predetermined.

We can easily see how changes in exogenous variables, the interest rate i, for example, affect P_1.

Alternatively, a graphic analysis of the three-equation system is straightforward. Figure 6.1 gives the demand for consumption D_c, the demand for storage D_I, and predetermined supply H_{01}, all on the understanding that exogenous variables are held constant. The horizontal summation of D_c and D_I yields aggregate demand D. The intersection of D and S gives the equilibrium price in period 1, that is, P_{1c}. Equilibrium consumption and storage equal, respectively, Q_{11} and I_{11}. (We defer until later what happens when $H_0 < H_0'$.) A ceteris paribus increase in β_0' expands the demand for storage and increases both P_1 and I_1; Q_1 falls. A ceteris paribus increase in β_0 (first-period demand expands) increases P_1 and Q_1, but I_1 falls. What happens if H_0 expands? What happens if the carrying charge (γ or i) increases? Importantly, the student should recognize that, once the equilibrium value for I_1 is determined, the solution for $E_1(P_2)$ can be determined using the expected price function.

Up to this point in the analysis, the harvested output at the end of the initial period ($t = 0$) is taken as predetermined or exogenous. To complete the model, we now consider the determination of output. We abstract from production uncertainty and assume that the producers of the output are risk neutral. Maximizing expected profit implies that the farmer optimizes output by equating the marginal cost of production to the price expected to prevail at harvest time. With these assumptions, the aggregate supply function is given by

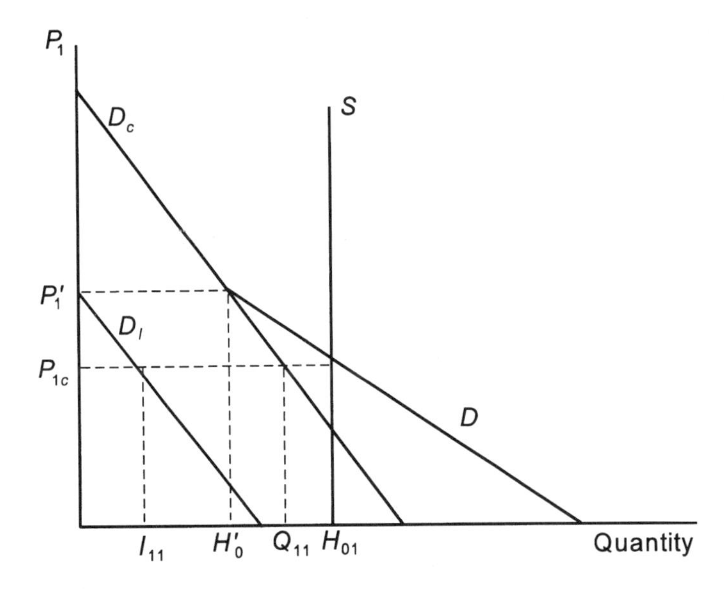

Figure 6.1 Demands for consumption and storage and total demand and predetermined supply.

$$H_0 = S[E_0(P_1)] \tag{6-14}$$

The value of e_1 in the first-period demand, although known in the first period, is not known in the initial period. This explains why, in period 0, P_1 must be viewed as a random variable. How, then, can producers compute $E_0(P_1)$? Go back to the system given by Eqs. (6-11a) through (6-11d). The system is linear with H_0 treated as exogenous. This means that the reduced form solutions are also linear. The reduced form for P_1, given by Eq. (6-13), is a case in point. On the assumption of rational expectations, we take the expectation of P_1 according to Eq. (6-13), which yields the expected price function for period 0:

$$E_0(P_1) = \frac{(2\gamma + N\beta_1')(\beta_0 - H_0) + N\beta_0'}{\beta_1(2\gamma + N\beta_1') + N(1+i)} \tag{6-15}$$

Equations (6-14) and (6-15) form a two-equation system that can be solved for the solutions or reduced forms for the two unknowns H_0 and $E_0(P_1)$.

Graphically, the relationship between H_0 and $E_0(P_1)$ according to the supply Eq. (6-14) is upward sloping; that for the expected price function, Eq. (6-15), is downward sloping. The intersection of the graphs of the supply and expected price functions yields the equilibrium. The effects of exogenous shocks can again be derived. If, for example, the interest rate rises, the graph of the expected price function falls. Both the expected price $E_0(P_1)$ and H_0 decrease. The intuition behind this result is simply that an increase in the interest rate makes storage more expensive, driving up prices in the second period and lowering consumption.

At this point in the analysis, the student should look back on what has been accomplished. A determinate structural model has been derived to explain market performance in the initial period. We can tell the world how the production of the commodity, late summer potatoes or whatever, is determined in the initial production period. We could not have done this, however, if we had not first derived and solved a complete model for the market in period 1, which led to a reduced form expression for P_1. But notice that we could not have derived and solved the market model for period 1 without having first derived the complete model for period 2. *In short, to model the market for the initial period, we first had to model the market for the last period!* This is an extremely important strategy in the development of dynamic theories in general and of storage and pricing theories in particular.

Our model of a market that allows for storage can be modified or generalized in various ways. We now consider, in turn, a simplification and a complication that are both important for understanding the findings of recent work on commodity storage and pricing. This model could be simplified by replacing the supply for storage given by Eq. (6-11c) with the following arbitrage condition:

$$E_1(P_2) = (1 + i)P_1 + C_1 \tag{6-16}$$

where C_1 is the per unit cost of bin space, which is assumed to be constant for all levels of I_1. This arbitrage condition may be justified by assuming a long-run model in which keen competition among arbitrageurs always drives expected profit to zero and where the storage industry is a constant-cost industry. (Neither the interest rate nor the per unit cost of bin space rise with increases in I_1.) Adopting this simplification leads to the following demand for storage:

$$I_1 = \frac{\beta_0' - C_1 - (1+i)P_1}{\beta_1'} \tag{6-17}$$

Graphic analysis based on a figure like Fig. 6.1 is straightforward and need not detain us here.

We turn next to a remarkable complication that held back the development of the theory of pricing and storage for many years. The complication is remarkable because on the surface it looks like a mere technicality, whereas, in practice, dealing effectively with it requires the use of a computer. The objective is to generalize the preceding model to allow for economic situations in which $E_1(P_2)$ is so low, less than $[(1 + i)P_1 + C_1]$, that storage in the first period does not occur (i.e., where $I_1 = 0$).

Consider the following numerical model in which an arbitrage condition, similar in some respects to Eq. (6-16), replaces the supply for storage:

$$\begin{aligned} &Q_1 = 10 - P_1 + e_1 && \text{first-period demand} \\ &Q_1 + I_1 = H_0 && \text{market clearing} \\ &E_1(P_2) \le P_1 + 4 && \text{arbitrage condition} \\ &E_1(P_2) = 10 - I_1 && \text{expected price function} \end{aligned} \tag{6-18}$$

For simplicity, the interest rate is set equal to zero and we let $C_1 = 4$. Furthermore, assume that e_1 takes on three values, $-4, 0, +4$, with equal probabilities. If $E_1(P_2) = P_1 + 4$, then either I_1 is positive or, as a corner solution, I_1 equals zero. Because keen competition drives expected profit from arbitrage to zero, we need never worry about the possibility that $E_1(P_2) > P_1 + 4$. If we could be sure in advance that $E_1(P_2) = P_1 + 4$, then simple algebra can be used to show that

$$P_1 = 8 - 0.5H_0 + 0.5e_1$$

and $E_0(P_1) = 8 - 0.5H_0$. The previous analysis, which assumed that I_1 was always positive, applies.

If, however, we want to allow for the possibility that $E_1(P_2) < P_1 + 4$, then the first-period price is no longer a linear function of H_0. With the inequality appearing in the arbitrage condition, the model given by Eqs. (6-18) is highly nonlinear and surprisingly difficult to analyze. Anyone who thinks otherwise should try to find the reduced form for P_1. Going back to Fig. 6.1, if H_0 is less than H_0', price is determined by S and D_c. If H_0 is larger than H_0', price is determined by S and D. Two alternative structural models are involved in the determination of price P_1, and in the initial period it cannot be determined in advance which model will apply because demand D_c is stochastic. A sufficiently strong first-period demand, for example, might drive P_1 up so high that no storage occurs.

If we cannot find the reduced form for P_1, how can we derive the counterpart to Eq. (6-15), which is needed to complete the model for the initial period? As it turns out, given models like Eqs. (6-18), numerical methods can be used to approximate the expected price function for period 0 as closely as desired. Here is where a computer comes in handy, but for our illustrative model a pencil and some scratch paper will do very well.

The objective is to approximate the relationship

$$E_0(P_1) = f(H_0) \tag{6-19}$$

in light of Eqs. (6-18). As noted, the first-period price is determined by one or the other of two structural models. If $E_1(P_2) < P_1 + 4$, then $I_1 = 0$, and $P_1 = 10 - H_0 + e_1$. We can forget about period 2. If $E_1(P_2) = P_1 + 4$, then $I_1 \geq 0$, and $P_1 = 8 - 0.5H_0 + 0.5e_1$. To estimate Eq. (6-19), we set H_0 equal to several different values and calculate $E_0(P_1)$ for each value using trial and error. If, for example, $H_0 = 2$, then $E_0(P_1|H_0 = 2) = 8.3333$. How is this figured? We estimate that price P_1 equals 12, 8, and 5 with equal probabilities depending on the value of e_1. By trial and error, we discover that we must use the equation $P_1 = 10 - H_0 + e_1$ when H_0 equals 2 and e_1 equals either 0 or +4. For these two values of e_1, the carryout I_1 would be negative if we let $P_1 = 8 - 0.5H_0 + 0.5e_1$ with H_0 set equal to 2. We use $P_1 = 8 - 0.5H_0 + e_1$, however, when $e_1 = -4$. The values of $E_0(P_1)$ conditional on H_0 set equal to 2, 4, 6, 8, and 10 are thus estimated using trial and error. The results are plotted in Fig. 6.2. When joined by straight lines, the resulting spline curve labeled $f(H_0)$ is the approximation that we seek. Its intersection with the supply curve S_0 gives market equilibrium. [Although the demands are linear, $f(H_0)$ is nonlinear and convex to the origin.]

Two further aspects of this numerical procedure may be noted. First, by choosing a large number of possible values for H_0, $f(H_0)$ can be approximated as closely as desired by a spline function, even though the actual function is smooth. As an alternative to a spline function, the analyst could quantify $f(H_0)$ using any one of several statistical techniques.

Second, not having an exogenous shifter in first-period demand saved a lot of work. To see this, suppose that we add shifter Z_1 to first-period demand, where Z_1 and e_1 both take on 10 different values with known probabilities. Then our procedure calls for estimating 100 possible prices for each value of H_0 in order to estimate $E_0(P_1)$, always being careful to discard prices associated with negative carryouts. If

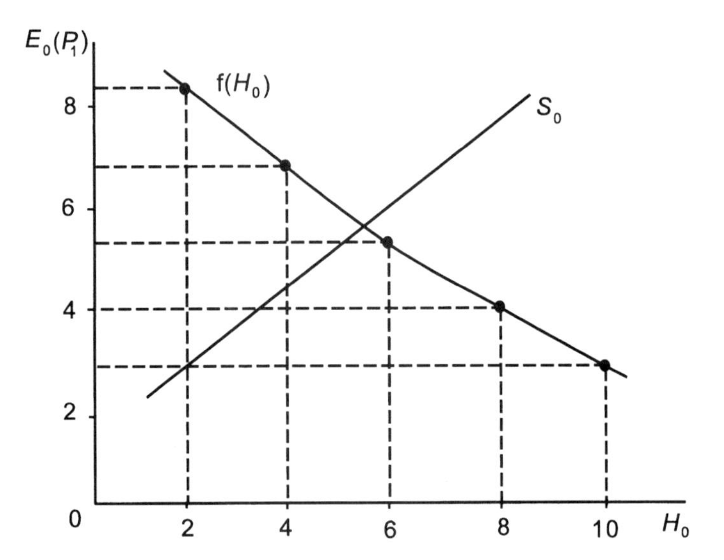

Figure 6.2 Expected supply and approximated expected price function.

we add still another shifter Y_1 that also takes on 10 different values with known probabilities, then 1000 prices would need to be calculated for each value of H_0. Suppose further that we wanted to distinguish between the stochastic demand for domestic consumption and the stochastic demand for exports and that we wanted to treat the interest rate and the cost of bin space as random variables. Clearly, the number of possible prices that would need to be computed for each value of H_0 could easily slip into the millions, and bear in mind that we are still considering a simple case. This is why computers are needed to solve the storage and pricing problems of major interest in studies of commodity markets, studies that we will consider in a moment.

Storage with Hedging

We now modify our analysis to allow for the role of hedging in decisions to store and to produce output. Futures markets exist for a wide range of commodities, such as corn, wheat, and cotton; hedging by arbitrageurs is commonplace.

The profit function for an arbitrageur who has the option of speculating in a futures market is given by

$$\pi_2 = [P_2 - (1 + i)P_1]s_1 - C(s_1) - TFC + [F_1(P_2)g_s - P_2 g_s] \tag{6-20}$$

where $F_1(P_2)$ is the price of futures and g_s is the quantity of futures sold in period 1 and bought in period 2. Expected profit and variance of profit are given by

$$\Theta = \Theta_P(s_1 - g_s) - (1 + i)P_1 s_1 - C(s_1) - TFC + F_1(P_2)g_s$$
$$\sigma^2 = \sigma_P^2(s_1 - g_s)^2 \tag{6-21}$$

Let the arbitrageur's utility function be given by $u_1 = \phi_0 + \phi_1\Theta - \phi_2\sigma^2$, where $\phi_1 > 0$ and $\phi_2 > 0$ under risk aversion. Then the first-order conditions for utility maximization are given by

$$\frac{\partial u_1}{\partial s_1} = \phi_1\left[\Theta_P - (1+i)P_1 - C'(s_1)\right] - 2\phi_2\sigma_P^2(s_1 - g_s) = 0 \tag{6-22a}$$

$$\frac{\partial u}{\partial g_s} = \phi_1\left[F_1(P_2) - \Theta_P\right] - 2\phi_2\sigma_P^2(g_s - s_1) = 0 \tag{6-22b}$$

Solving Eq. (6-22b) for g_s as a function of s_1 and using the resulting equation to get rid of g_s in Eq. (6-22a) lead to

$$F_1(P_2) - (1 + i)P_1 = C'(s_1) \tag{6-23}$$

which, when solved for s_1, yields the new supply function for storage. The optimal value for the level of speculation, as measured by the value of g_s, depends on Θ_P. If, for example, $F_1(P_2) = \Theta_P$, then the arbitrageur hedges completely ($s_1 = g_s$) and the optimal variance of profit equals zero. (We will not be concerned in what follows with the level of speculation on the part of arbitrageurs.) At this point, of course, the student will realize that the present analysis is but a slight variation of that given in Chapter 3, where the focus was on a farmer hedging his or her production decisions in the absence of production uncertainty.

On the assumption that $C(s_1) = \gamma s_1^2$, as before, our new aggregate supply function for storage is derived thus:

$$\begin{aligned} I_1 &= Ns_1 \\ &= \frac{N}{2\gamma}\left[F_1(P_2) - (1+i)P_1\right] \end{aligned} \tag{6-24}$$

The system given by Eqs. (6-11) is modified by replacing the old supply for storage with Eq. (6-24) and adding a new equation, which asserts that

$$F_1(P_2) = E_1(P_2) \tag{6-25}$$

This, too, goes back to Chapter 3 where it was argued that futures market processes may be viewed as an estimation procedure. Importantly, Eqs. (6-24) and (6-25) together with the last equation of Eqs. (6-11) constitute a three-equation system with the four unknowns: I_1, $F_1(P_2)$, P_1, and $E_1(P_2)$. Using two equations to rid the system of $F_1(P_2)$ and $E_1(P_2)$ yields the demand for storage, which is exactly the same as that given by Eq. (6-12). What this means is that the model given by Eqs. (6-11a), (6-11b), and (6-12) can be derived through assuming that arbitrageurs are risk neutral or, more realistically, that arbitrageurs are risk averse but hedge storage operations by speculating in the market for futures. This is an important result.

We may now proceed one step further by modifying the model for the initial production period. Allowing for a futures market and hedging, the supply function for output in the initial period is given by

$$H_0 = S[F_0(P_1)] \tag{6-26}$$

where $F_0(P_1)$ is the price of first-period futures as determined in the initial period. This assumes that production uncertainty does not exist (see Chapter 3). The expected price function for the initial period is given by Eq. (6-15) in the case when storage from the first to the second period is known always to occur *or* by Eq. (6-19) as approximated by a figure such as Fig. 6.2 in the case when

$$E_1(P_2) \leq (1 + i)P_1 + C_1$$

In either case, the model for the initial period may be completed by assuming that $F_0(P_1) = E_0(P_1)$.

This completes our exploration of the theoretical underpinnings of intrayear storage and pricing models. The assumption of rational expectations is basic to the analysis. In addition, two strategies from dynamic analysis are employed. One strategy involves developing a complete model for the current or initial period by developing and analyzing a model for the last period first and then working backward toward the current period one period at a time. The second strategy involves using trial and error methods to quantify expected price functions, methods that are unavoidable if the arbitrage condition involves an inequality. The payoff from the analysis consists of expected price and storage supply functions that can be used to derive the demand function for storage. The demands for consumption and storage together with predetermined output constitute important components of all the storage and pricing models considered later.

6.2 INTERYEAR STORAGE

The preceding models centered on intrayear storage. We turn next to the problem of interyear storage, where in a year of abundant supplies old crop might be carried over into the new crop year. Field crops provide many important examples. Intrayear storage is ignored, but this omission is attended to in Section 6.3, which integrates both intrayear and interyear storage. A steppingstone to Section 6.3, the interyear model is also basic to models of market stabilization schemes to be studied in Chapter 9.

The consumption demand for a field crop is given by

$$Q_t = D(P_t) \tag{6-27}$$

where Q_t and P_t equal, respectively, consumption and price in year t. To simplify, we assume that demand is not stochastic. Acreage planted in period t is an upward sloping function of the price expected to prevail at harvest time (i.e., in period $t + 1$). We have

$$A_t = S[E_t(P_{t+1})] \tag{6-28}$$

Although the crop is planted in period t, yield depends on subsequent weather. Production equals acreage planted the previous year, A_{t-1}, times current random yield per acre given by Y_t. We have

$$H_t = A_{t-1}\, Y_t \tag{6-29}$$

Market clearing requires

$$H_t + I_{t-1} = Q_t + I_t \tag{6-30}$$

where I_t is the carryout from one crop year to the next. Assume that yield is exogenous and stochastic, with a known probability distribution.

We next assume that keen competition among arbitrageurs does not allow positive expected profits from arbitrage. More particularly, we assume an arbitrage condition given by

$$E_t(P_{t+1}) \le P_t(1 + i) + C \tag{6-31}$$

where i equals the annual rate of interest and C is a constant, equaling average cost of bin space paid at the end of period t. If $E_t(P_{t+1}) < P_t(1 + i) + C$, as might happen in years of sufficiently small harvests, nothing is committed to storage by arbitrageurs; $I_t = 0$. If expected returns from arbitrage are normal, then the carryout is positive. We have $I_t > 0$ and $E_t(P_{t+1}) = P_t(1 + i) + C$. (This equality might also hold as a corner solution where $I_t = 0$.)

Where do we now stand in our modeling effort? If the inequality of condition (6-31) holds, we have four equations, Eqs. (6-27) through (6-30), and five unknowns Q_t, P_t, A_t, H_t, and $E_t(P_{t+1})$. Notice that $I_t = 0$. (Notice also that we could solve for H_t, Q_t, and P_t.) If, on the other hand, the equality of condition (6-31) holds, then I_t need not and very likely will not equal zero. In any event, we have five equations and six unknowns. Clearly, our model is not complete.

To close the model, we assert that the following equation can be derived and must be added to the model:

$$E_t(P_{t+1}) = f_t(A_t, I_t) \tag{6-32}$$

According to Eq. (6-32), the rational price expectation depends on the acreage planted in the current year and the level of the carryout of old crop. Increases in the carryout or in acreage planted tend to lower the expected price.

To justify including Eq. (6-32) in the model, we once again show how it can be derived using numerical methods. We envisage a distant period T such that arbitrageurs and farmers pay no attention to market events beyond period T as they form, in period t, their expectations with regard to period $t + 1$. Period T is a horizon beyond which no one can see or cares to see. Perhaps all market participants expect the end of the world at the close of period T. On this assumption, we equate A_T and I_T to zero. We need not bother with $E_T(P_{T+1})$. We cannot predict precisely what P_T will equal, but market clearing requires

$$A_{T-1}Y_T + I_{T-1} = D(P_T) \tag{6-33}$$

Solving for P_T, we find that

$$P_T = g_T(A_{T-1}Y_T + I_{T-1}) \tag{6-34}$$

We see that P_T depends on the total supply available in period T. Next we assume that Y_T takes on any one of several specified values with probabilities that are either known or estimated. With estimates of the structural parameters for Eq. (6-34) in hand, we can then estimate $E_{T-1}(P_T)$ for whatever sets of values for A_{T-1} and I_{T-1} we might care to choose. Following the procedure introduced in Section 6.1, choose a set of values for A_{T-1} and for I_{T-1}. For each value of Y_T, we then calculate P_T. Because the probabilities for the various values for Y_T are known, the probabilities for the corresponding values for P_T are known, too. This means that we can calculate the expected price in period T conditional on the chosen values for A_{T-1} and I_{T-1}. The vertical distance to point A in Fig. 6.3 measures $E_{T-1}(P_T)$ for $A_{T-1} = A'$ and $I_{T-1} = I'$. Other sets of values for A_{T-1} and I_{T-1} are given in Fig. 6.3: (A', I''), (A'', I'), and (A'', I''). The vertical distance above each point in the AI plane, shaded gray, measures the corresponding value for $E_{T-1}(P_T)$. The surface $ABCD$ is like the top of a table with uneven legs. One can use interpolation to estimate the distances to the top of the table, the value of $E_{T-1}(P_T)$, for all points in the shaded area directly beneath the table top. We may now imagine a surface above the entire AI plane consisting of a great many surfaces similar to the surface $ABCD$. This surface is our graphic repre-

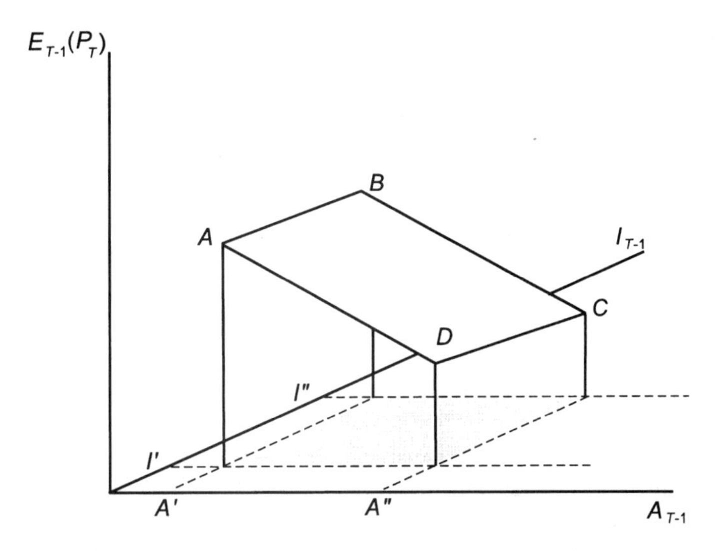

Figure 6.3 Four points on the expected price function.

sentation of Eq. (6-32) for $t = T - 1$. With this estimate in hand, the system for $T - 1$ is complete. This allows forming a graph showing $E_{T-2}(P_{T-1})$ for various values for A_{T-2} and I_{T-2} in the same manner used to generate Fig. 6.3. This completes the model for $T - 2$. The strategy is to work our way back from the horizon, period T, to the current period t. Experience with computational models suggests that a horizon of five years will often yield close approximations of the expected price function for period t.

To derive the demand for storage, we use Eq. (6-31) to get rid of $E_t(P_{t+1})$ in Eq. (6-32). We thus have

$$P_t \geq \frac{f_t(A_t, I_t) - C}{1 + i} \tag{6-35}$$

If the current price P_t is sufficiently high, I_t equals zero. Otherwise, the quantity stored increases with decreases in price. Notice that the demand for storage has acreage planted as one of its arguments. Increases in acreage are associated with contractions in the relationship between P_t and I_t. In a sense, planted acreage substitutes for carryout of old crop.

A step by step numerical example will help clarify model derivation. The student should use pencil and paper and follow the example step by step. Let the current period be period 1. The market model is given by

$$
\begin{aligned}
&P_1 = 10 - Q_1 && \text{consumption demand} \\
&I_1 + Q_1 = \overline{S} && \text{market clearing} \\
&A_1 = 0.9E_1(P_2) && \text{acreage response} \\
&E_1(P_2) \leq 1.04P_1 + 0.5 && \text{arbitrage} \\
&E_1(P_2) = f_1(A_1, I_1) && \text{expected price function}
\end{aligned}
\tag{6-36}
$$

where $\overline{S}$ equals the available supply given by the sum of newly harvested output plus any carryover of old crop into the current crop year. Notice that the interest rate i equals 0.04 and the per unit cost of bin space C equals 0.5. The objective of this example is to explain the theoretical underpinnings of the expected price function by showing how farmers and arbitrageurs could estimate it. For this purpose, we assume that market participants act as if the world will come to an end on the last day of year 3. In other words, we take year 3 as the horizon and assume that $A_3 = I_3 = 0$.

The model of the market in year 3 is given by

$$
\begin{aligned}
&P_3 = 10 - Q_3 && \text{consumption demand} \\
&Q_3 = Y_3A_2 + I_2 && \text{marketing clearing}
\end{aligned}
\tag{6-37}
$$

where Y_3 is random yield. It is assumed that Y_t equals 0.5, 1.0, and 1.5 with equal probabilities. It follows that $E_2(Y_3) = 1.0$. The reduced form or solution for P_3 is

$$P_3 = 10 - Y_3A_2 - I_2 \tag{6-38}$$

The model of the market for year 2 is given by

$$
\begin{aligned}
&P_2 = 10 - Q_2 && \text{consumption demand} \\
&I_2 + Q_2 = Y_2A_1 + I_1 && \text{market clearing} \\
&A_2 = 0.9E_2(P_3) && \text{acreage response} \\
&E_2(P_3) \leq 1.04P_2 + 0.5 && \text{arbitrage} \\
&E_2(P_3) = 10 - A_2 - I_2 && \text{expected price function}
\end{aligned}
\tag{6-39}
$$

It is easy to derive the expected price function given immediately above because the model for year 3 is linear. We need merely take the expectation of P_3 using Eq. (6-38) with $E_2(Y_3) = 1$.

Second-period yield Y_2 is not known in the first period. If Y_2 is low, it is possible that $E_2(P_3) < 1.04P_2' + 0.5$, in which case $I_2 = 0$. We give P_2 a prime to indicate a case of low yield. If Y_2 is high, then $E_2(P_3) = 1.04P_2'' + 0.5$, in which case $I_2 \geq 0$. We then give P_2 a double prime, $P_2 = P_2''$ to indicate a case of high yield.

If $E_2(P_3) < 1.04P_2' + 0.5$, the low-yield case, then

$$P_2' = 10 - Y_2A_1 - I_1 \tag{6-40}$$

using the consumption demand and market clearing equations of Eqs. (6-39), where I_2 is set equal to zero. Notice that in this case we omit the arbitrage condition, and the solution values for A_2 and $E_2(P_3)$ are found using the acreage response equation and the expected price function.

If $E_2(P_3) = 1.04P_2'' + 0.5$, then, using the last three equations of Eqs. (6-39), we see that

$$1.04P_2'' + 0.5 = 10 - I_2 - A_2$$

$$A_2 = 0.9(1.04P_2'' + 0.5)$$

which, using the second equation to eliminate A_2 from the first, yields

$$I_2 = 9.05 - 1.976P_2'' \tag{6-41}$$

Also, substituting for I_2 and Q_2 in the market clearing equation, we have

$$(9.05 - 1.976P_2'') + (10 - P_2'') = A_1Y_2 + I_1$$

which yields

$$P_2'' = 6.4012 - 0.336Y_2A_1 - 0.336I_1 \tag{6-42}$$

This is the solution for P_2 in the high-yield case.

In period 1, the market participants will not know whether Y_2 will be low, in which case $P_2 = P_2'$, price being determined by Eq. (6-40), or whether Y_2 will be high, in which case $P_2 = P_2''$, price being determined by Eq. (6-42). What to do? Here is where trial and error methods come into play. We want to experiment with many different sets of values for A_1 and I_1, calculating the value of $E_1(P_2)$ for each set. For a starter, we let $I_1 = 0$ and $A_1 = 5$. Recall that $Y_2 = 0.5, 1.0$, and 1.5 with equal probabilities.

If $Y_2 = 0.5$, then

$$P_2' = 7.5$$

$$P_2'' = 5.5612$$

but $P_2 = 5.5612$ implies that $I_2 = -1.9389$, using Eq. (6-41), which is clearly impossible. Hence $P_2 = 7.5$.

If $Y_2 = 1.0$, then

$$P_2' = 5$$

$$P_2'' = 4.7212$$

but $P_2 = 4.7212$ implies that $I_2 = -0.2791$, again using Eq. (6-41), which again is clearly impossible. Hence $P_2 = 5$.

If $Y_2 = 1.5$, then

$$P_2' = 2.5$$

$$P_2'' = 3.8812$$

but $P_2 = 3.8812$ implies that $I_2 = +1.3807$. If $Y_2 = 1.5$, then zero expected profit from arbitrage in the second year implies that $I_2 = +1.3807$. Hence $P_2 = 3.8812$. With these calculations in hand, we see that

$$E_1\left(P_2 \mid I_1 = 0,\ A_1 = 5\right) = \frac{1}{3}\left(7.5 + 5 + 3.8812\right) = 5.4604$$

We do not know the equation for the expected price function appearing in Eqs. (6-36), but we have calculated one point that would appear on the surface of its graph. Many such points can be calculated, and we may use these points to approximate the expected price function. Table 6.1, for example, provides the values for $E_1(P_2)$ conditional on a small grid of values for A_1 and I_1. A graphic approximation to the true graph for the expected price function, based on Table 6.1, is given in Fig. 6.4. Interpolation can be used to approximate $E_1(P_2)$ for intermediate values of A_1 and I_1. For example, using interpolation and the true values given in Table 6.1, we find that the approximate value for $E_1(P_2 \mid I_1 = 0, A_1 = 5)$ equals 5.5246, which is

TABLE 6.1 Expected Second-year Price, $E_1(P_2)$, for Alternative Values of First-year Acreage, A_1, and Carryout, I_1

	A_1				
I_1	0	2	4	6	8
0	10.0	8.0	6.1284	4.9208	4.0275
1	9.0	7.0	5.2568	4.3635	3.3772
2	8.0	5.9071	4.6995	3.7132	3.0412
3	7.0	5.0355	4.0492	3.3772	2.7052

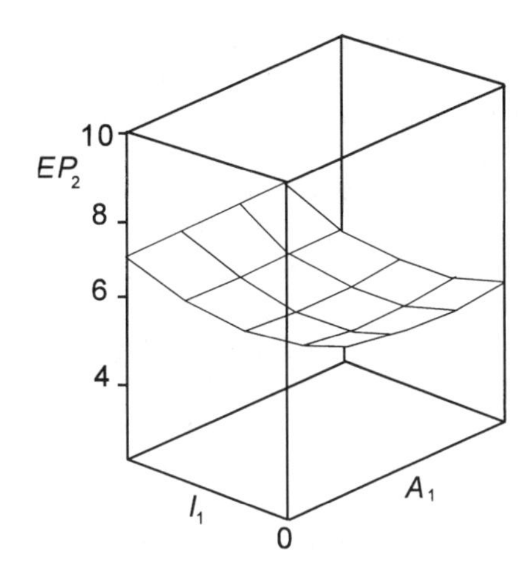

Figure 6.4 Approximated surface of the expected price function.

halfway between 6.1284 and 4.9208, exceeding the true value, calculated earlier, by 0.0642, a 1 percent error. Armed with the approximation given by Table 6.1 (or in Fig. 6.4), the system given by Eqs. (6-36) is complete and the market performance for the first period could be found, using a computer, for alternative values of initial supply $\overline{S}$. The analysis could be repeated for alternative values of the interest rate i and the cost of bin space C. In short, it would be possible, with the aid of a computer programmed to interpolate, to examine how the performance of the market in the first period is altered by changes in exogenous variables. The results of such an analysis are given in Section 6.3.

6.3 MODEL OF INTRAYEAR AND INTERYEAR STORAGE

The previous analysis centered on interyear storage, but ignored intrayear storage. The time has come to take the plunge into deeper waters, allowing for both intrayear and interyear storage. We have in mind commodities such as foodgrains, feedgrains, oil crops, cotton, processed fruits and vegetables, and other crops that are produced seasonally but consumed year round. Obviously, these commodities taken together constitute a large part of any nation's food supply.

To analyze the performance of markets for such commodities, we propose a quarterly model. This proposal is not made lightly. Annual models for field crops, like that considered in the previous section, do not allow for the proper modeling of the sequence of important decisions that market participants make during the year, as we shall soon see. Monthly models, although more complex, add little to what can be learned from quarterly models as to the role of storage in agricultural markets.

At the beginning of the first quarter (September, October, and November) of what will be called the marketing year, the total supply consists of newly harvested output plus, possibly, a carryover of old crop into the new marketing year. This total supply is available for consumption during the four quarters of the marketing year plus, possibly, a carryout at the end of the fourth quarter. Modeling the first two quarters is comparatively easy; given a fixed initial supply, price is determined by total demand, consisting of the demand for consumption and the demand for storage. In the third quarter, farmers plant acreage for a new crop. This complicates matters, partly because of the possible substitution of old crop for new crop in the next marketing year. The greater the acreage planted in the third quarter, for example, the smaller is the quantity of old crop likely to be carried over into the next marketing year. Modeling the fourth quarter is complicated by an arbitrage condition that allows for positive carryouts in some years and zero carryouts in others. In short, the arbitrage condition involves an inequality. Because theoretical analysis of the model is difficult, we will give considerable attention to an empirical application that relies on numerical methods. Our objective is to show how the performances of quarterly markets respond to changes in the interest rate and in the cost of bin space. The question of market stability is a central issue.

Many of the variables appearing in the models developed next have two subscripts; the first indicates year t; the previous year, $t-1$; or the subsequent year, $t+1$. The second subscript j, $j = 1, 2, 3, 4$, indicates quarter j.

The model for the first quarter is given by

$$
\begin{array}{llll}
(1) & Q_{t1} = D_1(P_{t1}, e_{t1}) & \text{consumption demand} & \\
(2) & E_{t1}(P_{t2}) = P_{t1}(1 + i) + C & \text{arbitrage} & \\
(3) & E_{t1}(P_{t2}) = f_1(I_{t1}) & \text{expected price function} & (6\text{-}43) \\
(4) & Q_{t1} + I_{t1} = S_{t1} & \text{market clearing} & \\
(5) & S_{t1} = Y_{t-1,4}A_{t-1,3} + I_{t-1,4} & \text{available supply} &
\end{array}
$$

The first function is the first-quarter consumption demand with the random term e_{t1} having zero expectation and finite variance. This demand may be interpreted as the sum of the demand for domestic consumption and the demand for exports. Although demand is stochastic, we assume that its parameters remain the same through time. Shift variables such as population or per capita income are ignored. The second function is the arbitrage condition assuming that the carryout I_{t1} is always positive. The quarterly rate of interest and the constant quarterly cost of bin space per unit of output are given by i and C, respectively. It is assumed that a futures market exists that allows for hedging and that the futures price for the second quarter, $F_{t1}(P_{t2})$, equals $E_{t1}(P_{t2})$. Because of hedging, we may ignore risk-averse behavior and risk premiums in the storage industry, which is assumed free of production uncertainty. The third function is the expected price function, justified on the basis of numerical rational expectations modeling, which entails the assumption of rational expectations and the numerical methods required to compute expected price functions, as explained in the previous sections.

The fourth function states a market clearing condition where S_{t1} equals the predetermined available supply. According to the last function, supply equals the sum of the new harvest plus any carryout of old crop $I_{t-1,4}$. The harvest is given by the product of acreage planted in the third quarter and yield determined in the fourth quarter, both quarters of the previous year. Notice that the endogenous variable $E_{t1}(P_{t2})$ can be eliminated through substituting the right-hand side of the arbitrage condition for the left-hand side of the expected price function. Thus simplified, the model consists of two demands, one for domestic consumption together with exports and the other for storage, plus an initial supply and a market clearing condition.

The model for the second quarter of year t is essentially the same as that for the first except that the initial supply S_{t2} equals nothing more than the first-quarter carryout I_{t1}. Writing out this model is an exercise left to the student.

Letting A_{t3} equal acreage planted currently, the model for the third quarter is given by

(1) $Q_{t3} = D_3(P_{t3}, e_{t3})$	consumption demand	
(2) $E_{t3}(P_{t4}) = P_{t3}(1 + i) + C$	arbitrage	
(3) $E_{t3}(P_{t4}) = f_3(I_{t3}, A_{t3})$	expected price (f) function	
(4) $A_{t3} = A_3[E_{t3}(P_{t+1,1}), W_{t3})]$	acreage response	(6-44)
(5) $E_{t3}(P_{t+1,1}) = g_3(I_{t3}, A_{t3})$	expected price (g) function	
(6) $C_{t3} = \alpha A_{t3}$	seed input	
(7) $Q_{t3} + I_{t3} + C_{t3} = I_{t2}$	market clearing	

Notice, first, that two expected price functions appear in this model. An explanation of the second of these will be given momentarily. The first expected price function, the f function, is akin to the expected price functions that appear in the models for the first two quarters *except* that A_{t3} is included as an additional endogenous variable. Why is that? Because the larger the acreage planted to a crop is, the less the need for the carryout of old crop. We therefore suppose that $\partial f_3/\partial A_{t3}$ is negative. Also, $E_{t3}(P_{t4})$ may easily be eliminated from the model, but it is interesting to note that the resulting demand for storage has A_{t3} as one of its arguments. In other words, in the demand for third quarter stocks, I_{t3} depends not only on P_{t3} but on A_{t3} as well.

According to the fourth function, acreage planted depends on the price that is expected to prevail at the time of harvest $E_{t3}(P_{t+1,1})$ and on the random term W_{t3} that measures planting conditions. This simple model assumes that expected yield is not affected by expected price. Expected yield is both exogenous and random.

Because acreage planted depends on the expected harvest price, the expected price function given by the fifth equation, *the g function*, is required to complete the model. Since the parameters of all future quarterly demands for consumption are fixed, the expected harvest price depends on (1) the current availability of old crop given by I_{t3}, some of which could be carried over into the new marketing year; and (2) the level of acreage planted, A_{t3}. The larger A_{t3} is, the lower is the expected harvest price. The same is true of I_{t3}, except that if I_{t3} is below some critical value then changes in I_{t3} will have no effect on expected harvest price. (If I_{t3} is sufficiently small, there is no chance that the carryout I_{t4} will be positive, but more on this later.)

The sixth function merely asserts that the amount of the crop required for seed is proportional to acreage planted.

The model for the fourth and final quarter is given by

$$
\begin{aligned}
&(1)\ Q_{t4} = D_4(P_{t4}, e_{t4}) && \text{consumption demand}\\
&(2)\ E_{t4}(P_{t+1,1}) \le P_{t4}(1+i) + C && \text{arbitrage}\\
&(3)\ E_{t4}(P_{t+1,1}) = f_4[(Y_{t4}A_{t3}) + I_{t4}] && \text{expected price function}\\
&(4)\ Q_{t4} + I_{t4} = I_{t3} && \text{market clearing}
\end{aligned}
\qquad (6\text{-}45)
$$

If the inequality of the arbitrage condition holds, the carryout I_{t4} equals zero. The price in the first quarter of the new marketing year depends on acreage planted in the third quarter and the fourth-quarter growing conditions, which determine crop yield. Good weather and high yields would encourage fourth-quarter consumption and a zero or at least small fourth-quarter carryout. If the equality of the arbitrage condition holds, then the carryout will almost certainly be positive, although there is the remote and unimportant possibility that I_{t4} equals zero. In and of itself, a low fourth-quarter yield would encourage carrying some of the old crop into the new marketing year, but it must be recognized that the current supply, I_{t3}, could also be very small. In this latter case, the stage is set for high prices during both period $t4$ and period $t+1, 1$. The student should recognize, drawing on the analysis in Section 6.2, that two structural models are implicit in the fourth-quarter model. Under low price conditions the fourth-quarter demand for storage is a relevant structural relationship. Under high price conditions, this demand is irrelevant and plays no role in determining market performance.

Having set forth a quarterly model of pricing that allows for uncertainty and storage, we turn next to the implications of the model for market performance and to the effects of changes in the interest rate i and in the cost of bin space C. Of particular interest is the stability of market performance. (What, for example, determines price variability?) More often than not in economics, the analyst constructs a structural model, derives the reduced forms for all endogenous variables, and then deduces, where possible, the signs of the partial derivatives of each endogenous variable with respect to the exogenous variables. This time-honored approach is not very promising as a means for analyzing the quarterly model at hand. The model is too large and too complicated.

We therefore propose in what follows to take advantage of a study of the U.S. soybean market by Lowry, Glauber, Miranda, and Helmberger (LGMH) (1987) that analyzes a quantitative version of the quarterly model given previously. Their estimates of the structural parameters are first considered for the sake of completeness. Then, to provide further necessary background, some attention is given to a research method called stochastic simulation and to the notion of stationary state in stochastic dynamic analysis. After that we consider in some detail the LGMH findings. What we are primarily interested in is how market performance is affected by changes in the interest rate and in the cost of storage.

Quantitative analysis of our quarterly model requires making assumptions with regard to the forms of functions and estimating all structural parameters. The probability distributions for all stochastic terms must also be estimated. Estimates of i and C are also required. The quarterly demands for soybeans were estimated by LGMH using a statistical procedure called two-stage least squares and quarterly observations for the years 1965 to 1979. The consumption demand for each quarter consisted of the domestic demand for crushing and the demand for exports. The functional form for demand was assumed to be multiplicative or log linear. (For example, $Q = \beta_0 P^{-\beta_1}$, where β_1 equals demand elasticity, constant for all values of Q.) Elasticities were estimated on the assumption that the domestic demand elasticities were the same across quarters and similarly for export demands. Estimated domestic and export elasticities equaled –0.56 and –0.63, respectively.

The acreage response function, also assumed to be multiplicative, was estimated using annual data for the period 1951 to 1979 and ordinary least squares. The average price of futures in March for delivery of soybeans in November was taken as the measure of the harvest price of soybeans expected at planting time. The estimated expected price elasticity of acreage response was 0.89.[3]

The quarterly interest rate i was set at 0.02. The quarterly cost of storage per bushel C was set at 10¢ per bushel.

[3]As to stochastic terms, yield was assumed log normally distributed with a mean of 29.7 bushels per acre. The estimated variances of the stochastic terms in the domestic demand, export demand, acreage response, and yield were 0.0051, 0.0150, 0.0016, and 0.0202, respectively. These four terms were assumed to have zero means.

Finally, numerical methods were used to estimate the five expected price functions appearing in the quarterly model. These functions were estimated on the basis of the demand and supply conditions prevailing in the marketing year commencing with quarter 1 of 1977. Figure 6.5 gives graphic representations of the expected price functions that link quarterly carryouts I_{tj} to the prices expected to prevail in the next quarter $E_{tj}(P_{t,j+1})$. (The t subscripts in the figure are dropped for simplicity.) As we might suppose, the expected price function for quarter $j+1$ lies to the left of that for quarter j, except that the fourth-quarter function, representing as it does the sum of new harvest plus any carryout of old crop, lies farthest to the right. These curves contain a property that is basic to understanding the role of storage in pricing. Notice that at a price just above $4.00 per bushel the expected price functions tend to become very elastic. This means that, when the expected price (or the spot price for that matter) is low, slight decreases prompt relatively enormous increases in the willingness of arbitrageurs to hold stocks. As we shall see more clearly in a moment, this behavior of arbitrageurs tends to put a floor on cash prices and tends to stabilize several but not all dimensions of market performance.

Figure 6.6 provides a graphic representation of the g function in the model for the third quarter, linking the expected harvest price with acreage planted and the third-quarter carryout. As expected, holding the carryout constant, the larger the

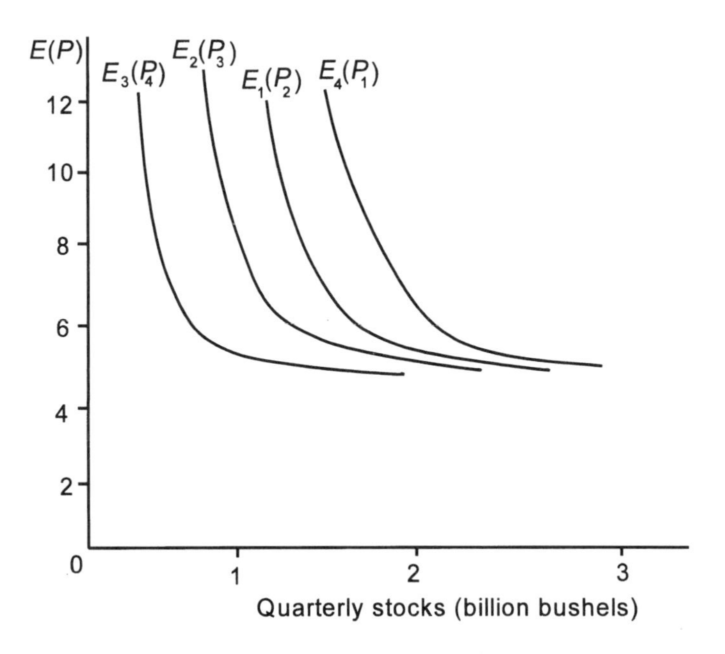

Figure 6.5 Quarterly expected price functions for U.S. soybean market with planted acreage set at 60 million acres. *Source:* Lowry, Glauber, Miranda, and Helmberger (1987).

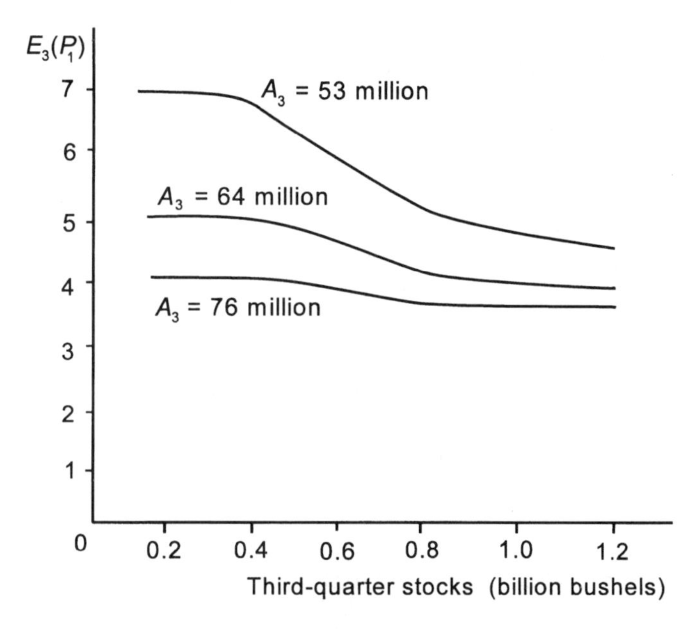

Figure 6.6 Third-quarter expected price function (the *g* function) for the U.S. soybean market for three levels of planted acreage. *Source:* Lowry, Glauber, Miranda, and Helmberger (1987).

acreage is, the lower the expected harvest price. Interestingly, as I_3 declines, the expectation of a positive fourth-quarter carryout declines and eventually equals zero for a critical value of I_{t3}. Third-quarter carryout below this critical value will, on average, be consumed entirely in the fourth quarter. For such low values of I_{t3}, expected harvest price depends only on acreage planted; decreases in I_{t3} below the critical value have no effect on $E_{t3}(P_{t+1,1})$.

Stochastic Simulation and Steady-state Equilibria

Given estimates of all structural equations, including the expected price functions, the LGMH study used a research method called stochastic simulation to analyze the implications of the quarterly model. To explain stochastic simulation, we consider briefly the following cobweb model:

$$D_t = 10 - 3\,P_t + e_t$$

$$S_t = 2\,P_{t-1} \tag{6-46}$$

$$D_t = S_t$$

where e_t is a random variable equaling 0.1 half the time and −0.1 half the time. Alternatively, if we let $e_t = 0$, the cobweb model is nonstochastic. We explore both possibilities next.

Suppose that we set t equal to 1 for the current period or year on the assumption that the price observed last year, P_0, equaled 4.0, bother how that price ever came to be. Now consider period $t = 1, 2, 3, \ldots, 10$. For the nonstochastic case, we use simulation to obtain the values for S_t, D_t, and P_t, which is a fancy way of saying that we use simple algebra to solve consecutively for S_1, D_1, and P_1, which then allows solving consecutively for S_2, D_2, and P_2, and so on and so forth. The resulting values are given in Table 6.2. Price oscillates beginning with the lowest value 0.6667. The oscillations are damped, however, and, as we might guess from the generated data, it can be shown that in time an equilibrium is reached, with price equaling 2 and output equaling 4. This is an example of a dynamic equilibrium. The equilibrium is stable in that P_t and D_t tend toward the values 2 and 4 for any plausible initial price.

Turning to the case when e_t does not equal zero, we toss a coin 10 times with the understanding that, when heads appear, $e_t = +0.1$, and when tails appear, $e_t = -0.1$. The actual results from such an experiment are given in Table 6.2. Once again, simple algebra can be used to solve for the endogenous variables consecutively through time. The procedure is an example of stochastic simulation. Now suppose that stochastic simulation is used to continue this experiment up to, say, $t = 50$ in order to rid the system of the effects of having arbitrarily set the initial price equal to 4.0. Suppose that we then repeat the experiment 10,000 times. In other words, using a random process we generate 10,000 different 50-year paths of values for e_t. Flipping coins and doing algebra would be plainly impractical, but a computer, correctly programmed, would carry out the random process and do the necessary calculations in no time at all. We would then have 10,000 independently determined observations on P_t for $t = 1, 2, 3, \ldots, 50$ and similarly for $S_t = D_t$. We could then estimate the means,

TABLE 6.2 Simulated Values for the Endogenous Variables of a Cobweb Model

	$e_t = 0$			$e_t \neq 0$		
t	$S_t = D_t$	P_t	e_t	$S_t = D_t$	P_t	$E_0(P_t \mid P_0 = 4)$
1	8	0.6667	0.1	8.0	0.7	0.6667
2	1.3333	2.8889	0.1	1.4	2.9	2.8889
3	5.7778	1.4074	−0.1	5.8	1.3667	1.4074
4	2.8148	2.3951	0.1	2.7333	2.4556	2.3951
5	4.7901	1.7366	−0.1	4.9111	1.6630	1.7366
6	3.4733	2.1756	−0.1	3.3259	2.1914	2.1756
7	4.3512	1.8829	−0.1	4.3827	1.8391	1.8829
8	3.7659	2.0780	0.1	3.6782	2.1406	2.0780
9	4.1561	1.9480	−0.1	4.2812	1.8729	1.9480
10	3.8960	2.0347	−0.1	3.7458	2.0514	2.0347

variances, and coefficients of variation for price (and quantity) for all 50 time periods on the basis of the 10,000 observations for each period. We might find that these estimated population parameters remain the same in successive periods beginning, say, with $t = 30$. In this case we would say that the stochastic dynamic system had reached steady-state equilibrium at $t = 30$. A stochastic dynamic system is in steady-state or long-run equilibrium when the probability distributions for the endogenous variables remain the same from one period to the next for a given set of initial conditions (for $P_0 = 4$ in the present example). Steady-state equilibria are the counterpart to dynamic equilibria in nonstochastic dynamic models and to long-run equilibria in comparative static models.

As an example, the expected values of price P_t, $t = 1, 2, 3, \ldots, 10$, are given in the last column of Table 6.2. These expectations, all formed in the initial period conditional on $P_0 = 4$, are parameters that could be estimated using simulated data. The average price of 10,000 independently distributed prices in the tenth year would equal, for example, approximately 2.0347. If the simulation experiment were continued beyond the tenth year, it would be discovered that the average price in the 27th year, $E_0(P_{27}|P_0 = 4)$, and in all subsequent years, equals 2.000. We would say that the stochastic dynamic system reaches steady-state or long-run equilibrium in period $t = 26$. In stochastic systems, endogenous variables fluctuate around and may never equal exactly the steady-state average values. (The student familiar with basic statistical theory will recognize how we could use the 10,000 deviations from the mean price in period $t = 26$, or in other periods as well, to estimate the variance of price.)

Effects of Changes in the Carrying Charge

Turning to the implications of storage for the performance of quarterly commodity markets, we first consider the question how commodity pricing would be affected if interyear commodity storage were disallowed. Alternatively, suppose that the cost of carrying fourth-quarter stocks over into the new marketing year is prohibitively high. Figure 6.7 provides a 50-year history of annual U.S. soybean prices simulated under 1974 economic conditions with and without interyear storage. (The graphs are taken from research that used a model and computational techniques similar to those described in connection with the LGMH study.) The main conclusion to be drawn from this figure, and it is of basic importance, is that a competitive storage industry tends to put a floor on prices. Notice that, with storage, price never falls below \$2.20 per bushel; without storage, price often falls below \$2.00 and nearly plummets to zero in the 49th year. A storage industry puts a floor on prices because the demand for stocks becomes nearly perfectly elastic at low prices. The stocks that accumulate in years of abundance tend to moderate upward price swings that would later occur because of bad weather and low yields. *Storage tends to dampen price variability*. High price peaks, on the other hand, are the result of consecutive poor harvests. With no carryin, a small harvest sends price soaring.

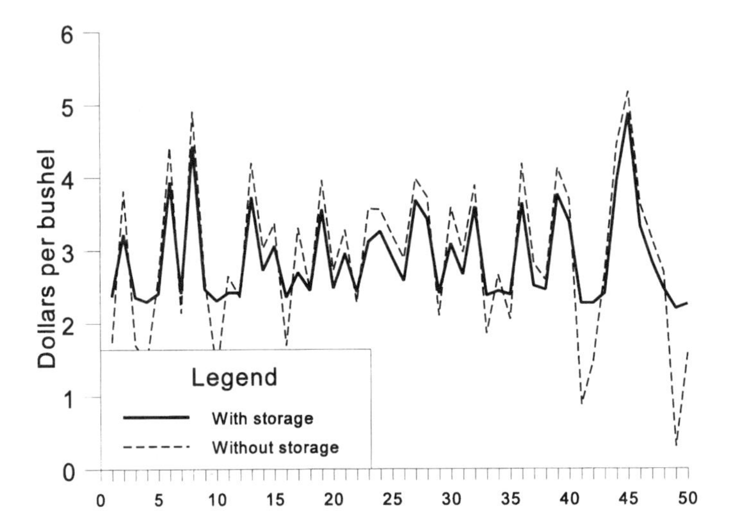

Figure 6.7 Simulated annual soybean prices with and without competitive storage. *Source:* Helmberger and Akinyosoye (1984).

Table 6.3 is drawn from the LGMH study referred to previously. All estimates are steady-state values based on stochastic simulation using the quantitative model described and assuming the economic conditions that prevailed in the 1977–1978 marketing year. All the findings are implicit in the model and should be viewed as hypotheses that might or might not be confirmed by further research.

Several hypotheses merit discussion. Consider the first column of numbers first. The average or mean quarterly prices (price $j, j = 1, 2, 3, 4$) rise from \$5.584 per bushel in the first quarter to \$6.239 in the fourth quarter, with the differences between consecutive quarters equal to the carrying charge. There is a tendency for commodity prices to rise seasonally in real-world markets, but such markets are continually bombarded by exogenous shocks; the clear-cut pattern seen in Table 6.3 is rarely as obvious in actual markets.

Turning to variability, we see that price variability as measured by the coefficient of variation (CV) remains about the same through the marketing year, but it picks up some in the last quarter. Probably the reason for this is that prices in the first three quarters reflect expected yield in the fourth quarter, which is constant. Yield gets determined in the fourth quarter, however, and low yields drive fourth-quarter prices up; high yields have the opposite effect. The mean quarterly carryouts (carry $j, j = 1, 2, 3, 4$) fall during the marketing year from 1.355 billion bushels in the first quarter to a mere 0.037 bushels in quarter 4, the latter level equaling about 2 percent of average production. The variability of quarterly carryouts rises modestly during the marketing year, but soars in the last quarter. The higher variability of ending stocks in the fourth quarter is likely associated with the increased fourth-quarter price variability and the highly elastic demand for stocks at low prices.

TABLE 6.3 Estimated Expected Values and Coefficients of Variation (in Parentheses) for the Endogenous Variables of a Quarterly Model for U.S. Soybeans for Alternative Levels of Interest Rates i and Storage Costs C[a]

	Alternative Levels of Interest Rates and Cost of Bin Space				
Variable	$C = 0.100$ $i = 0.020$	$C = 0.050$ $i = 0.020$	$C = 0.100$ $i = 0.010$	$C = 0.050$ $i = 0.010$	$C = 0.000$ $i = 0.005$
Price 1	5.584	5.606	5.609	5.630	5.659
	(15.35)	(13.86)	(13.89)	(11.99)	(7.22)
Price 2	5.798	5.769	5.766	5.736	5.687
	(15.27)	(13.92)	(13.83)	(12.05)	(7.33)
Price 3	6.016	5.936	5.926	5.845	5.716
	(15.53)	(14.30)	(14.09)	(12.40)	(7.63)
Price 4	6.239	6.106	6.086	5.953	5.745
	(18.09)	(17.07)	(16.87)	(15.25)	(9.68)
Annual price	5.902	5.849	5.841	5.788	5.701
	(15.41)	(14.14)	(14.00)	(12.26)	(7.50)
Domestic demand	0.926	0.929	0.930	0.933	0.937
	(8.35)	(7.64)	(7.57)	(6.64)	(4.45)
Export demand	0.838	0.841	0.842	0.844	0.849
	(9.62)	(8.85)	(8.78)	(7.85)	(5.84)
Supply	1.802	1.824	1.826	1.859	2.002
	(10.64)	(10.60)	(10.56)	(10.62)	(11.83)
Carry 1	1.355	1.379	1.381	1.416	1.562
	(11.44)	(11.70)	(11.64)	(12.08)	(14.25)
Carry 2	0.920	0.944	0.946	0.981	1.126
	(13.05)	(13.83)	(13.77)	(14.84)	(18.51)
Carry 3	0.448	0.469	0.471	0.503	0.643
	(18.78)	(21.06)	(21.02)	(23.73)	(30.10)
Carry 4	0.037	0.053	0.055	0.082	0.217
	(159.09)	(142.14)	(140.00)	(121.47)	(83.85)
Production	1.764	1.771	1.772	1.777	1.786
	(11.77)	(11.86)	(11.86)	(11.97)	(12.10)
Acres	59.630	59.835	59.868	60.066	60.352
	(2.38)	(2.77)	(2.80)	(3.22)	(3.68)
Producer revenue	9.690	9.782	9.794	9.890	10.053
	(5.82)	(5.91)	(5.96)	(6.66)	(9.37)
Cov(P, H)	−0.162	−0.142	−0.143	−0.116	−0.053
Producer quasi-rent	4.997	5.057	5.066	5.163	5.323
	(9.87)	(8.93)	(8.91)	(9.27)	(14.59)

[a] Prices are measured in dollars per bushel, demands and supplies are in billions of bushels, acres are in million acres, and producer revenue and quasi-rent are measured in billion dollars. Cov(P, H) is the covariance between price 1 and newly harvested production.

Source: Mark Lowry, Joseph Glauber, Mario Miranda, and Peter Helmberger, "Pricing and Storage of Field Crops: A Quarterly Model Applied to Soybeans," *American Journal of Agricultural Economics,* 68, no. 4 (November 1987), 740–749.

The negative covariance (or correlation) between first-quarter price and the new harvest, Cov(P, H), clearly reflects the negativity of demand. High (low) yields mean low (high) prices.

Columns 2 and 3 show what happens when the carrying charge is lowered either through holding i constant and decreasing C or holding C constant but decreasing i. The last two columns show what happens when both the interest rate and the storage cost are lowered by rather significant amounts. The results are qualitatively the same for all cases. Lowering the carrying charge decreases modestly mean annual prices and all mean quarterly prices except that for the first quarter; the latter increases. Lowering the carrying charge lowers the cost of supplying buyers during much of the marketing year, thus encouraging consumption. Increased average consumption could not occur, however, if average production did not increase. But the only way greater production can be elicited is through increased prices in the first quarter. Notice that, as the carrying charge is lowered, annual production and consumption both rise.

On the basis of Fig. 6.7, we indicated that competitive storage, relative to a no-storage regime, tends to put a floor on prices and to dampen some price peaks. The result is *more stable prices*. This hypothesis is consistent with the numbers given in Table 6.3. The variability of annual price, measured by the coefficient of variation, is cut by more than half as the interest rate and storage cost are lowered from the base levels (0.02 and 10 cents) to near-zero levels (0.005 and zero). Annual domestic consumption and exports are also stabilized through lowering the carrying charge. The average fourth-quarter carryout rises from 0.037 billion bushels (2 percent of average production) to 0.217 billion bushels (12 percent of average production).

Market performance has many dimensions, and it should not be supposed that, because lowering the carrying charge stabilizes some dimensions, it stabilizes all of them. Returning to Table 6.3, we note that lowering the carrying charge destabilizes production slightly. If interyear carryouts were disallowed, the expected first quarter price and expected production would not vary from year to year. Interyear storage allows for the substitution of old crop for new crop. A high yield in one year leads to a large stock that discourages planted acreage the next.

Perhaps of greater importance than destabilized production is the complex relationship between the carrying charge and the stability of producer quasi-rent. Table 6.3 indicates that, as the carrying charge is decreased from the base level, producer quasi-rent is at first stabilized slightly, but is eventually destabilized. The reasons for this complex relationship are not at present well understood. We note, however, that gross producer revenue is destabilized through lowering the carrying charge. It seems likely that total variable cost is stabilized in light of the stabilization of first-period price. Perhaps with small decreases in the carrying charge, but not for large decreases, the destabilization of revenue is offset by the increased stability of cost. It may also be noted that the destabilization of total revenue is closely associated with the remarkable decline in the covariance between first-period price and new production as the carrying charge is lowered to the near-zero level. An important point the student should bear in mind is that exogenous changes that tend to stabilize price need not stabilize farm income.

6.4 SHORT-RUN COMPETITIVE MODEL OF AGRICULTURAL CROP PRODUCTION, MARKETING, AND PRICING: AN OVERVIEW

The purpose of this section is to draw together in a synthesis some of the models discussed in this and in the foregoing chapters. The idea is to conceptualize in broad terms the competitive processes that determine farm prices and other dimensions of market performance, to provide a window, as it were, on the wondrous complexity of farm commodity pricing. We start with the first quarter of the marketing year. Rather than being upward sloping as in the typical textbook, the supply curve for farm output is perfectly inelastic, consisting of new crop production and any carryover of old crop into the new crop year. Total first-quarter demand consists of the demands for domestic consumption, exports, and stocks.

Population, per capita income, prices of marketing inputs, and technological change in the marketing sector are likely important shifters of the domestic demand for food. It may also be noted that these shifters or exogenous variables move rather predictably over time; changes in them are not likely a serious source of instability. If the model is to be applied to a single farm commodity (pork), as opposed to an aggregate such as food, then we need to be concerned about prices of related foods (beef). The question also arises whether we should enlarge and generalize the model through assuming multiple products. This question is particularly relevant for those involved in empirical research.

Important shifters of export demand include all the variables that shift domestic consumption demand except, of course, that such variables must be defined for other countries. Additionally, the factors that shift supply functions in foreign nations, yields, for example, must now be viewed as shifters of the demand for U.S. exports. The distinction between short- and long-run functions, of basic importance in supply analysis, is important as regards the demand for exports. Specifically, the demand for exports is less elastic in the short run than in the long run. Exchange rates are important shifters of export demand. The strengthening of the dollar against foreign currencies, for example, contracts demand. The demand for exports is also subject to the farm and trade policies of foreign countries. In light of the differing sets of demand shifters, as between domestic consumption and export demand, it is little wonder that the latter tends to be much less stable and less predictable than is the former.

The demand for stocks is derived from arbitrage conditions and expected price functions. Futures markets exist for major crops, which allow the demanders of commodities for stocks to hedge and to reduce the costs of incurring risk. Although the interest rate, storage cost per unit, and expected crop yield are the important shifters of expected price functions in a stationary world, as analyzed previously, these functions in a nonstationary world are also shifted by changes in the expected values of the shifters of the domestic consumption demands and export demands for future quarters, particularly, we would think, for the future quarters of the current marketing year. The factors that cause export demand to be unstable likely also lend instability to the expected price functions. Modeling the formation of price expectations is complex, but there is no escaping the central role that speculation and expected prices play in determining the demand for stocks.

Much of what was said with regard to the first-quarter model applies to the second-, third-, and fourth-quarter models as well. The perfectly inelastic supply for the second quarter, for example, is merely the carryout from the first.

The third quarter offers new challenges for the price analyst, however, because in this quarter farmers plant crops and make decisions that to a considerable extent set input levels and production costs. The third-quarter demands for consumption, exports, and stocks are, of course, as relevant as before. Price expectations are again critical, not only because of their role in storage decisions, but also as regards acreage planted. Modeling is complicated by the interdependence among price expectations, acreage planted, levels of storage, and other variables. The upward sloping supply function in theory textbooks is displaced by an acreage response function that links acreage planted to expected price or expected gross returns per acre. (The costliness of risk must also be taken into account if a banking system, diversification, and hedging or forward pricing cannot be used to avoid or substantially diminish risk.) The period t in the models of Chapter 2 must now be interpreted as the third quarter of the marketing year. The price expected to prevail in period $t + 1$ must now be interpreted as that for the first quarter of the new marketing year. Changes in technology and shifts of input supply functions are important shifters of the acreage response function. (The importance of farm programs as shifters of acreage response functions is taken up later.)

Unlike the models of Chapter 2, when storage was ignored, an expected price function, the g function, is required that links together expected price for the new harvest with acreage planted and ending third-quarter stocks. [See the fifth equation of Eqs. (6-44).] In a nonstationary setting, the g function is shifted by changes in the expected values of the shifters of future quarterly demands for domestic consumption and exports. Factors that shift expected yields also shift the g function. For example, the introduction of a new crop variety, which is expected to increase crop yields, would tend to lower the expected harvest price for any combination of acreage planted and ending carryout.

In the fourth quarter of the marketing year, crop yields get determined and, unlike previous quarters, the end-of-quarter carryouts equal zero in years of scarcity. In the latter case, the total demand is merely the aggregate of the demands for domestic consumption and export; its intersection with predetermined supply, given by the third-quarter carryout, determines the price. A more general treatment than that given in the previous section would also take into account what is often referred to as pipeline stocks that have little to do with temporal price relationships. Such stocks are carried mainly to assure continuity of processing and marketing operations over a period of time when the old crop is petering out and the new crop is becoming available. The student should not be misled into thinking that interyear storage is unimportant in commodity pricing merely because ending inventory tends to be, on average, a small percentage of average production. In periods of abundance the fourth-quarter demand for stocks puts a floor on price and plays an important role in stabilizing price and consumption.

Finally, it is important to recognize that although day to day conversations regarding prices often relate to the current prices relevant to actual or potential transactions, agricultural price analysis usually involves averages of prices over stated periods of time—weekly, monthly, quarterly, and annually. The importance of expectations and storage in conceptualizing how prices get determined likely diminishes as the periods over which prices are averaged are lengthened. Analyses designed to explain why farm and consumer prices are relatively low or high over periods of several years, for example, might well center on changes in demand and supply conditions that actually occurred over those periods. Expectations of future demand and supply conditions formed during periods of several years duration might easily be of secondary importance, and the movement of commodities into storage is essentially offset by the movement out. It is in short-run analysis and in attempting to understand questions of market stability that expectations and storage assume a role of central importance.

PROBLEMS

6.1. A commodity, harvested at some point in time, is consumed over three months. At the end of the third month, it turns to garbage. Monthly demands are given by

$$P_1 = 20 - 4Q_1$$

$$P_2 = 20 - 2Q_2 + e_2$$

$$P_3 = 20 - 0.5Q_3 + e_3$$

where e_2 and e_3 are random terms with zero expectations. Assuming a zero rate of interest and a cost of bin space per unit stored equal to 1, the arbitrage condition for storage is

$$E_t(P_{t+1}) = P_t + 1$$

where $t = 1$ and 2. The initial harvest is $H_0 = 30$. Calculate the competitive equilibrium values for P_1, Q_1, I_1, and $E_1(P_2)$.

6.2. Suppose that the third period demand in Problem 6.1 is given by

$$P_3 = \frac{16e_3}{Q_3^2}$$

where $e_3 = 0.5, 1.0$, and 1.5 with equal probabilities.

a. Find the functional relationship between $E_2(P_3)$ and I_2 and between $V_2(P_3)$ and I_2.

b. Derive the demand for stocks in the second period.

6.3. Consider the numerical example given at the end of Section 6.2 (see Table 6.1).

a. Calculate $E_1(P_2|I_1 = 0.1, A_1 = 5)$, $E_1(P_2|I_1 = 0, A_1 = 5.2)$, and $E_1(P_2|I_1 = 0.1, A_1 = 5.2)$. [*Hint:* Review Eqs. (6-39) and the derivation of Eqs. (6-40) and (6-42). Note that $i =$ 0.04 and $C = 0.5$ and that yield equals 0.5, 1.0, and 1.5 with equal probabilities. Calculate P_2' and P_2'' for each combination of I_1 and A_1. Use Eq. (6-41) to check nonnegativity of I_2 for all calculated values of P_2''.]

b. Suppose that the interest rate changes from 0.04 to 0.08. Calculate $E_1(P_2|I_1 = 0, A_1 =$ 5.2).

c. Suppose that the cost of bin space per unit rises from 0.5 to 0.7, keeping the interest rate at 0.4. Calculate $E_1(P_2|I_1 = 0, A_1 = 5.2)$.

d. From parts b and c, we see that increases in the carrying cost lower $E_1(P_2|I_1 = 0, A_1 =$ 5.2). Why?

REFERENCES

Helmberger, Peter G., and Vincent Akinyosoye, "Competitive Pricing and Storage under Uncertainty with an Application to the U.S. Soybean Market," *American Journal of Agricultural Economics,* 66, no. 2 (May 1984), 119–130.

Leuthold, Raymond M., Joan C. Junkus, and Jean E. Cordier, *The Theory and Practice of Futures Markets.* Lexington, Mass.: Lexington Books, 1989.

Lowry, Mark, Joseph Glauber, Mario Miranda, and Peter Helmberger, "Pricing and Storage of Field Crops: A Quarterly Model Applied to Soybeans," *American Journal of Agricultural Economics*, 69, no. 4 (November 1987), 740–749.

Williams, Jeffrey C., and Brian D. Wright, *Storage and Commodity Markets*. New York: Cambridge University Press, 1991.

7

Measuring Welfare Effects

Although this book is concerned mainly with the price and output and input quantity effects of exogenous shocks, its scope is now broadened to include welfare effects as well. There are three reasons for this. First, the effects of exogenous shocks on human welfare, happiness, well-being—call it what you will—are obviously important. It is, in fact, of ultimate importance. Second, in the chapters that follow, particularly in Chapters 9 and 10, we will analyze various forms of government intervention as an important source of exogenous shocks. Although government intervention can and does have significant impacts on pricing, the motivation for intervention is explicable largely in terms of intended welfare effects. Third, broadening the scope of our work to include welfare effects can be accomplished at modest cost: The marginal gain to the student exceeds, we believe, the marginal effort. But the marginal effort required is not trivial, and the major reason for this is that welfare effects are difficult to measure.

As dimensions of market performance, prices and output and input quantities are not only measurable, they are ordinarily set forth in contracts of exchange. For example, the shopper leaves the supermarket with a receipt showing the quantities of items purchased together with the prices paid. Market, sector, and nation-wide statistics, such as the consumer price index, are based on the measurable terms of trade that appear in contracts of exchange, terms that are tabulated across various aggregates of firms, consumers, and input suppliers. The levels of welfare or utility achieved through buying and selling are not stated in exchange contracts, however, nor can they be measured directly as, for example, we might measure the bushels of wheat in a bin. Fortunately, the question of how to measure or quantify welfare effects, including the effects of various forms of government intervention, is addressed in detail in the field of applied welfare economics often called benefit–cost analysis.

To incorporate welfare effects in our analyses, we pause in this chapter to develop and explain benefit–cost concepts that will be needed later. Sections 7.1 through 7.4 center on measuring changes in the welfare of consumers, farm input suppliers, farmers, and marketing firms. Section 7.5 applies benefit–cost analysis to technological change as an important illustration. Section 7.6 introduces the student to the distinction between the distributive and efficiency effects of government intervention. (For proofs of several results discussed in this chapter, see Appendix E.)

7.1 MEASURING CHANGES IN CONSUMER WELFARE

We start by posing a simple question. The demand of an individual consumer for food is given in Fig. 7.1. An exogenous shock of some kind raises the price from P_0 to P_1. For concreteness, we assume that the exogenous shock is a sales tax on food. Is the consumer's welfare diminished by the price increase? The answer is yes, at least according to the *principle of consumer sovereignty*. This principle states that the consumer knows best. The consumer is held to be the ultimate judge as to how much a price change affects his or her welfare. Exceptions are not hard to find. The market for hard drugs is driven underground by legal prohibitions and law enforcement, presumably to increase public welfare. Here the principle at work seems to be that the government knows best, not the consumer. Doctors urge consumers to stop smoking. Even commodities such as butter, eggs, and red meats pose a problem. Decreased consumption following price increases of these products might help people to control their blood cholesterol levels. The principle of consumer sovereignty, it seems, should be used cautiously.

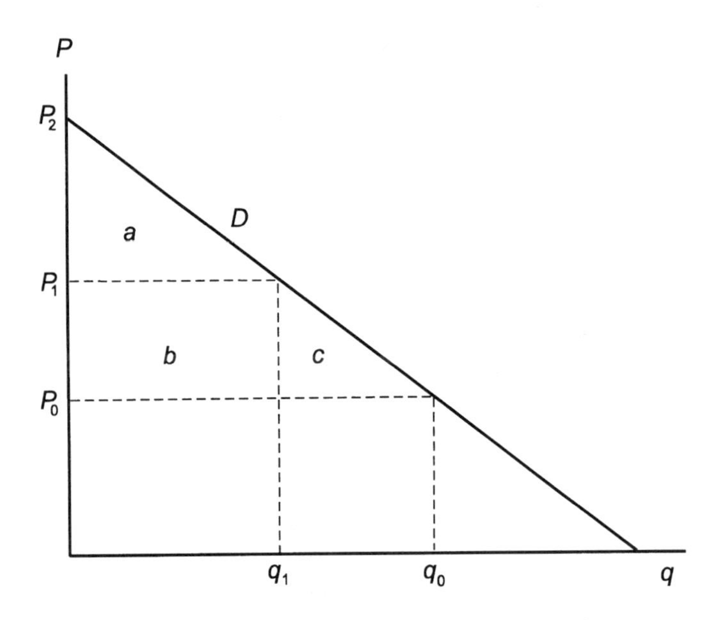

Figure 7.1 Consumer demand and consumer surplus.

Another possible objection to the conclusion that a price increase lowers consumer welfare involves the distinction between what might be called selfish losses (or benefits) and selfless losses (or benefits). This distinction can be seen if we make the further assumption that the revenues raised by the government through sales taxes on food are to be used to build shelters for the homeless. Two very different questions might be put to our hypothetical consumer. What is the amount of the welfare loss due to the price increase assuming that other people are not affected by the program? Alternatively, what is the loss if the program helps the homeless?

These questions might elicit very different replies. A consumer loss apart from the consumer's regard for the welfare of others is referred to as a *selfish loss*. A consumer loss that takes into account the concern for the welfare of others is called a *selfless loss*. The distinction between selfish and selfless welfare gains is left to the reader. In our example, the consumer might be willing to pay a higher price for food to finance shelters for the homeless. The consumer then suffers both a selfish loss and a selfless gain from the program. Although this discussion centers on a price increase, welfare changes associated with a price decrease can be handled in the same way.

Because selfless gains and losses are at best difficult to measure, we concentrate on selfish welfare changes. Returning to Fig. 7.1, our problem is to figure out a way of measuring or quantifying the consumer's selfish welfare loss caused by a price increase. The key concept advanced in the modern treatment of problems of this kind is willingness-to-pay money to avoid welfare losses (or to pay money for benefits received). We could ask the consumer, What is the maximum amount of money you would be willing to pay to avoid having to shop at price P_1 instead of at the price P_0? In the case of a potential price decrease, we might ask the consumer, What is the maximum amount of money you would be willing to pay for the privilege of shopping at price P_0 instead of at the price P_1? In practice, the typical consumer might find it difficult to answer such questions. In addition, the consumer might have an incentive to misrepresent his or her estimate of the welfare gain (or loss) to influence policy choices (whether, for example, to have a sales tax on food).

Fortunately, there is an alternative way of eliciting from the consumer the desired information without having to ask dubious questions. In the literature on benefit–cost analysis, it has been shown that under a wide range of real-world circumstances the area under the consumer's demand curve and between the two price lines, area $(b + c)$ in Fig. 7.1, closely approximates the consumer's willingness-to-pay money to avoid the decreased welfare caused by a price hike. (For a price decrease, from P_1 to P_0, the area $(b + c)$ measures approximately the willingness-to-pay for the welfare gain.) The area under the demand curve and between the two price lines, area $(b + c)$, is called the change in *consumer surplus*. One or the other or both of two conditions must be met if the change in consumer surplus is to approximate closely willingness-to-pay. (Again we call attention to Appendix E for a rigorous treatment.) First, the quantity demanded by the consumer is insensitive to changes in his or her income; that is, the consumer's income elasticity for the good in question is close to zero. Second, the consumer spends a small share of his or her income on the good. In this and in succeeding chapters we will assume that at least one or the other or both of these two conditions hold such that the change in consumer surplus closely approximates willingness-to-pay. It may be noted, for example, that the income elasticity of the farm-level demand for food is thought to be positive but fairly small.

As a special case of great importance, consider the welfare gain to the consumer for the privilege of shopping at a competitive price, which we assume equals P_1 in Fig. 7.1, as opposed to not being allowed to buy the commodity at all. In other words, we might inquire as to the value to the consumer of having the opportunity of shopping for a good in a competitive market. To answer this question, we compute the change in consumer surplus for a price decline from P_2 to P_1. This gives area (a), which we interpret as the *total consumer surplus* associated with the privilege of participating in a competitive market. Clearly, the lower the competitive price is, the greater the value of the privilege. This concept will be of considerable use in this and later chapters.

At this juncture we turn attention from the individual consumer to the aggregate of all consumers. Willingness-to-pay is measured in dollars, which means we can add together the changes in consumer surplus for all consumers taken together. Thus, imagine that the demand curve in Fig. 7.1 is the aggregate demand for food, with total quantity demanded Q replacing quantity demanded by a single consumer q on the horizontal axis. Then area $(b + c)$ would measure approximately the total amount of money consumers taken together would be willing to pay to avoid the welfare losses associated with a price increase. What this means, importantly, is that the welfare change on the part of all consumers can be measured in principle by first estimating the aggregate demand for food and then calculating the area under the demand and between the two price lines. As noted, these results are valid on the assumption that the share of income spent on food is small or, what is more likely, that the income elasticity for food is small.

Before continuing, however, we consider still another limitation of benefit–cost analysis, a limitation in addition to those noted with regard to the principle of consumer sovereignty and the distinction between selfless and selfish gains (losses). It is sometimes argued that in adding up the dollars of willingness-to-pay across all consumers, the dollars of any one consumer are given the same weight as the dollars of any other, even though some consumers—because of age, poverty, poor health, or whatever—might be much more in need than others. The importance of this objection can be easily exaggerated. All agents (not just consumers) that participate in the market may be grouped according to whatever demographic factors are judged relevant. The benefits or costs to each group may be measured using willingness-to-pay, leaving to the policy maker the question of what weights should be given to the various groups. In fact, a major objective of benefit–cost analysis is to identify programs that channel benefits to various societal groups, including those deemed most in need (or those with the most political clout) and thus targeted by government programs.

7.2 MEASURING CHANGES IN THE WELFARE OF FARM INPUT SUPPLIERS

As in the case of consumers, the welfare of farm input suppliers is also affected by changes in prices. Considered in turn are land, labor, and other inputs lumped together in the category we have previously called producer goods. For the present, we assume that an increase in product demand increases the demands for inputs. (None of the inputs is inferior.) In contrast to the study of consumers, the distinction between short and long run is of great importance in the study of farm input suppliers.

Suppliers of Land Input

As before, we take the total amount of land available for farming as given and fixed. Both the short- and long-run supply curves for land are perfectly inelastic, as in Fig. 7.2. Although land may be bought and sold freely in the long run, in the short run, land ownership is assumed fixed.

Suppose that an increase in the demand for land input elevates annual rent from R_0 to R_1. We take area (x) as our measure of the short-run benefits to initial land owners. With a perfectly inelastic supply curve, area (x) is nothing more than the increase in total rents received, equaling $(R_1 - R_0)A_0$. In the study of the land market, however, it is important to distinguish between a resource and resource use,

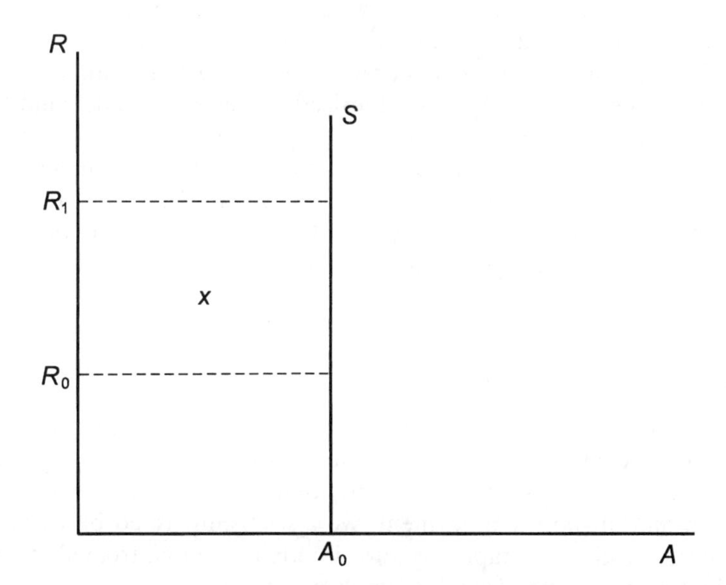

Figure 7.2 Supply for land input and land owner benefits from increased rent.

as we saw in Chapter 4. If, in a world of certainty, all buyers and sellers of land had assumed initially that the world was stationary, with no prospective changes in product demand, technology, input supply functions, and the like, then the initial value of land V under competition would have been given by

$$V_0 = \frac{R_0 A_0}{i} \tag{7-1}$$

where i equals the annual interest rate. On the assumption that all buyers and sellers assume a once and for all increase in the demand for land, caused by an increase in product demand, the value of land would increase in the long run, with land being negotiable, from V_0 to V_1, where

$$V_1 = \frac{R_1 A_0}{i} \tag{7-2}$$

Based on these arguments, one long-run effect of the increase in the demand for land is the creation of an aggregate capital gain to initial land owners given by

$$CG = \frac{(R_1 - R_0)A_0}{i} \tag{7-3}$$

This formula for measuring capital gains rests, of course, on restrictive assumptions and must be viewed as providing crude first approximations.

That land can be bought and sold obviously has important implications for benefit–cost analysis. The hypothesized increase in the demand for land increases the short-run flow of benefits or rental payments to land owners by $(R_1 - R_0)A_0$, as noted. Under our simple assumptions, the future flow of rent increases becomes capitalized in the price of land. Land owners find their short-run rental incomes increased; upon selling their land in the long run, they capture the present value of the future flow of increased rents as well. Future land owners receive no benefits whatever.

Suppliers of Farm Labor

It was assumed in previous chapters that the labor of a family farm is a fixed input. To avoid complications and to limit the scope of our analysis, we abstracted from choices between hours worked and hours of leisure and between hours devoted to farm and off-farm employment. We also distinguished between family and hired labor, showing in Chapter 4 how hired labor can be treated much as any variable input with an upward sloping supply curve.

The long-run effects of wage changes on the welfare of family farm labor are now considered, assuming the number of families is free to vary. Many of our results will be shown to apply to hired labor as well. The short-run case, with the number of farm families held constant, is considered in Section 7.3. There the student will be reminded that the quasi-rent to farmers in the short run represents a payment to all fixed inputs, not just to fixed family labor.

Letting h_0 equal the representative family's labor input in farming, panel a of Fig. 7.3 gives an aggregate, long-run stepped labor supply curve for three households. The transfer earnings for the first family, TE_0, are the lowest of the three, where $TE_0 = W_0h_0$. (Recall from Chapter 4 that the long-run implicit wage per hour of family labor W equals the total return to family labor divided by hours worked.) The transfer earnings for the second and third families equal, respectively, $TE_1 = W_1h_0$ and $TE_2 = W_2h_0$.

Now suppose that, in an initial long-run competitive equilibrium, the total return to family labor is such that the imputed wage equals W_0. The first farm is then a marginal farm in that actual returns to labor just equal the household's transfer earnings. If an increase in product demand increases the demand for farm labor, causing W to increase from W_0 to W_2, then the second and third households enter farming; the third farm becomes the new marginal farm. In the new equilibrium, the actual return to the first family's labor *exceeds* its transfer earnings by the area $(p + s)$. Similarly, the actual return to the second family's labor exceeds its transfer

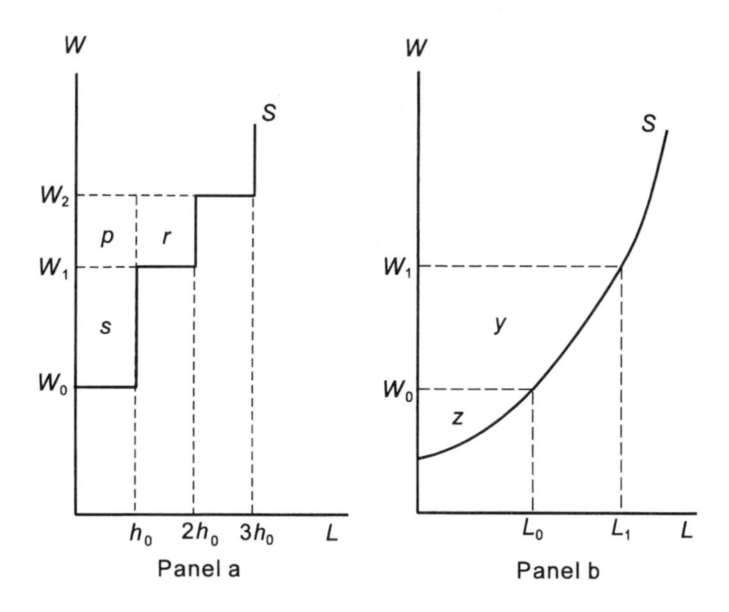

Figure 7.3 Labor supply curves and labor surplus.

earnings by area (r). We take, as our measure of willingness to pay for labor benefits generated by the hypothetical increase in labor demand, area ($p + s$) for the first family and area (r) for the second. Notice that the total willingness-to-pay for labor benefits equals area ($p + s + r$), which is the area above the stepped supply curve and between the wage lines given by W_0 and W_2.

A smooth and continuous supply curve for family labor, as in panel b, Fig. 7.3, is appropriate when there are thousands of family farms, each contributing a relatively small amount of labor (h_0) to total farm labor given by L. Area (z) measures the aggregate willingness-to-pay for the privilege of working in the farm industry if W_0 is the long-run equilibrium imputed wage. Similarly, area ($y + z$) measures willingness-to-pay if the wage equals W_1. Therefore, the willingness-to-pay for the family labor benefits generated by an increase in the imputed wage from W_0 to W_1 is given by area (y), the area above the supply curve and between the two wage lines. Borrowing from the terminology in the literature on benefit–cost analysis, we will refer to area (z) for $W = W_0$ [or area ($y + z$) for $W = W_1$] as the total *family labor surplus*. The area (y) equals the increase in family labor surplus associated with the increase in the wage from W_0 to W_1.

Having analyzed the long-run welfare implications of an increase in the returns to family labor, we analyze hired farm labor in much the same way. More particularly, assume that when a person joins the farm work force he or she puts in a fixed hour day (8 hours, say) without regard to the wage rate. Furthermore, hours of hired work are not divided between farm and nonfarm employment. In contrast to family labor, however, it is assumed that workers are free to enter or exit the farm sector in both the short and long run.

Referring once again to panel b, Fig. 7.3, we now reinterpret the variable L to equal aggregate hired labor input as opposed to family labor input. Accordingly, the variable W is to be taken as the hired wage rate. (The supply curves for hired and family labor need not, of course, be the same.) Based on the arguments set forth, area (y) measures the increase in welfare, based on willingness-to-pay, generated by an increase in the hired wage rate from W_0 to W_1. Following previous terminology, area (y) is the increase in *hired labor surplus*.

There is one significant difference, however, in the welfare analysis of hired and family labor. Little is gained in terms of analytical convenience by assuming that the level of hired labor is fixed in the short run, that short-run entry and exit of hired labor are not possible. There is, moreover, considerable evidence that the movement of hired labor between the farm and nonfarm sectors is rather fluid. Thus, in the remainder of this book, we will often assume that the supply curve for hired labor is upward sloping in both the short and long run. Changes in hired labor surplus may be generated by either short- or long-run changes in the hired wage rate. One reason why this distinction should be born in mind is that the long-run elasticity of the hired labor supply function likely exceeds the short-run elasticity. Indeed, for small farm industries, the long-run supply for hired labor might be perfectly elastic.

Suppliers of Producer Goods

We define producer goods as inputs manufactured in the nonfarm sector. In the short run, with a fixed number of manufacturers, each with a fixed plant, we must expect the law of diminishing returns to be in full force. What this means is that the short-run supply curve for producer goods is upward sloping. An increase in demand will surely increase the profits of suppliers, and we might be content simply to take the increase in profits as the measure of the increase in the welfare of producer goods suppliers. Although this is a good starting point, some complications are conveniently considered later, after we have had the chance to examine the welfare implications of product price increases for farmers themselves.

The long-run supply curve for producer goods might also be upward sloping, but as much to state an important conclusion as to analyze actual cases, we assume that the curve is flat; that is, the long-run supply of producer goods is perfectly elastic. In this case, the long-run price of producer goods is exogenous; it will not change in response to a change in demand. As a consequence, there will be no long-run benefits to the suppliers of producer goods. Bear in mind that, under competitive conditions, input suppliers can sell as much as desired at the going price. A supplier of diesel fuel, for example, enjoys no benefit simply from selling more fuel to farmers and less to the trucking industry.

7.3 MEASURING CHANGES IN THE WELFARE OF FARMERS

Measuring the increase in the welfare of farmers resulting from an increase in output demand, for example, and the associated increase in output price is a difficult business. This is largely because of the interdependence among farm output and input markets. Farm output and input prices might rise in tandem, which complicates measuring the change in net farm income. Whether farm input prices are exogenous or endogenous becomes an important question, which turns on whether the farm industry is a small or large buyer of inputs relative to the total quantities of inputs sold in input markets. The farm industry is a relatively small buyer of diesel fuel, for example. We do not anticipate that diesel fuel prices vary significantly in response to changes in farm purchases. The farm industry or sector is a relatively big user of land, on the other hand, and, in the long run at least, land rent and output price are jointly dependent. To figure out how changes in product prices affect farmer income and welfare in the long run, we must surely take account of the jointly determined changes in land rent.

Four different models are analyzed next depending on (1) whether the model is short run or long run and (2) whether at least some input prices are endogenous or none are. The models do not allow for uncertainty. We start by assuming that all farm input prices are exogenous, that is, independent of the level of farm output, focusing, in turn, on short- and long-run analysis.

Figure 7.4 gives the marginal cost (*MC*) curve for a representative farmer with a fixed plant. In this simple linear case, the marginal cost curve is also the farmer's supply curve. If price equals P_0, for example, the optimal output equals q_0. With constant input prices, the area under the *MC* curve over the range of output from zero to q_0 equals the total variable cost. Since total revenue equals area $(u + v)$, quasi-rent equals area (u). Because family labor and other inputs (land, for example) are fixed in the short run, quasi-rent is often interpreted as the implicit return to all fixed factors, including family labor. This explains why our welfare analysis of family labor was limited to the long run.

Getting back to the main stream of our analysis, however, if price rises from P_0 to P_1 in Fig. 7.4, total variable cost rises from area (v) to area $(v + w)$ and total revenue rises from area $(u + v)$ to area $(u + v + t + w)$. Quasi-rent rises to area $(u + t)$. Therefore, the increase in quasi-rent equals area (t), the area above the farmer's supply curve and between the two price lines. We take this as our measure of the farmer's willingness-to-pay for the privilege of selling output at P_1 instead of at P_0.

In the literature on benefit–cost analysis, the area (u), assuming that price equals P_0, is called the *total producer surplus*. The area above the producer's supply curve and between the two price lines, area (t), equals the increase in producer surplus associated with the price increase. The motivation for introducing the term total producer surplus—to measure what in this simple case is nothing more than quasi-

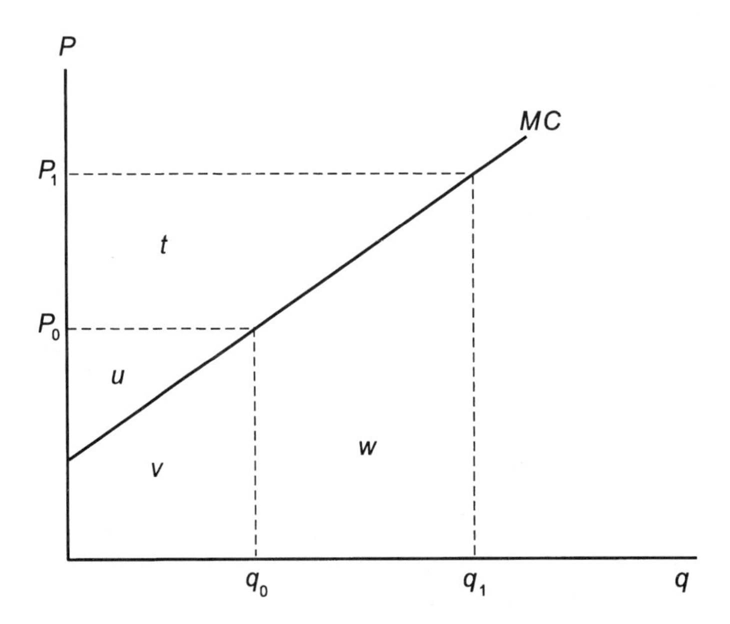

Figure 7.4 Marginal cost curve and producer surplus.

rent—will become clear when we consider cases involving endogenous input prices. Also, our analysis, which assumes a price increase, could be easily recast for a price decrease.

Aggregative analysis is straightforward. Figure 7.5 gives the short-run supply curve for the farm industry as a whole. With aggregate farm-level demand given by the curve labeled D_0, aggregate producer surplus (quasi-rent) is given by area (c). The increase in aggregate producer surplus (quasi-rent) generated by the increase in demand, which causes price to rise from P_0 to P_1, is given by area ($a + b$).

The long-run model under the assumption of exogenous (constant) input prices, which assumes that all farm families (established and potential) have the same transfer earnings, is not of much interest here. With exogenous input prices, the long-run supply curves for output and all inputs, including family labor, are perfectly flat. Demand shifts do not change long-run prices. Under competitive conditions, only one price will prevail, and the question of how long-run price changes affect farmer welfare is beside the point. The analysis here is the same as for the suppliers of producer goods with a perfectly elastic supply.

Models that allow for the endogeneity of the prices of at least some variable inputs are now considered. In other words, the farm industry is assumed to be a relatively big buyer of at least some variable inputs. A fundamental point to bear in mind in both short- and long-run analysis is that, for every equilibrium product price, there will be a set of equilibrium variable input prices. If the product price rises because of

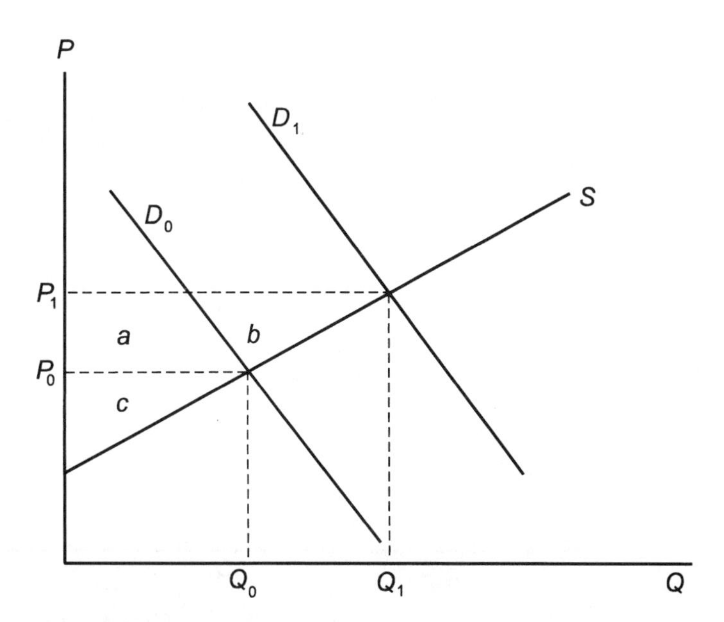

Figure 7.5 Aggregate demand curves, supply curve, and producer surplus.

an increase in product demand, for example, the prices of those variable inputs with upward sloping supply curves (short and long run) will rise as well. Of course, some variable input prices, those that are exogenous, might remain constant.

We let the curve labeled MC' in Fig. 7.6 equal the aggregate or horizontal sum of the short-run marginal cost curves of a fixed number of farmers, drawn on the assumption that the set of input prices is the equilibrium set associated with the producer price P_0 and the optimal output Q_0. In other words, the point (Q_0, P_0) is on the industry's short-run supply curve.

For purpose of illustration, we may suppose that the only two variable inputs are hired labor L_h and producer goods K. The supply curves for these two inputs are given in Fig. 7.7. The equilibrium in the product market (Q_0, P_0) is linked to the equilibrium points (L_{h0}, W_{h0}) and (K_0, J_0), respectively, in the markets for hired labor and producer goods. Now let the equilibrium product price rise from P_0 to P_1 in Fig. 7.6 in response, say, to an increase in product demand, not shown to avoid clutter. In response to an increase in product price, farmers expand output. They would move along the MC' curve *if* all input prices remained the same. But under present assumptions, input prices rise, which causes the aggregate marginal cost curves to rise as well. The price P_1 is a new equilibrium price because at this price the quantity produced equals the quantity demanded (Q_1) and because the equilibrium set of input prices associated with P_1 is, by assumption, the set of input prices for which the aggregate marginal cost curve MC'' is drawn. The new equilibrium in the output mar-

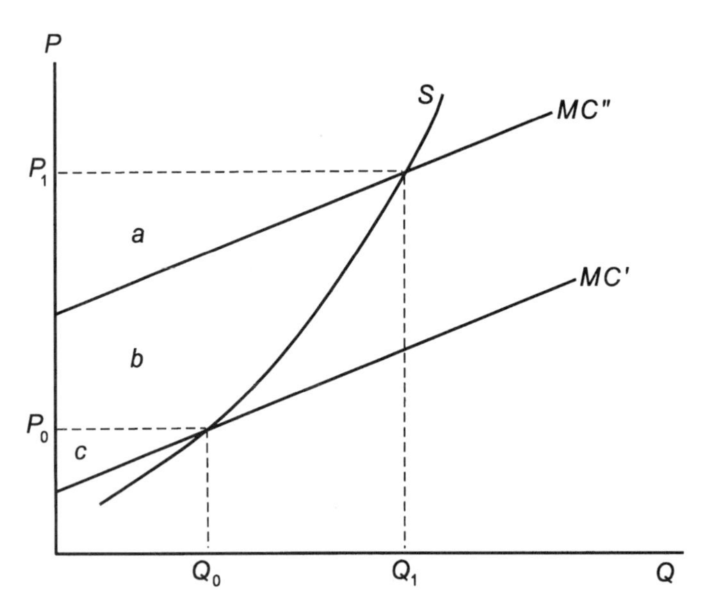

Figure 7.6 Aggregate marginal cost curves, market equilibrium supply curve, and producer surplus.

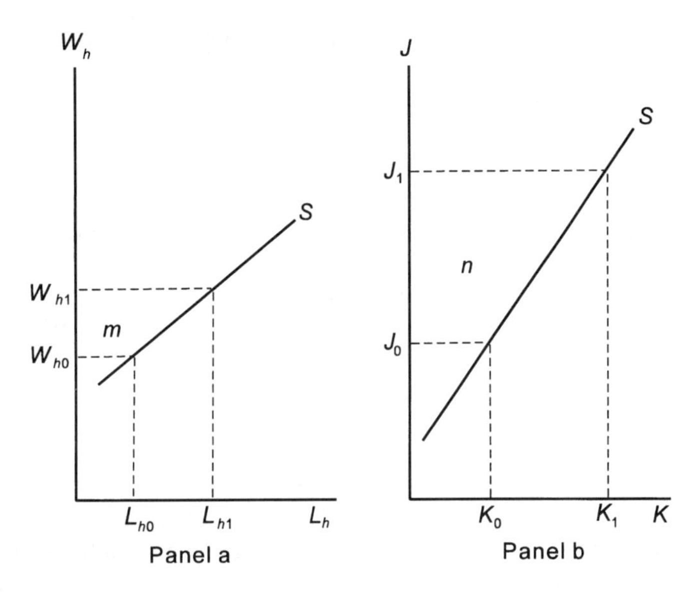

Figure 7.7 Supply curves for labor and producer goods and labor and producer surplus.

ket (Q_1, P_1) is linked with the new equilibrium points (L_{h1}, W_{h1}) and (K_1, J_1), respectively, in the markets for hired labor and producer goods.[1] Many points such as (Q_0, P_0) and (Q_1, P_1) may be hypothesized to exist, and the locus of such points is the short-run supply curve that shows how much farmers are willing to produce at alternative product prices allowing for the associated endogenous changes in input prices.

An important result is now stated without proof, but a numerical example is given as an illustration. (For a proof, see Appendix E.) *Assuming that product price and at least some input prices are interdependent, the area above the market equilibrium supply curve and between the two price lines equals the sum of the increase in quasi-rent to farmers plus the increases in the surpluses to variable input suppliers.* This result is applicable in both the short run, where several inputs may be fixed, and in the long run, where family labor, although fixed to each established farm, is variable through entry and exit of farms. Returning to Figs. 7.6 and 7.7, we suppose that product price rises from P_0 to P_1. The change in producer surplus given by area $(a + b)$ in Fig. 7.6 equals the sum of area (m) and area (n) in Fig. 7.7, plus the increase in quasi-rent to farmers. The latter quantity, it may be noted, is given by area $(a - c)$ in Fig. 7.6.

[1]At this point the student might like to review the material covered in Chapter 4. There it is shown that, given a production model, a market equilibrium supply function is a reduced form function (or solution) for output with market price taken as exogenous. Some input prices may be endogenous.

We now take up a simple numerical example. Consider the following model:

$$Q = L^{0.6} \quad \text{aggregate production function}$$

$$L = 0.5W \quad \text{labor supply function} \qquad (7\text{-}4)$$

$$W = 0.6L^{-0.4}P \quad \text{profit maximizing condition}$$

The student should recognize that, according to the third equation, the price of the input W is equated to its value of marginal product $(\partial Q/\partial L)P$. Treating output price P as exogenous, we find the reduced form for Q:

$$Q = 0.5969P^{0.4286} \qquad (7\text{-}5)$$

This, of course, is the market equilibrium supply function. As price rises from 1 to 2, equilibrium output rises from 0.5969 to 0.8034. The area above the supply curve, with P plotted on the vertical axis, and between the two price lines equals approximately 0.707. This is the change in producer surplus. The equilibrium values for L and W when product price equals 1 are, respectively, $L = 0.4232$ and $W = 0.8463$. Similarly, when product price equals 2, $L = 0.6943$ and $W = 1.3886$. The change in worker surplus, given by the area above the labor supply curve and between the two wage lines, equals 0.303. By computing the total revenue and the total variable cost for both prices, it is easy to show that aggregate quasi-rent rises from 0.2387 when price equals 1 to 0.6427 when price equals 2. Hence the increase in quasi-rent (0.404) plus the change in worker surplus (0.303) sums to the change in producer surplus (0.707).

It is now convenient to tie a loose end left from our discussion of the welfare of the suppliers of producer goods. Suppose that the S curve in panel b, Fig. 7.7, is the sum of the short-run marginal cost curves of the firms that manufacture producer goods, assuming that the prices paid by these firms for their variable inputs are exogenous. Then area (n) measures the increase in quasi-rents to producer goods manufacturers if price J rises from J_0 to J_1. If some variable input prices in the producer goods industry are endogenous, however, and rise with increases in the input of K, then area (n) measures the sum of the increase in quasi-rent to the producer goods manufacturers and the increases in surpluses to their input suppliers. The student should recognize that the welfare changes caused by a price change in one market, the farm output market, for example, may spread out over a wide sector of the economy, like the ripples from a stone falling in a lake.

7.4 BENEFITS AND COSTS IN THE MARKETING SECTOR

Our welfare analysis up to this point has assumed that farmers sell their outputs directly to consumers. This assumption is needlessly restrictive, and we now explore various interpretations that can be given to areas under a farm-level demand curve and above a price line or between two price lines. For this purpose, we assume, as in Chapter 5, that retail output equals farm output (broilers produced equal broilers sold at retail) and that the prices of all inputs in the marketing sector are exogenous. In the short run, we assume that the marketing margin increases with increases in output because of the law of diminishing returns to fixed plants. We show in what follows that the area under the farm-level demand D and above the price line measures the sum of total consumer surplus and aggregate quasi-rent to the marketing sector.

Assuming linear functions allows an easy algebraic demonstration. The retail demand, supply for marketing services embodied in retail output, and the resulting farm-level demand are given by

$$P^r = a - bQ$$

$$MM = c + dQ \qquad (7\text{-}6)$$

$$P = (a - c) - (b + d)Q$$

The last equation follows from the arbitrage condition: $P^r - MM = P$. For $Q = Q_0$, we have $P^r = P_0^r$, $MM = MM_0$, and $P = P_0$. Aggregate quasi-rent equals $(MM_0 - c)Q_0(\frac{1}{2})$. Total consumer surplus equals $(a - P_0^r)Q_0(\frac{1}{2})$. The area under the farm-level demand and above the price line, which we will call *total buyer surplus*, equals $(a - c - P_0)Q_0(\frac{1}{2})$. Since $P_0^r - MM_0 = P_0$, buyer surplus equals the sum of consumer surplus and aggregate quasi-rents to the marketing sector. It is left to the student to show that the area under the farm-level demand and between two price lines equals the sum of the change in consumer surplus and the change in quasi-rents. With a constant marketing margin, as in the long-run case, the total buyer surplus equals the total consumer surplus. This can easily be seen by drawing the appropriate graph or letting $d = 0$ (in which case, $MM_0 = c$) in the preceding algebraic analysis.

7.5 BENEFIT–COST IMPLICATIONS OF TECHNOLOGICAL PROGRESS

In Chapter 4, the effects of exogenous shocks consisted of changes in prices and output and input quantities. Little was said regarding the welfare of market participants. We now show how benefit–cost concepts can be used to augment the analysis of Chapter 4 by analyzing the welfare effects of technological change.

Let the demand curve and the initial supply curve be given by D and S_0, respectively, in Fig. 7.8. Suppose that technological change shifts the supply curve to S_1, decreasing price from P_0 to P_1 and increasing output from Q_0 to Q_1. A decrease in

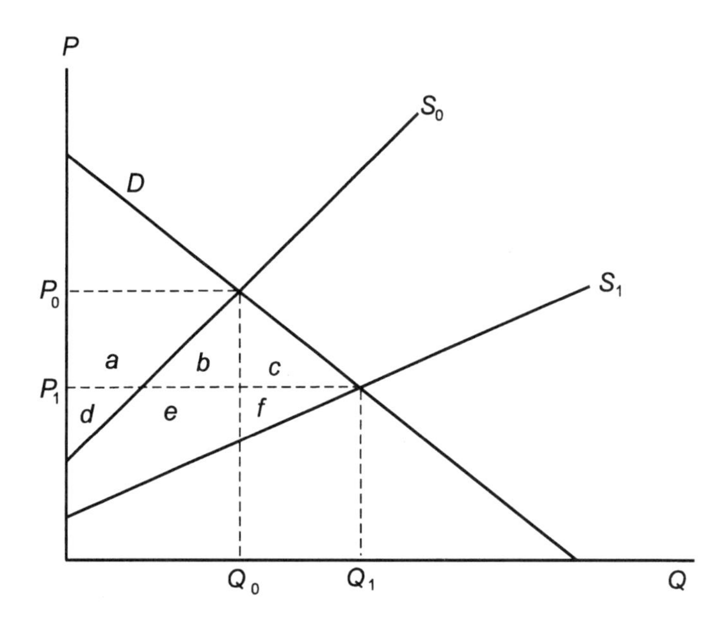

Figure 7.8 Aggregate demand and supply curves and consumer and producer surplus.

price and an increase in output are the inevitable consequences of technological progress. It is thus clear that consumer surplus is increased by technological change. Their willingness-to-pay for such gains equals approximately area $(a + b + c)$. Subtracting the total producer surplus before the technological change, area $(a + d)$, from total producer surplus after the change, area $(d + e + f)$, leaves the gain to producers of area $(e + f - a)$. (The student should always bear in mind that the interpretation to be given to the term producer surplus depends on the nature of the model being analyzed, as explained in Section 7.3.) Through experimentation with different graphs, the student will quickly see that the producer gain can be positive, negative, or zero. In any event, adding area $(a + b + c)$ and area $(e + f - a)$ yields a net gain, area $(b + c + e + f)$. In this simple model, society gains from technological change. What this means is that society as a whole would be willing to pay to have technological change occur. We hasten to add two qualifications. First, technological change does not simply drop from the skies like manna from heaven. It is rather the payoff from investment in research and development. This raises the question of whether the net gains depicted in Fig. 7.8 are large enough to justify the previous investment. Although important, this question is not analyzed here.

The second qualification concerns external costs. The introduction of a dangerous farm chemical, for example, might lower the costs incurred by farmers in the production of food, shifting supply to the right, but at the same time contaminating the nation's groundwater. The net gain given by area $(b + c + e + f)$ might easily be offset by the external cost of pollution.

Under what conditions will the change in producer surplus brought about by technological progress be either positive or negative? This question is not easily answered. In any event, changes in total producer surplus may mask important distributional consequences. Suppose, for example, that demand is inelastic. Technological change biased against labor but in favor of other factors lowers both total producer surplus and the imputed returns to family labor. Whether such change lowers returns to other factors, such as land, cannot be determined theoretically. This example calls attention to the aggregative nature of the concept of producer surplus. Input markets must be studied one by one if the distributional consequences of exogenous shocks are of interest.

7.6 DISTRIBUTIVE AND EFFICIENCY EFFECTS OF GOVERNMENT INTERVENTION

Much of the remaining material in this book centers on the effects of government intervention designed to alter market performance. Chapter 8 deals with monopoly power and the ways in which governments act either to destroy or protect it. Chapters 9 and 10 deal with farm commodity programs intended by governments to elevate farm prices and incomes, sometimes with results that are the opposite of those intended. For the present, however, we are interested in an example of government intervention that illustrates the application of benefit–cost concepts to policy analysis.

The redistributive effects of technological change have led some writers to argue that the nation needs a policy to regulate such change, a policy that presumably would require measuring the effects of potential changes prior to their being allowed to occur. In the analysis of such proposals, of government proposals that affect market performance, it is important to distinguish between distributive and economic efficiency effects. The performance of a market is said to *be efficient* if the total welfare of all market participants, as measured by willingness-to-pay, is maximized. Participants in a farm commodity market must be defined broadly to include the suppliers of farm inputs. If the willingness-to-pay on the part of beneficiaries exceeds the loss to program losers, the latter measured by willingness-to-pay to avoid the program, we say the program increases economic efficiency. In addition, the increase in efficiency can be measured by net benefits, that is, by the excess of benefits over costs. The definition of a decrease in economic efficiency follows accordingly. If program losers are willing to pay more to avoid a program than gainers are willing to pay to secure the program, we say the program causes inefficiency. The extent of the efficiency loss can be measured by the net loss of willingness-to-pay.

We may now apply these welfare concepts to the suggestion that technological change should be regulated. Returning to Fig. 7.8, we suppose that the market is initially in equilibrium with price equal to P_0. Allowing the technological change to occur would lower producer surplus on the assumption that demand is inelastic. In other words, suppose that area $(e + f - a)$ is negative. To protect farm land owners, family, and hired labor, and perhaps other farm input suppliers as well, the government blocks the technological change by, for example, making the sales of some new kind of seed or chemical illegal. The policy thus protects the welfare of farm input suppliers at the expense of consumers, but the distributive effects come at a cost in terms of economic efficiency. More particularly, the efficiency loss caused by blocking technological change equals area $(b + c + e + f)$.

Importantly, it does not follow that a policy decision is a bad one merely because of an efficiency loss. Governments are mightily interested in programs designed to redistribute national income (or, better still, the welfare benefits generated by the economic system) in order to help some people (the poor, say) at the expense of others (the rich). Efficiency losses are very often the unwanted side effects of such programs. Important trade-offs must be made between the perceived equitable distribution of the national income and the maximization of its size. Many examples will be provided in the chapters that follow.

Aside from environmental effects, which everyone agrees must be taken into consideration before technological change is allowed to go forward, is the regulation of technological change an appropriate means of an income redistribution perceived by policy makers to be desirable? Three points need to be made in this connection. First, except perhaps in special cases, it is not possible statistically to measure with acceptable accuracy the effects of technological change *before* such change occurs. It is in fact often difficult to measure statistically such effects *after* the change has occurred, when the researcher is able to draw on historical experience (e.g., time series data) to reach conclusions. This makes the distributional and efficiency effects of blocking technological change extremely uncertain.

Second, several policy options exist for ameliorating any redistributive effects of technological change judged to be unfair, aside from blocking the change itself. Labor training comes to mind as an example. Regulating technological change need not be an efficient means for effecting the desired redistribution relative to other means.

Third, technological progress is an important source of general economic growth and development. New agricultural technologies lower food prices and allow consumers to spend more of their incomes on nonfood products. This, in turn, increases the demand for the labor and other inputs needed to produce nonfood products, providing employment for the very labor and other inputs freed up by agricultural technological growth. New agricultural technologies have accounted historically for a massive worldwide reallocation of scarce resources, particularly labor, away from the production of the basic necessity food to the production of automobiles, rock concerts, novels, and all manner of luxuries that contribute to the good life.

PROBLEMS

7.1. In perfectly inelastic supply, farmland equals 1000. All land is rented and rent per acre equals 10. As a total surprise to everyone, the government places a property tax on farmland equal to 5 per acre.

a. Assuming fixed land ownership in the short run, calculate the net efficiency loss.

b. Letting the interest rate equal 0.08, calculate the long-run capital loss to initial land owners.

c. Will rent be affected? Explain.

7.2. Retail demand and the short-run supply for food marketing services are given by $P^r = 20 - Q$ and $MM = 2 + Q$. The prices of marketing inputs are exogenous. Because of technological change at the farm level, equilibrium output increases from 6 to 8. Calculate the increases in consumer surplus and in the quasi-rent to the marketing sector. Also calculate the area under the farm-level demand and between the two equilibrium price lines.

7.3. You are given the aggregate production function and labor supply function:

$$Q = L^{0.4}A^{0.6}$$

$$L = 0.1W^3$$

Acreage A is fixed at 100. The output price P rises from $P = 1$ to $P = 2$.

a. Derive the output supply function.

b. Find the equilibrium levels for Q, L, W, and rent R for the two output prices.

c. Calculate the change in the producer surplus and the sum of the increase in the aggregate expenditure on rents and the increase in worker surplus.

7.4. Demand and initial supply are given by $P = 20 - Q$ and $P = 2 + 2Q$. Under a new technology, supply becomes $P = 2 + Q$. Calculate the changes in the consumer and producer surpluses and the efficiency gain.

REFERENCES

Helmberger, Peter G., *Economic Analysis of Farm Programs*. New York: McGraw-Hill Book Co., 1991.

Just, Richard E., Darrell L. Hueth, and Andrew Schmitz, *Applied Welfare Analysis and Public Policy*. Upper Saddle River, N.J.: Prentice Hall, 1982.

Willig, Robert D., "Consumer Surplus without Apology," *American Economic Review,* 66, no. 4 (September 1976), 589–597.

8

Industrial Organization of Agricultural Markets

Industrial organization is one of the most important fields of economics if for no other reason than its breadth: it seeks to apply economic theory to virtually all markets. Indeed, the subject matter of the previous chapters readily falls within the purview of industrial organization. When economists speak of industrial organization, however, they are more often than not concerned with problems associated with market power, not with the kind of markets modeled heretofore in this book. The reason for this is that actual markets, particularly in the manufacturing sector, are not characterized very accurately by the assumptions of perfect competition, by which firms take prices as given. Many industries are dominated by a few large firms. Entry of new competitors is often difficult. Product differentiation, at least in retail markets, is ubiquitous. Since such conditions are frequently found in the marketing channels for farm outputs as well, the previous chapters on agricultural marketing, which adhered closely to competitive models, must be viewed as first approximations that are in need of modification and reexamination.

Accordingly, in this chapter, attention centers on the analysis of markets characterized by the presence of firms that do not take prices as given, that is, as parameters beyond their control or influence. More particularly, we analyze markets in which big sellers or buyers or both use market power to generate profit and have some latitude in setting prices. For such markets, the theories of monopoly and oligopoly come into play.

8.1 COMPETITION AND MONOPOLY: A COMPARATIVE ANALYSIS

The objective of this section is to compare the welfare implications of competition and monopoly. We start with some simple definitions. In this chapter, we say a product is homogeneous if every consumer or buyer is indifferent between any two units of it, without regard to whether the units are produced by the same or different firms.[1] The set of firms selling a homogeneous product constitutes a selling industry. The set of buyers to whom the firms sell their output constitutes a buying industry. Unless otherwise noted, the buyers considered in this chapter are the ultimate consumers. The buyers and sellers taken together constitute a market. Market definition in the real world is complicated by the presence of spatial and temporal dimensions and by product heterogeneity, but such complications are mainly of interest to researchers and they will receive limited attention in what follows.

Perfectly Competitive Markets

Perfectly competitive and monopoly markets differ in several important respects. A competitive market consists of many consumers (or buyers) and sellers, with each agent's purchases or sales accounting for a minute proportion of the total output traded. There are, in addition, no barriers to entry, meaning that the established producer has no advantage over the potential entrant in terms of the ability to buy inputs, organize production, sell output, and, in general, make a profit. The product is ordinarily assumed to be homogeneous, although we have seen in Chapter 5 how this assumption can be relaxed. Market agents have all the information required to maximize their objectives in market equilibrium, subject, of course, to given prices and constraints, such as production functions in the case of firms and utility functions in the case of consumers.

Although an inclusive definition of market structure will be given later, we note the following: In the literature on industrial organization, a market characterized by a large number of buyers and sellers, free entry of potential buyers and sellers, homogeneous product, and perfect information is said to have a *competitive structure*. If a market has a competitive structure, each seller makes choices with regard to levels of input and output independently of all other sellers, both established and potential. That is, each seller ignores completely the impacts of his or her choices on the decisions of other sellers. Similarly, each consumer's choices are made independently of the choices of other consumers. The structure of the market affects the way that participants make decisions.

Monopolized Markets

Turning to monopoly, if a firm is the only seller of the product in question, if entry of new sellers is not possible, and if the buying industry is perfectly competitive, the firm is said to be a monopolist. We note here for use later that a market character-

[1]This definition of homogeneity differs somewhat from that given in Chapter 5. Here we want to allow for product differences that are either real, as defined in Chapter 5, or fancied by consumers who might not have the ability to judge quality and are therefore subject to the influence of advertising.

ized by a monopolist is said to have a *monopoly structure*. A market with a single buyer (a monopsonist) and a perfectly competitive selling industry is said to have a *monopsonistic structure*. *Bilateral monopoly*, on the other hand, refers to a market with a single seller and a single buyer. Monopsony and bilateral monopoly will be considered briefly later.

For the present, however, we are interested in comparing the welfare implications (benefits and costs) of competition and monopoly. It is convenient to begin by reviewing the elementary theory of monopoly.

Let the long-run profit function for a monopolist be given by

$$\pi = PQ - C(Q) \tag{8-1a}$$

$$\pi = D(Q)Q - C(Q) \tag{8-1b}$$

where $C(Q)$ is the long-run total cost function and $P = D(Q)$ is the inverse demand for output. For simplicity, we let the demand be linear: $P = a - bQ$. Assume that the monopolist seeks to maximize profit. According to the first-order condition for maximum profit, the monopolist equates the marginal revenue and marginal cost thus:

$$a - 2bQ = C'(Q) \tag{8-2}$$

where the term on the left equals marginal revenue (MR) and the term on the right is the marginal cost (MC). [To derive this result, substitute the linear demand for $D(Q)$ in Eq. (8-1b) and differentiate with respect to Q.] The structural model that determines the four dimensions of market performance—market price and output and monopoly cost and profit—consists of four equations: the market demand, $P = D(Q)$; the cost function, $C = C(Q)$; and Eqs. (8-1b) and (8-2). Exogenous or shift variables could easily be included in demand and in the total cost curve were we interested in explicitly deriving hypotheses on how market performance is affected by changes in variables such as population, consumer income, and the prices of inputs. It is easy to show, for example, that if the demand parameter a in the linear case rises because population increases, then price, output, total cost, and monopoly profit will all rise as well.

Profit maximization implies conditions in addition to that given by Eq. (8-2). These conditions are derived with the aid of Fig. 8.1, where the demand and marginal revenue curves are given by D and MR and the marginal cost and average total cost curves are given by MC and AC. The first-order condition, marginal revenue ($MR = a - 2bQ$) equals marginal cost [$MC = C'(Q)$], is satisfied for $Q = Q_M$, with the corresponding price equal to P_M. The second-order condition for profit maximization is satisfied if $d^2\pi/dQ_2 < 0$, which holds if the marginal cost curve is upward sloping in the relevant region, as in Fig. 8.1. It must also be true, however, that the price associated with optimal output is not less than average cost. This condition is also

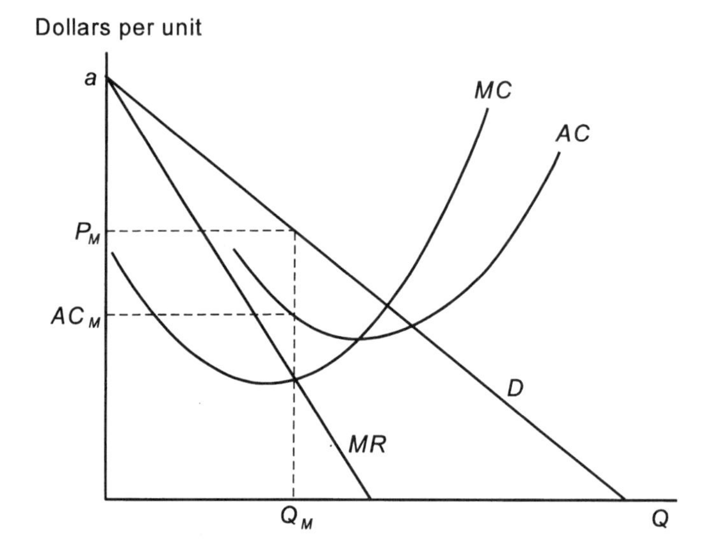

Figure 8.1 Cost curves and demand and marginal revenue curves for a monopolist.

satisfied in Fig. 8.1, since average cost for output Q_M equals AC_M. Maximum profit is given by the rectangular area $(P_M - AC_M)Q_M$.

Welfare Implications of Monopoly and Competition

We now consider a new maximizing problem. Following the developments of Chapter 7, we take total consumer surplus TCS (the area below the demand curve and above the price line) as an approximate measure of the willingness-to-pay on the part of consumers for the privilege of buying the commodity, rather than doing without the commodity completely. Importantly, TCS is inversely related to price and positively related to quantity purchased. Taking advantage of the linearity of the demand assumed previously and the formula for the area of a rectangle, total consumer surplus may be expressed as a function of output as follows:

$$\begin{aligned} TCS &= (a - P)Q/2 \\ &= bQ^2/2 \end{aligned} \tag{8-3}$$

where the parameter a is the vertical intercept of the demand curve (as in Fig. 8.1). To get the second expression for TCS, substitute $(a - bQ)$ for P in the first expression. For $Q = Q_M$, for example, TCS equals $\frac{1}{2}(a - P_M)Q_M$.

If we take profit as the measure of the monopolist's willingness-to-pay for the privilege of selling output as opposed to being shut down, the sum of total consumer surplus and profit equals the total willingness-to-pay for the privilege of participating in a market on the part of consumers and the monopolist taken together. Obviously, total willingness-to-pay depends on whatever quantity of output is chosen by the monopolist and the associated price. Letting TWP equal total willingness-to-pay, we have

$$\begin{aligned} TWP &= PQ - C(Q) + bQ^2/2 \\ &= aQ - bQ^2/2 - C(Q) \end{aligned} \tag{8-4}$$

We now suppose the monopolist seeks to maximize the value to society of having the market in question, where value is measured by total willingness-to-pay. The first-order condition for maximizing TWP is

$$a - bQ = C'(Q) \tag{8-5}$$

Since $P = a - bQ$, Eq. (8-5) states that the value of a market is maximized if market price is equated to the marginal cost of production. The second-order condition for a maximum is satisfied if the marginal cost curve is upward sloping, which we assume it is. We let Q_e denote the quantity that maximizes the social value of the market, that is, the quantity satisfying Eq. (8-5).

The result given by Eq. (8-5) is apparent from graphic analysis. The demand and marginal cost curves from Fig. 8.1 are reproduced in Fig. 8.2. The profit-maximizing level of output is again given by Q_M. Total consumer surplus plus monopoly profit for any level of output equals the area under the demand and above the marginal cost curve. (Recall that the area below the marginal cost curve measures total cost.) It is clear that the value of the market is increased if output is allowed to go beyond Q_M and that, in fact, the maximized value is associated with output equal to Q_e, given by the intersection of D and MC. Market value under profit maximization is less than the socially optimum level by the triangular area (y). This triangular area plays an important role in economic theory and is sometimes referred to as the welfare loss (the efficiency or deadweight loss) due to monopoly. Given monopoly and profit maximization, the welfare loss triangle depends on basic economic data (the values of exogenous variables) that determine the positions and shapes of the demand and cost curves.

This brings us to a fundamental objection to the market performance generated under monopoly. A monopolist produces insufficient output (and charges too high a price) relative to the output that maximizes the benefits of all the participants in the market taken together. This objection to monopoly raises an important question for policy. What can be done (should be done?) to expand output beyond Q_M,

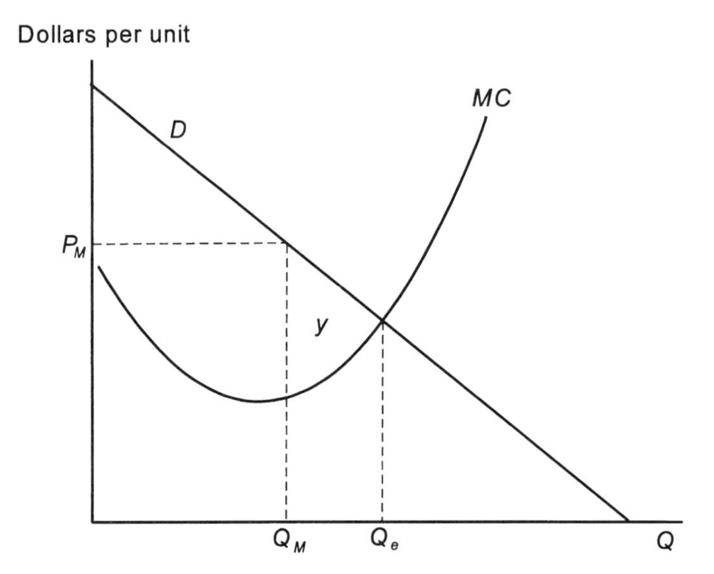

Figure 8.2 Marginal cost and demand curves for a monopolist showing efficiency loss.

maybe even to Q_e? A small menu of alternatives will be considered later, but we want first to consider the welfare implications of competitive performance.

To assess the welfare implications of perfect competition, we assume that the government converts by decree an otherwise perfectly competitive industry into a cartel. Without worrying about administrative costs, we assume that the government fixes both the number of firms and the output of each in such manner as to assure that, whatever output is chosen, it is produced at least cost. We wish to show that if the objective is to maximize the value of the market, as determined by willingness-to-pay, then the government will choose the competitive level of output. This means, of course, that if the market has a competitive structure the government need do nothing at all. Perfect competition maximizes the value of the market.

Demand and long-run supply curves are given by D and LRS, respectively, in Fig. 8.3. The competitive output is given by Q_c. If the government fixes the number of firms and the output of each such that Q_0 is produced, then price equals P_0. We assume that the government confiscates all excess profit for the benefit of taxpayers. In this manner, taxpayers are made participants in the market. If $Q = Q_0$ and $P = P_0$, for example, the total consumer surplus equals area (a), total producer surplus equals area (c), and the taxpayer benefit equals area (b), the latter given by $(P_0 - AC_0)Q_0$, where AC_0 is the competitive average cost of producing Q_0. If the chosen output is so large that price exceeds average total cost, we then assume that taxpayers must pay producers the balance to keep them in production.

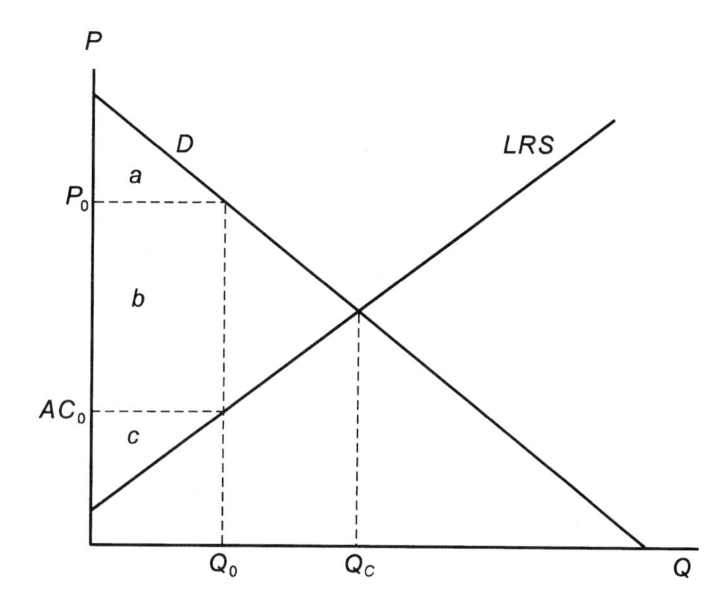

Figure 8.3 Demand and long-run supply curves and consumer, producer, and taxpayer benefits.

The total willingness-to-pay for the existence of a cartel, including the benefits to taxpayers, equals the area under demand, above supply, and to the left of the vertical dashed line that specifies the level of output. Clearly, to maximize this area, the government must set Q equal to the competitive output Q_c. (See, for example, Problem 8.3.) This shows that, in the long run, the value of a market is maximized by competitive performance. A government cartel dedicated to efficiency is not needed, all the more so, it may be noted, if managing a cartel is costly.

Whether efficient or otherwise, market performance is said to be *Pareto optimal* if it is impossible, through changing performance, to make one person better off without making another person worse off. In the absence of external costs, like environmental pollution, perfect competition is Pareto optimal. Starting at $Q = Q_c$ in Fig. 8.3, for example, the government could certainly increase benefits to taxpayers by decreasing output, but only at the expense of consumers and input suppliers. Perfect competition is not only efficient (i.e., maximizes total willingness-to-pay), it is also Pareto optimal. By comparison, a profit-maximizing monopoly is generally inefficient, although it might be Pareto optimal, as we will now see.

Monopoly and Collective Action

Is there some form of collective action that might increase the value of a monopolized market as measured by the willingness-to-pay of all market participants, including taxpayers? Two forms of collective action come to mind: consumer unions (cooperatives) and government policy.

Suppose that consumers form a union or cooperative in order to bargain with the monopolist. A numerical example is useful in understanding the principles involved. The aggregate demand for 100 identical consumers is given by $P = 10 - Q$. The monopolist's total variable cost is given by $TVC = \frac{3}{2}Q^2$. The marginal revenue and marginal cost functions are, respectively, $MR = 10 - 2Q$ and $MC = 3Q$. The monopoly output Q_M equals 2; the monopoly price P_M equals 8. Maximized quasi-rent equals 10, and the consumer surplus for the monopoly solution equals 2. Total willingness-to-pay for the privilege of participating in this monopolized market equals 12.

Equating price and marginal cost, on the other hand, yields output $Q_e = 2.5$ and $P_e = 7.5$. Monopoly quasi-rent falls by 0.625, but total consumer surplus rises by 1.125. Seeing that this is the case, the 100 consumers form a union, each chipping in 0.00625 to build a fund equal to 0.625. They then make a deal with the monopolist. The monopolist agrees to produce 2.5 units, earning a quasi-rent of 9.375, in return for a payment from the union equal to 0.625. The monopolist is then as well off as under the monopoly solution. The gain to the consumers is 0.5. If we assume that the formation of the union is costless, the monopoly solution is not only inefficient, it is also not Pareto optimal; it is possible to make some people (the consumers) better off without making anyone (the monopolist) worse off.

This discussion of a consumer union raises several questions, however. First, if the organization of a consumer union is costless, we would never have expected to find a monopoly solution in the first place. More generally, the market mechanism would include a union that assures performance that is both efficient and Pareto optimal: there would be no need for government action. Second, it is restrictive to assume that forming a consumer union is costless, and the assumption is less realistic in the presence of heterogeneous consumers. Organizational costs might exceed the 0.5 net benefit in our example. In this case, monopoly might be Pareto optimal. Third, once a consumer union is formed, why should we rule out the possibility of a consumer strike? Consumers might seek a price such that profit equals zero, where D intersects AC in Fig. 8.1. The formation of the union would alter the structure of the market and lead logically to the need to model a bargaining process in which consumers seek to maximize total consumer surplus and the monopolist seeks to maximize profit. Suddenly, the issues become complicated.

Another form of collective action is government policy. Three alternatives come to mind. First, the pope, a TV preacher, or a business school professor might be called on to talk ethics to our hypothetical monopolist, extolling the virtues of charity and emphasizing that God punishes the greedy. The idea is to change the monopolist's utility function and to get him or her to maximize total willingness-to-pay instead of profit. Given the likely efficacy of this approach, it would seem prudent to consider other alternatives.

A second alternative is important when the optimum size of the firm is large relative to the size of the market, as in Fig. 8.1, where competitive market structure is incompatible with economies of size. Here, we can forget about policies designed to assure a competitive structure, such as breaking up a large firm into many small ones. As an alternative, the government might embark on a program of regulation. A commission is appointed to collect information on demand and production costs and to order the monopolist to produce at the point where price equals marginal cost, at $Q = Q_e$ in Fig. 8.2. Regulated public utilities come readily to mind. It is easy to assume that the cost of effective regulation is negligible in a theoretical model, but the hypotheses derived from such a model may be of little use in the real world. If the regulatory commission is expensive to operate and/or is dominated by the monopolist, the whole exercise might be a fiasco or a charade. It is entirely possible that monopoly performance is the best that can be obtained.

A third alternative is a set of policies designed to forestall sellers from conspiring to behave like a monopolist or to disallow their using mergers and acquisitions to become a monopolist or, when there are no economies of size in the case of a multiple-plant monopolist, to break up a monopolist into several firms. This grab-bag alternative is best discussed after we have considered markets that are neither perfectly competitive nor purely monopolistic.

8.2 MONOPSONY AND BILATERAL MONOPOLY

A market made up of a perfectly competitive selling industry and a buying industry consisting of a single firm, with entry of new buyers being impossible, is called *monopsony*. A market consisting of a single buyer and a single seller with blocked entry in both the buying and selling industries is called *bilateral monopoly*. These two cases are analyzed briefly next.

Monopsony

Let the long-run profit function for a monopsonist be given by

$$\pi = f(Q) - PQ \tag{8-6}$$

where $TRP = f(Q)$ is the total revenue product function as developed in Chapter 4. To review briefly, this function shows, for every possible level of Q, the maximized difference between total revenue from sales and the outlays on all inputs other than the input Q. The reader may think of Q as a raw material purchased from farmers. The average revenue product (ARP) function, defined by $ARP = f(Q)/Q$, is displayed graphically in Fig. 8.4.

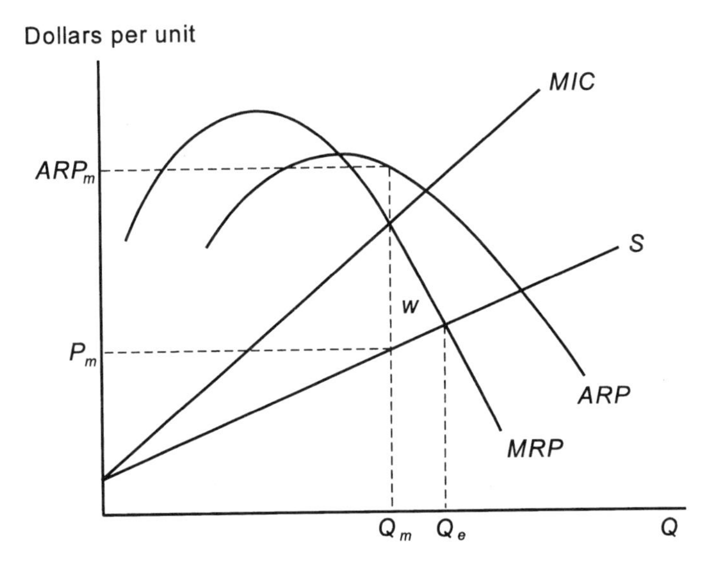

Figure 8.4 Revenue product curves and supply and marginal input cost curves for a monopsonist showing efficiency loss.

Because the monopsonist is the only buyer, the price of the raw material P must not be viewed as exogenous. In fact, price varies directly with the quantity purchased according to the farm supply function $Q = S'(P)$ or its inverse $P = S(Q)$. Equation (8-6) may therefore be rewritten as

$$\pi = f(Q) - S(Q)Q \tag{8-7}$$

Maximizing π, we have as the first-order condition

$$\frac{\partial f(Q)}{\partial Q} = \frac{\partial P}{\partial Q}Q + P \tag{8-8}$$

The expression on the left is, by definition (see Chapter 4), marginal revenue product (MRP). The expression on the right is the marginal input cost (MIC). If we assume, for example, that $P = c + dQ$ is the supply for Q, then the input cost is

$$S(Q)Q = cQ + dQ^2$$

and MIC is given by

$$MIC = c + 2dQ$$

At this point, a graphic analysis is convenient. Figure 8.4 gives the curves for the average revenue product (ARP), marginal revenue product (MRP), supply of farm output (S), and marginal input cost of farm output (MIC). To maximize profit, the monopsonist finds the optimum output Q_m by locating the intersection of the MRP and MIC curves. If Q_m is purchased, the farm price equals P_m. The second-order condition for a maximum is satisfied if the MRP and MIC curves are downward and upward sloping, respectively. In addition, we must be mindful that profit (or quasi-rent in a short-run formulation) is not negative. This latter economic condition for a maximum is certainly satisfied, since $ARP_m > P_m$. Maximum profit is given by the rectangle corresponding to $(ARP_m - P_m)Q_m$. As a useful exercise, it is left to the student to show (1) that the value of the market, measured by willingness-to-pay, is maximized by setting Q equal to Q_e (see Fig. 8.4) and (2) that the deadweight or efficiency loss due to monopsony is given by the triangular area (w). (*Hint:* Total willingness-to-pay equals the sum of the monopsonist's profit plus the total producer surplus, the latter given by the area above the farm supply function and below the price line.)

Bilateral Monopoly

The study of bilateral monopoly provides an introduction to bargaining theory. The profit functions for the single seller and buyer in the bilateral monopoly model are given by

$$\pi_s = PQ - C(Q) \tag{8-9a}$$

$$\pi_b = f(Q) - PQ \tag{8-9b}$$

where the subscripts s and b identify, respectively, the seller and the buyer. Here we encounter a conceptual problem that is new to this book, but which students will likely recall from their previous studies of oligopoly. (Also, see Section 8.3.) The profit of one firm cannot be maximized without that firm's dictating the terms of exchange to the other. But, clearly, it cannot be supposed in general that both firms dictate the terms of exchange. We could proceed to maximize the profit of the seller π_s, for example, if we could reasonably assert that $P = D(Q)$ is a demand function showing how much a price-taking buyer would be willing to purchase at the alternative prices set by the seller. In this case, the seller dictates the price to the buyer. The maximizing problem becomes identical to that for a monopolist. [See Eq. (8-1).] But why should a single buyer be so timid as to accept whatever price is offered by the seller?

Similarly, we could proceed to maximize the profit of the buyer π_b if we could reasonably assert that $P = S(Q)$ is a supply function showing how much a price-taking seller would be willing to supply at alternative prices set by the buyer. The maximizing problem becomes essentially the same as that for the monopsonist. [See

Eq. (8-7).] But why should a seller be so timid as to accept whatever price is offered by the buyer?

Confronted with this problem, many writers have argued that whatever deal is struck by the seller and buyer (whatever terms of trade are agreed on) it should be Pareto optimal. In other words, we should look for a solution to what is essentially a bargaining problem such that one party cannot be made better off without making the other party worse off. This means that the equilibrium output should maximize joint profit. Following this line of attack, we have

$$\begin{aligned} \pi_j &= \pi_s + \pi_b \\ &= f(Q) - C(Q) \end{aligned} \tag{8-10}$$

The first-order condition for a maximum of joint profits π_j is given by

$$f'(Q) = C'(Q) \tag{8-11}$$

In other words, the seller and buyer should agree to trade a level of output such that the marginal revenue product of the buyer (*MRP*) equals the marginal cost of the seller (*MIC*).

A graphic analysis is again illuminating. The relevant revenue product curves of the buyer and the cost curves of the seller are given in Fig. 8.5. Maximizing joint

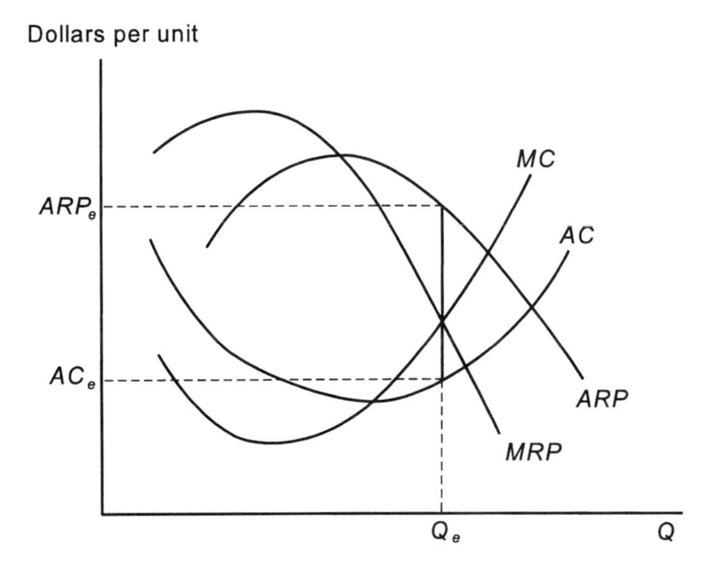

Figure 8.5 Revenue product and cost curves for bilateral monopoly with a price bargaining range.

profit implies choosing that level of output, Q_e, where the marginal revenue product curve (MRP) of the buyer intersects the marginal cost curve (MC) of the seller. This suggests that Q_e is the optimal output. The second-order condition for joint profit maximization is satisfied if the MRP and MC curves are downward sloping and upward sloping, respectively, as in the figure.

Notice, however, that nothing has been said about the price. Clearly, the long-run equilibrium price must be such as to disallow a negative profit to either party. This puts constraints on what might be called the *price bargaining range*, the set of prices that contains the solution. Specifically, $AC_e \leq P \leq ARP_e$. This leaves a range of indeterminacy in the absence of further assumptions. For example, an assumption might be made as to the psychology of the two combatants. Perhaps one is much tougher than the other. Alternatively, we might assume that one party has a great deal of financial resources, a deep pocket as is sometimes said in the literature. This might allow one firm to afford long, drawn-out strikes that would drive the other firm into bankruptcy, perhaps allowing the stronger firm to acquire the weaker firm. This suggests the interesting possibility of a merger. The terms of a merger would need to be negotiated, but the possibility of a merger points to a once and for all bargaining process that alters the structure of the market.

Clearly, setting output equal to Q_e maximizes the willingness-to-pay of the two firms for the privilege of entering a bargaining process. If, indeed, the seller buys his or her inputs in perfectly competitive markets, if the buyer sells the finished product to consumers in a perfectly competitive market, and if the two firms agree on Q_e, then the allocation of resources will satisfy a critical condition of perfect competition; that is, the price paid by the ultimate consumer equals the marginal cost of production. The result is an efficient allocation of resources. This has led some writers to argue that mergers that combine powerful buyers and sellers will likely entail efficiency gains for society. A detailed analysis of this proposition is beyond the scope of this book, but one serious objection arises in cases where entry in both the buying and selling industries is difficult but not impossible. The most favored entrants into the buying industry might be powerful sellers; the most favored entrants into the selling industry might be powerful buyers. Four sellers selling to four buyers could, if mergers are disallowed, lead to eight sellers of the finished product—sellers who are vertically integrated, but who do not buy a semifinished product in a bilateral oligopoly market.

8.3 OLIGOPOLY

Having reviewed briefly the theory of markets with market structures that stand at polar extremes, we now center attention on all the interesting in-between cases where fewness of sellers or fewness of buyers is the crucial market characteristic. More particularly, we take up the case of oligopoly, or, perhaps we should say, the many cases of oligopoly (few sellers) and oligopsony (few buyers). Oligopoly refers to a class of markets that is broad indeed. The number of sellers ranges from only two sellers to many sellers, some of whom are relatively large. Barriers to entry range from high to low. Outputs may be extremely homogeneous or so heterogeneous that it becomes hard to say whether the sellers of them should be included in the same market. The same considerations apply to oligopsony, for which fewness on the buying side of the market is the distinguishing characteristic.

The study of oligopoly is complex not only because oligopoly (or oligopsony) encompasses markets with widely ranging characteristics, but also because of what is called oligopolistic interdependency. Perhaps the easiest way to see the meaning of oligopolistic interdependency and to appreciate the difficulties it poses for theoretical analysis is to consider the classic case of duopoly. Consider a market in which the buying industry has a competitive structure, but the selling industry is characterized by the existence of two sellers, homogeneous outputs, and blockaded entry. If the product is perfectly homogeneous, only one price will prevail in the absence of consumer rationing. Letting q_i equal the output of the ith duopolist, $i = 1, 2$, such that $Q = q_1 + q_2$, and letting $P = D(Q)$ be market demand, we have

$$\begin{aligned} \pi_1 &= D(q_1 + q_2)q_1 - c_1(q_1) \\ \pi_2 &= D(q_1 + q_2)q_2 - c_2(q_2) \end{aligned} \tag{8-12}$$

The first firm cannot maximize π_1 without knowing the value for q_2 or how, exactly, q_2 will vary with q_1. The same considerations apply to the second firm. One way to proceed is to suppose that the entrepreneurs know the following two functions: $q_2 = f_1(q_1)$ and $q_1 = f_2(q_2)$. Through substitutions we could then get rid of q_2 and q_1 in the respective expressions for π_1 and π_2. This procedure for obtaining solutions is devoid of empirical relevance unless it can be shown how $f_1(q_1)$ and $f_2(q_2)$ can be derived. It is clear that what the first firm does affects significantly the profits of the second, and it cannot therefore reasonably be supposed that the second firm's decisions will not vary with those of the first. Modeling in a meaningful way the decision making of the two firms in this case must recognize oligopolistic interdependence.

The problem is to explain how output choices are made in this model. Standard price theory contains many solutions, some more plausible than others. We will not pause here to review in any detail the available models except to note the following: Instead of a unique solution implied by the assumption of maximizing behavior, we are confronted with several alternative patterns of behavior that are more or less plausible. First, instead of living together amicably, the two duopolists might decide to fight it out through price wars, each trying to drive the other out of business and to become a monopolist. It seems unlikely that such a mode of behavior would continue indefinitely, but the possibility of price wars does call attention to the likelihood that oligopolists need not always take market structure as given; they might try to change the structure in ways that facilitate making greater profits. Second, a distant cousin of a price war is product differentiation. The duopolists might be in a position to alter the outputs that they sell and to advertise and in other ways promote their outputs. It is not necessary for the physical characteristics of products to be greatly different when buyers are ignorant and can be convinced that one brand of perfume, say, leads to a better sex life than does another. Here, as in the case of a price war, the duopolists strive to change the structure of their market through product differentiation and market segmentation. Third, the duopolists might get together and agree on a price and a set of production quotas that maximize joint profit. The result is then equivalent to that for a multiplant monopolist. Establishing production quotas determines how the aggregate profit gets distributed between the two firms. [See Eqs. (8-12).] It seems likely that the more similar the two firms are in terms of costs, nature of products, financial resources, and the like, the greater the ease with which a collective agreement might be achieved. Finally, the duopolists might simply decide to merge, thus forming a monopoly.

An express agreement that sets price and assigns production quotas is not the only way to achieve concerted action, however. Verbal or written agreements might not be necessary. Two ballroom dancers move effortlessly across the floor without agreeing verbally when to spin and dip and without ever treading on each other's toes. They are able to coordinate their behavior without explicit forms of communication. A particularly important form of tacit collusion (no explicit agreement) among firms in a market is price leadership. Two types of price leadership may be distinguished. The first is monopolistic in nature; the second is competitive and reflects the fact that, with a limited number of firms, independent pricing is impossible.

In collusive price leadership, an announced price increase on the part of the first duopolist, to take effect next month, say, is read by the second duopolist as an invitation to increase his or her price as well. If the products are homogeneous or nearly so and the second duopolist does not follow the lead of the first, then almost certainly the price increase announced by the first duopolist will be rescinded, and price will remain where it was in the first place. In this situation, price change announcements to take effect later are substitutes for explicit agreements. Announced price changes signal one firm's belief to the other(s) that, with changes in demand and costs, a price change will increase profits. In a wide variety of situations, price changes that increase profits will always exist if industry marginal revenue is not equated to industry marginal cost. Proposed changes in price invite market analysis and cooperation. In actual market situations, a relatively large oligopolist, recognized for its efficiency and good management, might emerge in time as a price leader. The coordination of the activities of firms in this market setting might or might not lead to market performance similar to that expected in the case of monopoly. There is no question, however, but that such behavior could lead to noncompetitive results, with the firms enjoying excess profit over prolonged periods of time.

In competitive price leadership, unlikely in the case of duopoly with blockaded entry, prices vary over time because of cost and demand changes, but with a tendency for prices to equal the marginal costs of sellers. Competitive theory asserts that price is determined by supply and demand; in the real world, prices are determined by buyers and sellers when they enter contracts. We normally give little attention to the identities of buyers and sellers and to the slight differences that may occur in the hourly or daily prices paid and received in highly organized exchanges or auctions for farm commodity markets that often involve third parties, such as commission firms or auctioneers. A telephone call is sufficient to find out what the price for No. 2 corn is in Chicago. Absent organized exchanges, however, we must expect frequently to observe markets with several sellers and low barriers to entry where (1) announcements of price changes on the part of a leader reflect perceived changes in demand or cost conditions, (2) the identity of the would-be leader might change through time, and (3) announced changes are frequently canceled because price followers refuse to be led. Sellers might give the appearance of having considerable leeway in setting prices, when in fact the demand and cost constraints are the real but unseen (invisible) movers of price.

8.4 INDUSTRIAL ORGANIZATION RESEARCH

Because of complications caused by interdependency, economists have evolved models of oligopolistic (oligopsonistic) markets for special cases only. In general, however, it is difficult to specify consistent sets of structural equations that explain the operations of markets for products such as automobiles, television sets, cigarettes, or corn flakes. In practice, researchers have often framed reduced form relationships based on theoretical hunches, rather than on formal models, and have tried to assess the empirical validity of these relationships using real-world experiences. In this section, we will discuss the concepts, hypotheses, and research approaches that have been used in this endeavor. In Section 8.5, we will briefly consider examples.

Market Structure–Performance Paradigm

Drawing on the preceding discussion of competition, monopoly, and oligopoly, we begin by considering three important concepts. The first of these is the performance of a market, which consists of the level of inputs and outputs together with the levels of product prices and endogenous input prices. In Chapters 2 through 6, we referred to market performance as the solution values of all the endogenous variables entering the structural model of a market. This definition encompasses a good deal more than we might first suppose. Dimensions of performance of interest in industrial organization include (1) research and development input levels and the payoff in terms of new products and processes; (2) diversity of products produced in terms of quality, styles, and colors in light of the demand for diversity; (3) inputs used in product advertising and promotion and the effects of such inputs on sales, prices, education, and entertainment; (4) generation of waste products that despoil the environment; and (5) profitability of firms and the concentration of national income among those who amass market power, selling their products at prices above costs and, possibly, using the resulting economic power to indulge in extravagant consumption and to curry favors in the political arena.

The second concept of interest is market conduct. *Market conduct* refers to the manner in which the decisions of sellers are coordinated and made mutually compatible in satisfying the demands of buyers. (The student should modify this concept for a buying industry.) In perfect competition, for example, firms make their decisions independently, given observed market prices, and supply and demand coordinate the decisions of sellers and buyers in an impersonal manner. Many forms of conduct are possible in oligopolies, as we have seen, including conspiracies to fix prices, splitting up markets, predatory pricing to bankrupt a rival, and various forms of price leadership.

The third concept of interest is *market structure*, which consists of the characteristics of a market that have a strategic influence on the nature of market conduct. Some of the major dimensions of market structure are apparent from previous discussion. These include the number and size distribution of buyers and sellers, height of barriers to entry, and the extent of product heterogeneity or differentiation. Barriers to entry give established sellers advantages over potential entrants. A patent that awards a firm the exclusive right to produce a new product for 17 years, as in the United States, is a very high barrier. Economies of size may dictate the existence of at most a few large sellers in a market; a new firm would, to be of efficient size, bring a great deal of excess capacity to the industry, thus lowering profit for everyone. (The seventh survivor of a shipwreck who hauls himself into a raft designed for six risks the lives of all seven.) Product differentiation through the offering of differing designs, colors, and styles coupled with a large advertising campaign might protect a firm from the competition of established and potential rivals.

Other market characteristics may be important depending on the market. Labor unions are obviously important in many labor markets. Agricultural cooperatives organized on the principle of service at cost are important in many farm output and input markets. As another example, the elasticity of demand is an important dimension of structure when there are few firms and significant interdependency exists. Cheating on collusive agreements, whether tacit or explicit, is harder to detect the more elastic is total demand. Severe inelasticity is a stern disciplinarian for an oligopolist who would cut price as a means for enlarging market share.[2]

Importantly, however, factors other than market structure may influence conduct, and the most important of these, at least in the United States, is antitrust pol-

[2]It may be noted that the more narrowly we define a market, excluding producers of all but the closest substitutes, the more elastic is the demand for the remaining output. Narrower definitions of a market mean higher levels of concentration, but they also mean lower levels of demand elasticity.

icy. The key elements of this policy are briefly noted here, leaving detailed treatments to the references given at the end of the chapter. Section 1 of the Sherman Act of 1890 makes illegal per se all agreements among competing firms to fix prices or otherwise limit competition. Section 2 prohibits the monopolization of a market, that is, the amassing of sufficient power to control price and exclude competitors, except when the power has arisen from development of a superior product or out of historic accident. The Clayton Act of 1914, as amended, prohibits price discrimination and mergers that substantially lessen competition and specifies other illegal activities as well. The Federal Trade Commission Act of 1914 established five full-time commissioners, investing them with substantial quasi-judicial powers and charging them to eliminate unfair methods of competition. Antitrust policy has had a profound impact on the conduct of U.S. companies and on the performance of U.S. markets. Perhaps the best evidence of this has been the extent to which cartels and price-fixing agreements have flourished in foreign nations where such agreements have been legal. The dozens of violations of the prohibition of price-fixing agreements that are prosecuted each year in the United States should also be emphasized. (See Scherer, 1980.)

The basic hypothesis of the field of industrial organization is that market structure determines market conduct, in light of a given legal environment, and that market conduct determines market performance, in light of given values of exogenous variables. Competitive market structure, for example, means that firms take prices as given and behave independently, particularly if campaigns to establish cartels are illegal. Independent behavior means that the forces of supply and demand determine market performance for given exogenous variables (population, technology, etc.). A monopolist need not worry about competitors, but actual performance (price and output) depends on the shapes and positions of demand and cost curves. A few oligopolists, prohibited from entering a contract to fix prices, might rely on price leadership to coordinate their behavior. Where concentration and barriers to entry are high enough, conspiracies might not be needed to generate a near-monopoly solution.

Estimating Structure–Performance Relationships

The work in industrial organization includes a considerable effort to examine and quantify the nature of the associations between market structure and market performance. To see what is involved in this work, we propose a hypothetical experiment. We consider a market for a homogeneous product and assume that exogenous variables are held constant.

On the vertical axis in Fig. 8.6, we measure price, with P_c and P_M equaling the perfectly competitive price and the monopoly price, respectively. For present purposes, we interpret P_c as the price given by the intersection of demand and the aggregate marginal cost curves of the sellers. (All output levels are assumed to be efficiently produced.) The price P_M would prevail if industry marginal revenue and industry marginal cost are equated, bother how this could be accomplished.

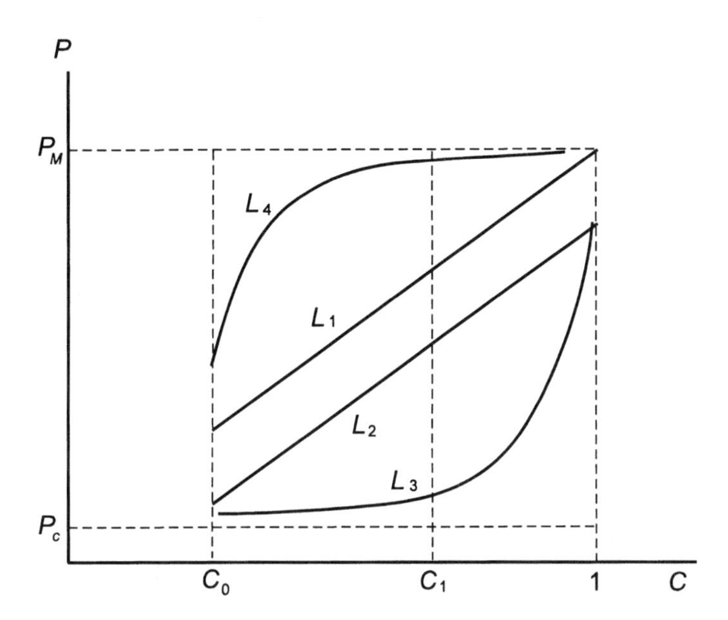

Figure 8.6 Hypothetical relationships between price and Herfindahl measures of concentration for two levels of barriers to entry.

On the horizontal axis we measure the concentration of sales among competing sellers, assuming that buyers are perfectly competitive. Concentration, given by the number and size distribution of sellers, is hard to measure with precision using a single number. To circumvent this difficulty, we use the *Herfindahl measure of concentration*, which equals the sum of the squared market shares of all the firms in the industry. This index equals 1 in the case of monopoly, for example, and 0.5 in the case of equal-sized duopolists. The index approaches 0 as the number of equal-sized firms becomes very large. The range of values for the level of concentration from $C = C_0$ to $C = 1$ in Fig. 8.6 is assumed to be consistent with exogenous variables, including economies of size. No firm is of suboptimum size, for example. This explains why we do not allow levels of concentration less than C_0 in Fig. 8.6. (The levels of concentration consistent with economies of size would presumably vary from one market to another.) Barriers to entry are also hard to measure using a single variable, and we will not try. Instead, we choose two hypothetical situations, one in which entry is impossible because the government will not allow it and the other in which entry is merely difficult because established firms have secret formulas that are hard to discover by outsiders. With blockaded entry, the hypothetical relationship between price P and concentration C is given by the line labeled L_1; with lower barriers, the relationship is given by L_2. (For the moment, ignore lines L_3 and L_4.) The relationship between an element of performance, price in this case, and structural characteristics, such as concentration and barriers to entry, is an example of a structure–

performance relationship. Such relationships are of fundamental importance in the field of industrial organization. The relationship hypothesized in Fig. 8.6 is rather like the reduced form relationship associated with a structural model, *except that we have not specified what the structural model is*. What may be offered instead of such a model is an informal discussion of the reasons why we believe such hypothesized relationships are plausible.

The profit to be had from monopolization is a potential in virtually every market. It must be supposed that sellers are aware of this and that they will often seek to turn potential gains into real ones through elevating price and restricting output to the point where industry marginal revenue and marginal cost are equal. There may, however, be obstacles that prevent firms from achieving monopoly gains, depending on the market. It is generally believed that the larger the number of sellers is, the more difficult it will be to achieve the monopoly price: The larger the number of sellers, the greater the potential for diverse operations, the greater the difficulty in agreeing on the optimal price, and the more likely that individual sellers will cheat on an agreement once settled.[3]

It may be argued, in addition, that the lower the barriers to entry to new sellers are, the more difficult will it be for established sellers to charge monopoly prices. In an uncertain world, success invites imitation. If an industry is profitable, word gets around and new entrepreneurs will strive to enter the industry. Successful entry will likely expand output and lower prices. To summarize, the hypotheses given in Fig. 8.6 are (1) holding barriers to entry and exogenous variables constant, the higher the level of concentration, the higher the market price and (2) holding the level of concentration constant, the higher the barriers to entry, the higher the market price.

The relationships given by L_1 and L_2 are linear, but linearity need not be assumed. The lines given by L_3 and L_4 are alternative representations drawn on the assumption of blockaded entry in both cases. The curvature of structure–performance relationships is important to policy formulation and is also a source of enormous controversy in the literature. If the relationship is like L_3 in the real world, then very high levels of concentration are compatible with prices that are close to the competitive price. Those in charge of antitrust policy might take a relaxed view if concentration is less, say, than C_1. For example, mergers that increase concentration to a level less than C_1 are nothing to worry about. If, on the other hand, the relationship is correctly depicted by L_4, even slight departures from competitive or atomistic structure lead to near-monopoly prices. Here the government might be very concerned about a merger that would elevate C close to C_1.

As it turns out, the imaginary experiment that undergirds Fig. 8.6 is not very useful in research aimed at quantifying structure–performance relationships, because the

[3]Interestingly, the history of U.S. agriculture is replete with instances in which farmers banded together, destroyed output, and withheld output from markets in an effort to raise farm prices. The failure of these movements to affect the performance of farm markets significantly is important evidence that large numbers of sellers and collusive agreements are often incompatible.

structures of markets tend to be very stable over time. We do not often see market concentration increasing quickly through time, for example, allowing the relationship between concentration and performance to be observed absent of significant changes in disturbing influences, such as changes in technology. Researchers trying to measure structure–performance relationships have rarely used time series data, relying instead on approaches based on cross-section data. That is, researchers have examined samples of markets with alternative structures. This latter approach has problems of its own. For one thing, prices and quantities of different products are often not directly comparable. In addition, exogenous variables differ among markets. At any one point in time, exogenous variables might encourage high profits in some markets and low profits in others. It is rarely easy to separate out the effects due to differing structures across markets and the effects due to other differing exogenous variables.

Both of these difficulties could be avoided or at least mitigated if, instead of plotting price on the vertical axis in Fig. 8.6, we plotted $(P - P_c)/P_c$, where P is the observed price. This idea is often impractical, however, because the competitive price will ordinarily be neither known nor estimable. Confronted with this problem, researchers have often centered attention on the ratio of pure profit to sales or the ratio of pure profit to net worth, the idea being that under keen competition such ratios should be close to zero.

Figure 8.7 is useful in understanding the difficulties associated with using cross-section data in the estimation of structure–performance relationships. The

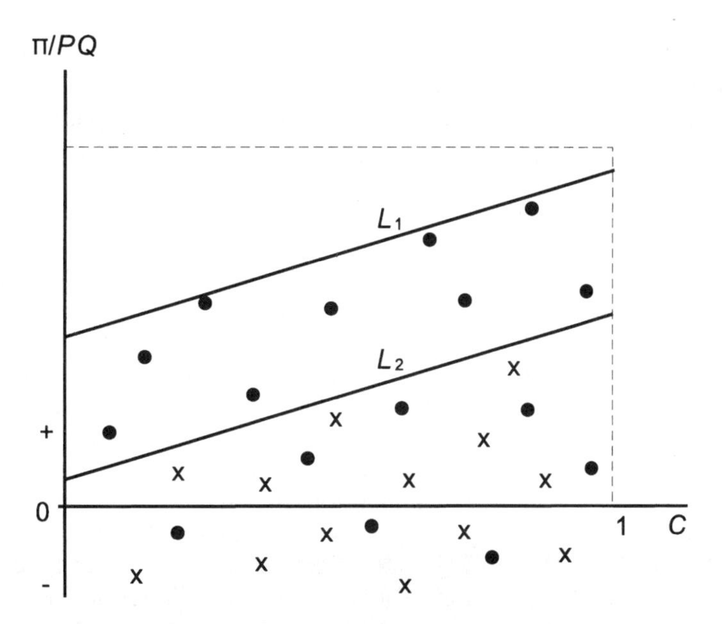

Figure 8.7 Hypothetical combinations of profit ratios and levels of concentration with two boundaries for different levels of barriers to entry.

Herfindahl measure of concentration (C) is again plotted on the horizontal axis. Again we may think of L_1 as being associated with high barriers to entry and L_2 with low barriers to entry. On the vertical axis, we measure the ratio of aggregate pure profit to aggregate sales, π/PQ, for all the firms held to be in the same industry. Dots represent hypothetical ratios for industries with the high barriers to entry. Crosses represent ratios for industries with low barriers. (The ratio of profit to net worth could just as well be plotted on the vertical axis.) The scatter of points given in Fig. 8.7 is intended to convey the idea that exogenous variables tend to make some industries profitable relative to others at any one point in time (or over a small number of years) *quite apart from differences in market structure*. A technological change that is in the process of destroying an industry, for example, might leave in its wake the last pathetic firm, a monopolistic seller with positive quasi-rent and negative profit, headed inexorably for extinction. Figure 8.7 suggests the hypothesis that the upper boundary on profit rates of observed industries is positively associated with the level of concentration, with the boundary being higher for industries with high barriers to entry and lower for those with low barriers to entry. High concentration and high barriers to entry might allow but need not assure high rates of profit.

Our discussion of hypothetical relationships between structure and performance and of some of the problems encountered in estimating such relationships raises questions with regard to what in fact industrial organization researchers have discovered. As noted, there is a great deal of controversy in this field, but it appears that by and large research findings support the basic hypothesis: Market structure does appear to affect market performance. The greater the departure of observed structure from the competitive extreme is, the greater the likelihood that market performance tends toward that associated with monopoly. The nature of this relationship, whether it is like L_3 or L_4 in Fig. 8.6 or something in between, is hotly disputed, however, and there are other bones of contention in the literature that we must now note.

Our attention up to this point has centered on the relationship between market structure and price (or profit ratios) and whether increases in concentration and barriers to entry increase price. Researchers must, of course, be cognizant of structural dimensions other than concentration and barriers to entry and mindful that measurement of structural dimensions depends on how we define a market. To measure concentration, for example, we might define a selling industry as a set of firms selling close substitutes. But how close is close? Is a Cadillac a good substitute for a Honda Civic? Are videos good substitutes for movie tickets? Are the supermarkets in Minneapolis and St. Paul in the same market?

There are, moreover, dimensions to market performance other than price. For example, the preceding analysis could be reworked on the basis of market output. We could plot $(Q - Q_c)/Q_c$ on the vertical axis of Fig. 8.6, where Q equals observed output and Q_c equals competitive output. We will not take up this idea here, since price and output are functionally related through market demand. The structure–quantity relationship is an inverted mirror image of the structure–price relationship.

The nature of the structure–price relationship might tell us very little, however, about the manner in which market structure is related to performance in such dimensions as technological progress, investment in new plant and equipment, the per unit cost of production, levels of advertising and promotional costs, the provision of products of varying qualities and designs, environmental pollution, and the concentration of national income among the wealthy few. It seems possible, for example, that although high concentration might lead to a high price it might also lead to technological progress. Indeed, some economists have argued that this is the case.

In summary, an important objective of industrial organization research is the estimation of relationships that quantify the effects of several carefully measured dimensions of market structure on each of several dimensions of market performance, with the disturbing influences of exogenous shocks eliminated. In the absence of a convincing theory of oligopoly, a common research strategy is to estimate reduced form relationships without specifying structural models in detail. Although this strategy is widely employed in industrial organization research—reflecting, as it does, needed guidelines for public policy, rather than an elegant recipe for scientific exploration—the fact remains that our knowledge of structure–performance relationships falls far short of the ideal.

Policy Implications of Structure–Performance Relationships

Aside from scientific curiosity, researchers seek to understand structure–performance relationships for practical policy considerations. A useful idea in policy analysis is the with-and-without government principle. On this principle, researchers seek to quantify the performance of markets with and without whatever policy is of interest. Of interest in industrial organization is a panoply of antitrust policies intended to foster competition, some of which have been identified. Go back to Fig. 8.6. How would relationships such as L_1 and L_2 (or L_3 and L_4, for that matter) be affected by a policy that outlaws price fixing agreements or decreases tariffs to foster competition from abroad?

Making sensible public choices requires more, however, than a knowledge of market structure–performance relationships with and without government policies of various kinds. There is also the question of what kind of performance we want.

The question of what performance effects are desirable is addressed in the literature on the evaluation of performance. Evaluation of performance, it must be noted, is very different from measurement of performance. The performance of a student on an examination may be defined as the number of questions answered correctly. Evaluation based on standards of comparison or norms is a prerequisite for assigning a grade. For example, is answering correctly 60 percent of the questions on an examination good enough to merit a B? Many graders would look to see how this score stacks up against the scores of others. Similarly, the question whether observed market performance is in some sense satisfactory or workable also requires norms. Establishing norms in theoretical analysis is relatively easy; establishing operational norms under real-world conditions is extremely difficult.

Following Jesse Markham, we may say that market performance is satisfactory (acceptable or workable) if there is no clearly indicated change in market structure and conduct that can be effected through public policy that increases social gains more than social losses. In other words, *performance is satisfactory if there is no policy that increases total willingness-to-pay for the privilege of participating in a market.* Taxpayers must, of course, be included along with consumers, producers, and input suppliers. The advantage of this prescript is that it calls attention to the need for remedial action, including government costs, and it warns against useless hand-wringing.

The difficulties encountered in the evaluation of actual market performance may be suggested by considering short-run allocative efficiency. Is actual output close to the level that equates price to marginal cost, thus maximizing the value of the market? To answer this question, the researcher would need to obtain sound estimates of market demand and the horizontal sum of the marginal cost curves of all sellers. How are these estimates to be obtained? There is, in addition, the question of what to do in situations where output can be shown to be well below that which equates price and marginal cost, where there lurks a problem of dead-weight efficiency loss. Are there policies that will change market structure, market conduct or both such that pricing improves? How will these policies affect other aspects of performance?

More specifically, with regard to the workability or adequacy of performance, we might ask (1) How progressive is an industry in terms of the development of new products and processes? We can seek to measure, at least approximately, the levels of inputs committed to research and development, but are the inputs used wisely? How can we tell what the optimum levels of research and development inputs are without some idea as to what is discoverable? (2) Is the observed level of output produced efficiently, bother if it falls short of some ideal where price equals marginal cost. We know the conditions of efficient production in theoretical models, but the real world is complicated by uncertain technological changes, uncertain input prices, uncertain demand, and fixed plants that may last for many years. What does cost efficiency mean in a world beset by uncertainty? (3) Do companies squander money on advertising and promotional costs in the fight for market share, without passing along useful information to consumers? What should one say of a highly misleading advertisement that gets consumers to try a new product that proves to be of great value to them? (4) Do companies offer consumers a suitable array of products of alternative designs in light of heterogeneous consumer preferences? (5) Do firms pollute the atmosphere, poison groundwater, and otherwise damage the environment in their choices of production processes? And, finally, (6) Do sellers make substantial excess profits that concentrate national income and wealth among the few, with possibly dire implications for the working of democratic processes that are supposed to be based on the one-person, one-vote principle? These questions are illustrative of those often asked, but rarely answered in a definitive way.

8.5 EXAMPLES OF INDUSTRIAL ORGANIZATION RESEARCH

Having discussed in rather general terms the objectives, methods, and difficulties encountered in industrial organization research, we now consider several examples. These examples are chosen not only to illustrate industrial organization research, but to show that the study of market structure and conduct is important to the understanding of agricultural pricing broadly defined to include food prices throughout the marketing channel.

Our first example centers on a conspiracy to raise bread prices in the state of Washington. The second centers on price leadership (tacit collusion) as a substitute for an explicit conspiracy in the rigging of cigarette prices. A third example provides estimates of the relationship between market structure and retail food prices in Vermont. A final example provides estimates of the structure–profit rate relationship in the U.S. food manufacturing sector.

Fixing Bread Prices in the State of Washington

Bread markets tend to be separated spatially because of the high cost of transporting freshly baked products. In addition, because of economies of size in both bread baking and retailing, a few relatively large bread bakers typically sell their outputs to a few relatively large grocery retailers. Large retailers often integrate backward into bread baking and sell their own private labels. This means they are often in a position to curtail sharply the market power of dominant bread bakers. The structure of the typical bread market tends, in other words, toward what might be called bilateral oligopoly.

In 1964, the Federal Trade Commission (FTC) issued a decision charging that the leading bread bakers in the state of Washington had conspired through their trade association, Bakers of Washington, Inc., to fix prices beginning in the 1950s and continuing through 1964. (See Federal Trade Commission, 1968.) There was considerable evidence and testimony that, through meetings, telephone calls, and memoranda, bakers often conspired to raise bread prices. Strict pricing discipline was achieved in part through threatening drastic price reductions to any firm that would not keep its prices in line. The success of the association rested in no small part on the cooperation of Safeway Stores, the dominant grocery chain in the Seattle area with its own bread-baking facilities. According to the FTC (page 1126), "While all the other bakers, wholesale and retail, were pressured to retail their bread for the same price that Continental (the largest wholesale bakery in Seattle) got for its 'Wonder Bread,' Safeway was permitted to sell for 1¢ less."

What was the impact of the conspiracy on bread prices in Seattle? Plotted in Fig. 8.8 are the annual average retail prices of white bread (cents per pound) for Seattle and the United States for the period 1950–1978. (Seattle prices for 1965 and 1966 were not available for this study.) The data suggest that the Washington conspirators were indeed successful in elevating Seattle bread prices. As noted, during

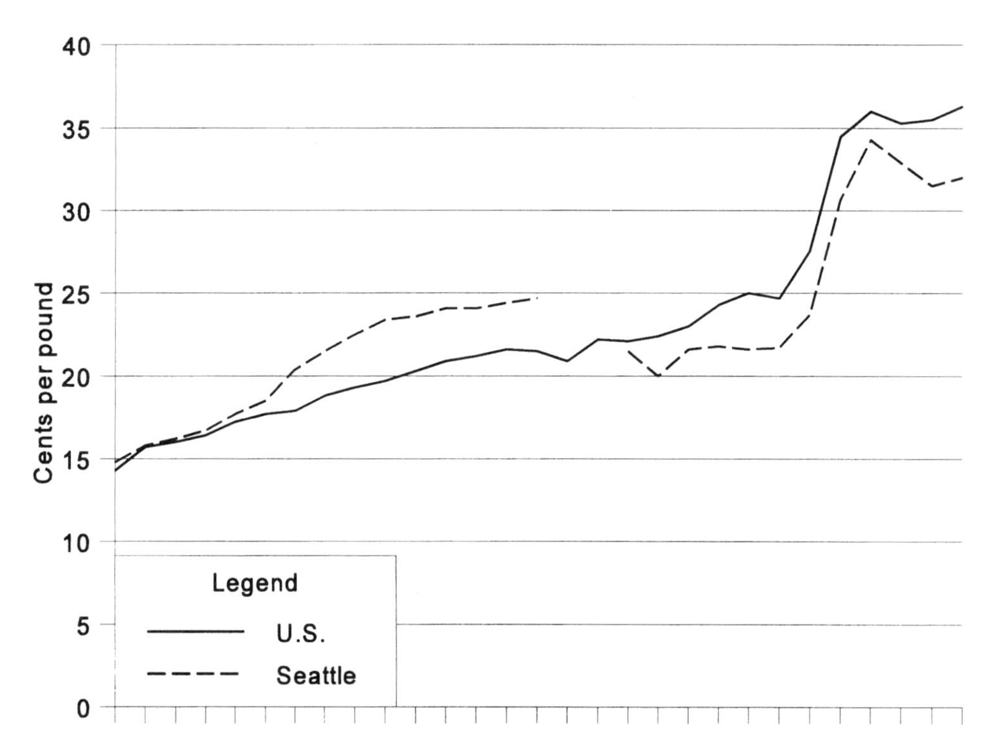

Figure 8.8 Average retail prices for white bread, Seattle and United States, 1950–1978. *Source:* U.S. Deparment of Labor, *Estimated Retail Food Prices by City,* Bureau of Labor Statistics, Washington, D.C., Annual Averages for Various Years.

the time of the conspiracy, Safeway Stores sold its brand at only 1¢ per loaf below the price of the wholesale baker brands. By the fall of 1966, however, after the conspiracy had been broken up, Safeway's price for its brand was 7¢ per loaf less than that of wholesale bakers.

The analysis based on Fig. 8.8 is rather simple and skeptics might easily ask embarrassing questions. Presumably, many factors affect Seattle bread prices, and the research problem is to separate out the effects of disturbing influences to see the pure effect of the conspiracy. Comparing the Seattle price to the average price for the nation is a crude way of correcting for changes in exogenous variables that affect the demand and supply for bread. The U.S. bread price may be viewed as a proxy for numerous factors that would likely affect the Seattle bread price. For example, if the price of flour jumps sharply, ceteris paribus, then bread prices can be expected to rise nationwide, including Seattle. Alternatively, if new bread-baking technologies lower per unit cost of production, we would expect bread prices to fall nationwide. But what of factors peculiar to the Seattle market? It is possible, after all, that all retail food prices in Seattle were high, relative to food prices nationwide, during the alleged period of the conspiracy.

These considerations led us to estimate the effect of the conspiracy on the *real* Seattle price of bread, that is, the price of bread relative to the prices of other foods. Using annual data for 1950–1964 and 1967–1978, we estimate the following equation:[4]

$$Y = 0.085 + 0.9164X_1 + 0.1111X_2$$

$$(5.12) \qquad (6.20)$$

where Y equals the Seattle price of white bread in cents per pound divided by the Seattle consumer price index (CPI) for food (1982–1984 = 100), X_1 equals the U.S. price of white bread in cents per pound divided by the U.S. CPI for food, and X_2 is a binary variable that equals 1 for all years during the conspiracy (1954–1964) and 0 otherwise. (The numbers given in parentheses below the estimated coefficients for X_1 and X_2 are t ratios.) Broadly speaking, bread prices are subject to some exogenous shocks that affect all food products, such as, for example, wage rates in food manufacturing and distribution, interest rates, and general food scarcity relative to demand (as occurred during 1972–1974). By deflating bread prices, we hope to correct for changes in these factors. Since these shocks (wage rate changes, for example) need not vary over time in Seattle as they do nationwide, we deflate the Seattle bread price by the Seattle CPI for food and the U.S. bread price by the U.S. CPI for food.

We hypothesize that the exogenous shocks affecting real bread prices nationwide, which tend to be unique to bread, also affect Seattle bread prices. It is further hypothesized that the binary variable X_2 allows the relationship between the real Seattle bread price and the real U.S. bread price to shift upward during the period of the conspiracy. That is, X_2 allows an upward shift in the constant term of the preceding equation during the period of the conspiracy.

Turning to the statistical results, the estimated elasticity of the real Seattle bread price with respect to the U.S. bread price equals 0.98 evaluated at the means of Y and X_1 *excluding* the years of the conspiracy. Absent the conspiracy, a 10 percent increase in the real U.S. bread price would be associated with a 9.8 percent increase in the Seattle price. Absent a price-fixing conspiracy, the percentage changes in the real prices of bread in Seattle were essentially the same as for the nation.

As to the conspiracy, we note that the coefficient for X_2 equals 0.1111. This means that, on average, during the time of the conspiracy the real Seattle bread price

[4]This equation was estimated using multiple regression analysis. The squared coefficient of correlation equaled 0.909.

was elevated 11¢ per loaf. This amounts to 17.8 percent of the real U.S. bread price over the sample period. (The high *t* ratios suggest that the estimated coefficients are statistically reliable.) The conclusion to be reached from these estimates is at once both obvious and important: Demand and supply analysis, based on competitive price theory, is unreliable when price-fixing conspiracies are in effect.

The 1946 Tobacco Decision

In 1946, the U.S. Supreme Court found the three leading tobacco companies, American Tobacco (Lucky Strike cigarettes), Liggett & Myers (Chesterfield), and Reynolds (Camel), guilty of an illegal conspiracy to fix cigarette prices. There was no hard evidence of an explicit agreement—letters, memoranda, or recorded conversations. Rather, the evidence was circumstantial, centering on the market shares, advertising expenditures, profits, and, above all, the concerted and parallel pricing policies of the big three tobacco companies. As regards announced changes in cigarette prices, they marched in lock step.

Briefly, the facts are these: the Big Three sold their highly advertised brands of cigarettes to jobbers and to selected dealers at list prices minus discounts. From at least 1923 on, the prices charged and discounts allowed were virtually identical. Following 1928, Reynolds was the first to announce price increases, and American and Liggett always followed suit, increasing their list prices in identical amounts. Importantly, with high levels of concentration and barriers to entry, the latter created by advertising, once the companies came to understand that their prices would always be identical, an announced price change by any firm, particularly by a firm that had led the way before, could only be interpreted as an invitation to follow suit. There is simply no need for skilled oligopolists to meet in a smoke-filled room when concentration is high.

On June 23, 1931, Reynolds increased the list price of Camel cigarettes from $6.40 to $6.85 per thousand. American and Liggett immediately increased the prices for Lucky Strikes and Chesterfields by the same amounts. These price increases—announced in the face of the Great Depression, at a time when both demand and costs were falling—led to expanded sales of the 10-cent brands, the little-advertised cigarettes that were sold by a fringe of small companies for a nickel per pack less than the cigarettes of the Big Three. The market share of the 10-cent brands rose from less than 1 percent in June 1931 to nearly 23 percent in November 1932. In response to loss of share, the Big Three lowered their price to $5.50 per thousand, but when the share of 10-cent cigarettes had fallen to 6.4 percent in May 1933, the price of the leading brands was increased to $6.10 per thousand. The 10-cent cigarettes maintained a modest market share until the 1940s, when they disappeared.

We have cited the *Tobacco* case for two reasons, the first of which has already been noted. When concentration and barriers to entry in a selling industry are high and unopposed by power on the buying side, express collusion is not required to avoid competitive pricing. A second reason for thinking about the *Tobacco* case is that it points up vividly a serious problem for government agencies committed to fostering keen competition. To some considerable extent, the Supreme Court punished the big cigarette companies for behaving in a way that was more or less preordained by the structure of their industry. It is as if the Creator were to say to the lioness, "I know your pride likes meat, but you simply must stop killing my other animals." When asked why his company raised cigarette prices at the outset of a depression, the president of American had the audacity to indicate a desire to make money. Well, of course. The quest for profit is the engine of a market economy.

The *Tobacco* case is illustrative of a large body of evidence that high-order oligopoly bestows market power, and we must suppose that the oligopolists in such markets will behave accordingly. This has led many economists to argue that the government should consider structural remedies in the case of high-order oligopolies—breaking up large companies, disallowing anticompetitive mergers and acquisitions, lowering barriers to international trade—as a means for assuring vigorous competition in the U.S. economy. As in so many areas in the literature on industrial organization, this suggestion has sparked considerable controversy.

Market Power in Vermont Retail Grocery Markets

A study by Cotterill (1986) has examined the relationship between market structure and the price levels of 35 supermarkets in 18 Vermont retail grocery markets. Since a supermarket can offer as many as 16,000 products, Cotterill used as his measure of performance in the price dimension an aggregate price index based on 121 representative product prices for each sample supermarket during August 1981. (The price levels of sample supermarkets were indexed relative to the lowest-price supermarket, which was assigned a value of 100.) What might account for variation in prices among supermarkets at any one point in time? On arguments advanced earlier in this chapter, we might expect that the higher the level of concentration is, the higher the prices. Several exogenous factors likely affect prices as well. First, according to Cotterill, cost studies suggest that at least up to some level (16,000 to 20,000 square feet of selling space), the average cost of sales declines with size. On the other hand, large size increases the range of items offered (drugs and/or clothing), which might allow a supermarket to charge relatively high prices to pay for the convenience of one-stop shopping. Second, the greater the distance between a supermarket and its wholesale distribution center is, the greater the transportation costs and the higher the prices that must be charged to cover long-run costs. Third, wage rates vary across markets, and we would expect that higher wages mean higher supermarket prices. Fourth, population growth tends to increase demand, which might imply that supermarkets in high-growth areas charge higher prices. Fifth, higher levels of per capita income might imply a greater demand for a "pleasant shopping experience," which might explain higher prices. Finally, since the unit of observation is a supermarket's price level, instead of the price level of a market, Cotterill argued that whether the supermarket is part of a chain (11 or more stores) or an independent

(10 or fewer stores) might also be a relevant exogenous variable. He hypothesized that independents tend to charge higher prices, but that they survive through doing a better job of catering to the particular demands of the local population.

Drawing on the evidence generated by the 35 sample supermarkets, Cotterill estimated the structure–price level relationship as follows:[5]

$$Y = 107.1 + \underset{(5.37)}{7.779X_1} + \underset{(2.51)}{2.068X_2} - \underset{(2.37)}{0.730X_3} + \underset{(2.35)}{0.027X_4} + \underset{(0.30)}{0.002X_5} - \underset{(0.87)}{0.041X_6} + \underset{(0.82)}{0.175X_7}$$

where Y equals the price level of a supermarket; X_1 equals market concentration measured by the Herfindahl index (defined in Section 8.4); X_2 is a binary variable to identify independent supermarkets (as opposed to chains), equaling 1 if the supermarket is an independent and 0 otherwise; X_3 equals size of the supermarket measured in square feet of sales space; X_4 equals X_3 squared; X_5 equals distance between the supermarket and its warehouse distribution center; X_6 equals market population growth between 1970 and 1980; and X_7 equals 1980 per capita income of market consumers. (The numbers appearing in parentheses below the estimated coefficients for X_1 through X_7 are t ratios.) What this equation suggests is that the higher the level of concentration in a Vermont grocery retail market, ceteris paribus, the higher were grocery prices in August 1981. We estimate that the elasticity of price (Y) with respect to concentration (X_1) equals 0.031. If, for example, we consider a ceteris paribus structural change in which three equal-sized supermarkets are replaced by two equal-sized supermarkets, the index of prices charged would rise by about 1.6 percent. (The t ratio for the coefficient of X_1 suggests significant statistical reliability.)

As Cotterill hypothesized, independent supermarkets appeared to have charged higher prices than chains. He also reasons, on the basis of the estimated coefficients for X_2 and X_3, that increased size lowered prices up to the size of 13,600 square feet. Beyond this size, prices rose as supermarkets offered greater convenience. The remaining variables (X_4 through X_7) appeared to have had little effect on price levels in that their estimated coefficients were not statistically reliable. This does not mean that these variables are irrelevant, only that their effects could not be measured with much reliability, given the sample data and the estimation procedures that were employed.

Market Power in Food Processing

A study by Imel and Helmberger (IH) (1971) examined the variation among profit rates of 99 large food and tobacco manufacturing companies over the period

[5]This equation was estimated using multiple regression analysis. The squared coefficient of correlation equaled 0.639.

1959–1967. The sample companies accounted for roughly 40 percent of the food manufacturing sector sales for that period. Using secondary data and primary data collected through interviews with company executives, IH estimated the following structure–profit rate relationship:[6]

$$Z = 0.0164 + \underset{(3.01)}{0.0533U_1} + \underset{(1.86)}{0.1433U_2} + \underset{(2.14)}{0.5511U_3}$$

where Z equals the ratio of a company's after-tax profits to sales averaged for the period 1959–1967. A normal return on year-end equity (0.05 times year-end equity) was subtracted from reported profits to estimate pure profit. (Again, the t ratios are given in parentheses below estimated coefficients.)

Market concentration was measured by the ratio of the market sales of the four largest sellers to total market sales in 1963. The variable U_1 equals a weighted concentration ratio for a food company, with the weights equaling the ratios of company sales in various markets to total company sales. Suppose, for example, that a company's sales are divided evenly between two markets, the first with a concentration ratio of 0.9 and the second with a concentration ratio of 0.3. The weighted concentration ratio equals 0.6, given by $(0.5)0.9 + (0.5)0.3$.

A widely held belief among researchers in industrial organization is that product differentiation is successfully employed by food manufacturers to protect themselves from the competition of extant and potential sellers. An important way a company can differentiate its product is to convince buyers (consumers, in the case of food products) through advertising that its product is better than those of its competitors. Real as opposed to fancied product differences may have little to do with the success of advertising. The variable U_2 equals a company's total advertising expenditure in the mass media divided by total company sales over the period 1959–1967. (Mass media includes TV, radio, magazines, newspapers, and billboards.)

For an explanation of U_3, IH argue that research and development expenditures also likely support a company's effort to differentiate its products through creating real product differences, that is, through introducing superior products that other companies might not be able to readily duplicate. In the preceding equation, U_3 equals a company's total research and development expenditures divided by total company sales for the nine-year period from 1959 to 1967.

Our equation is consistent with standard industrial organization hypotheses: as concentration and product differentiation increase, the ability of a company to increase profits through elevating prices rises as well. If, for example, we consider a ce-

[6]This equation was estimated using generalized least squares to allow for heteroskedasticity. The evidence of heteroskedasticity reported by IH supports the hypotheses implicit in Fig. 8.7.

teris paribus structural change in which eight equal-sized firms are replaced with four equal-sized firms, the concentration ratio rises from 0.5 to 1.0 and the ratio of profit (excess profit after tax) to sales rises by 2.7 percent. (The t ratios for the variables, particularly for U_1, suggest that the estimated coefficients are statistically reliable.)

Concluding Remarks

This book emphasizes the development of structural models to explain the operation of markets and the determination of prices. The evaluation of the corresponding reduced form models often yields important hypotheses regarding the effects of exogenous shocks. When applied to the food manufacturing and distribution sector, however, this approach is often impeded by a serious obstacle: It is very difficult to specify meaningful structural models in the presence of oligopolistic interdependency. Confronted with this problem, researchers have had to rely on informal theorizing—armchair speculation, the cynic might say—in the development, evaluation, and estimation of reduced form models. The estimated equations discussed in this section are illustrative examples.

Several conclusions seem warranted. First, the theory of demand and supply, based as it is on the assumption of a competitive structure, must not be applied indiscriminantly to real-world markets. It is important to study market structure and conduct to see if there is much likelihood of market power and collusive conduct, such as monopolistic price leadership or price conspiracies.

Second, in the presence of market power, price need not move quickly and freely in response to what we might think of as competitive price shocks, that is, to changes in the exogenous variables of demand and supply models. Collusive behavior may be an important factor affecting market price and performance, a factor that must be added to the long list of price-affecting factors predicted by competitive price theory. Collusive behavior may take a variety of forms, and it may be supported by noncompetitive market structures and agreements such as price-fixing conspiracies. The effectiveness of collusive behavior might vary considerably over time. In any event, when firms have market power, prices might move in ways that are completely at odds with what would have been predicted on the basis of supply and demand.[7]

Finally, when high levels of concentration, barriers to entry, and product differentiation are present, we should probably expect that prices will exceed average production costs. Excess profit that persists through time is a distinct possibility. The

[7]Price-fixing conspiracies may be broken up by government action or because of a falling out among the conspirators. Many students will no doubt be aware of the Organization of Petroleum Exporting Countries (OPEC) as an example of the latter. OPEC was able through concerted action to raise crude oil prices to extraordinary levels over the period 1973–1986. With the collapse of the cartel, prices have plummeted. For an interesting and informative discussion, see Carlton and Perloff (1990).

problems this poses for government policy, together with the research results on the nature of structure–performance relationships and the controversies over the relative importance of deadweight monopoly losses, technological progress, production efficiency, and other aspects of performance, are outside the scope of this book and are left to the literature on industrial organization.

PROBLEMS

8.1. You are given the following information: Suppose that a firm has a monopoly because of patent rights. The firm's long-run cost function is $C = Q^2/2$. Demand is $P = 12 - Q$.

a. Calculate the monopoly solutions and find the corresponding total willingness-to-pay for the market at the monopolistic price.

b. Find the level of Q that maximizes the total willingness-to-pay for the market, and calculate the deadweight loss due to monopoly.

8.2. The long-run supply is given by $P = 2 + 0.5Q$. The long-run average revenue product (ARP) function for a monopsonist is given by $ARP = 4 + 10Q - 0.5Q^2$. Find the level of input that maximizes the monopsonist's profit. Calculate marginal revenue product (MRP), ARP, price P, and π for the profit-maximizing input.

8.3. Long-run demand and supply are given by $P = 10 - 0.05Q$ and $P = 0.1 + 0.25Q$. The government forms a cartel and regulates the entry and exit of farmers and the output of each farmer. Let AC equal the minimized average cost of production. (Note that for the industry $AC = 0.1 + 0.25Q$.) Define the cartel residual CR as $CR = (P - AC)Q$. If $CR > 0$, the government collects taxes from farmers equal to $(P - AC)Q$. This lowers taxed paid by the general taxpayers. If $CR < 0$, the government reimburses farmers for the losses that they incur. This means that the general taxpayer picks up the tab. Farmers are not allowed to earn long-run profit nor are they forced to incur long-run losses.

a. Suppose that the government sets $Q = 36$. Calculate the total consumer surplus, the total producer surplus, and the gain (or loss) to the general taxpayers relative to zero output.

b. Find the level of Q that maximizes the willingness-to-pay of consumers, producers, and general taxpayers all taken together. For this optimum Q, compute total consumer surplus, total producer surplus, and taxes paid by the general taxpayers.

8.4. Consider duopoly with blockaded entry. Demand is given by $Q = 60 - 2.5P$. The total cost functions of the duopolists are given by

$$C_1 = 2 + q_1^2$$

$$C_2 = 1 + 0.5q_2^2$$

Find P, Q, q_1, q_2, π_1, and π_2 under each of the following regimes.

a. Following firm 2, firm 1 always sets its price equal to that set by firm 2.

b. The firms agree that firm 1 will always set its output equal to the output produced by firm 2.

c. The firms agree to maximize joint profit using output quotas.

REFERENCES

American Tobacco Co. et al. v. *United States,* 328, U.S. 781 (1946).

Carlton, Dennis W., and Jeffrey M. Perloff, *Modern Industrial Organization*. New York: Harper Collins Publishers, 1990.

Cotterill, Ronald W., "Market Power in the Retail Food Industry: Evidence from Vermont," *Review of Economics and Statistics,* 68, no. 3 (August 1986), 379–386.

Federal Trade Commission, *Federal Trade Commission Decisions: Findings, Opinions, Orders, Jan. 1, 1964 to March 31, 1964,* vol. 64. Washington, D.C.: U.S. Government Printing Office, 1968.

Imel, Blake, and Peter Helmberger, "Estimation of Structure–Profit Relationships with Application to the Food Processing Sector," *American Economic Review,* 61, no. 4 (September 1971), 614–627.

Markham, Jesse W., "An Alternative Approach to the Concept of Workable Competition," *American Economic Review,* 40, no. 3 (June 1950), 349–361.

Scherer, F. M., *Industrial Market Structure and Economic Performance,* 2nd ed. Chicago: Rand McNally College Publishing Co., 1980.

PART III

Farm Commodity Programs

9

Analysis of Farm Programs: Part I

Although the models of farm markets in previous chapters draw heavily on competitive price theory, the fact remains that government intervention in farm markets is pervasive in most of the developed and developing countries of the world. Competitive processes are still at work, but government intervention causes these processes to generate noncompetitive market results. The gun still works, but the hand of government points it at a different target. The analysis of farm programs is therefore imperative if we are to understand how agricultural prices and other performance dimensions are often determined.

The broad objectives of this and the next chapter are to describe the major policies governments use to solve farm price and income problems and to analyze their likely economic effects. In this chapter, we first examine the effects of a direct payment program under certainty and in the absence of international trade. Then, in Section 9.2, we center on both direct payments and market price supports, allowing for agricultural exports. Models that allow for imports will be considered in Chapter 10. Drawing on Chapter 6, Section 9.3 analyzes direct payments and market stabilization schemes under conditions of uncertainty. It will be seen that proper analysis of stabilization schemes must be based on a theory that takes explicit account of dynamics, uncertainty, price expectations, and commodity storage.

9.1 DIRECT PAYMENTS UNDER CERTAINTY

A direct payment program is important not so much because of its widespread use, but because it is often recommended by economists as having several advantages over other approaches, a recommendation politicians have been loath to accept. In a direct payment program, the government announces a target price P^* for a farm commodity, food, say, and makes direct payments to farmers to assure that the real prices they receive will not fall below P^*. More specifically, under a direct payment program, the farmer's output is sold at the market price P. If the market price equals or exceeds the target price P^*, the government does nothing. If the market price falls below P^*, the government sends a check to the farmer equaling $(P^* - P)q$, where q equals the farmer's output. The direct payment per unit of output equals $(P^* - P)$. In real-world applications, it is important that the direct payment be based on the market price, an average for the nation, say, as opposed to the actual price a farmer receives. This encourages the farmer to search for the highest price.

Price and Quantity Effects

We begin by focusing on the domestic market, abstracting from international trade and uncertainty, complications that will be taken up later. The domestic demand curve DD and supply curve SS are given in Fig. 9.1. They can be interpreted as either short- or long-run curves. In either case, the intersection of these curves gives the competitive price P_c and the competitive output Q_c. (The student should ignore for the time being the areas in the diagram denoted by lowercase letters.)

To find the market equilibrium with a direct payment program in effect, we note that, for any market price less than the target price P_0^* in Fig. 9.1, farmers receive a per-unit direct payment that brings the price they effectively receive up to P_0^*. For all such market prices the effective farm price equals P_0^*, and, in the aggregate, farmers produce Q_g. For market prices in excess of P_0^*, the market price be-

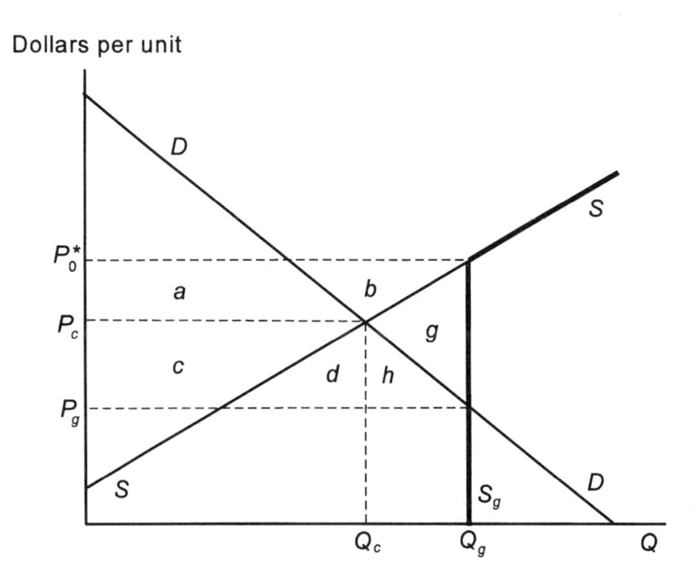

Figure 9.1 Demand and supply curves under perfect competition and a direct payment program, with program costs and benefits.

comes the effective price. Thus the supply curve that shows how much farmers would be willing to produce at alternative market prices is given by the darkened curve SS_g, kinked at P_0^*, in Fig. 9.1. The new equilibrium price P_g and output Q_g are given by the intersection of DD and SS_g. In equilibrium, buyers are able to buy what they want at the market price P_g; farmers sell as much as desired at the effective farm price P_0^*. Market participants are in equilibrium because they could not further their goals by changing their decisions. The farmer receives a government check equal to $(P_0^* - P_g)q$, with the direct payment per unit equaling $(P_0^* - P_g)$. Government expenditure equals $(P_0^* - P_g)Q_g$.

Equilibrium in the product market is linked to equilibria in the input markets. By increasing the revenue per unit of output produced, direct payments increase the demands for factors of production. (This assumes that none of the inputs is inferior.) In both the short and long run, increased input demands increase the levels of inputs (except for exogenous land), which drives up prices of all inputs with upward sloping supply curves.

Welfare Effects

Because a direct payment program lowers the market price, consumers are program beneficiaries. The aggregate gain in consumer welfare is given by area $(c + d + h)$. The increase in producer surplus is given by area $(a + b)$, but, again, we must be careful how this surplus is interpreted. With constant input prices in the short run, area $(a + b)$ measures the aggregate increase in quasi-rents to farmers. With endogenous input prices in the short run, area $(a + b)$ measures the increased quasi-rents to farmers plus the increases in surpluses to input suppliers. With endogenous input prices in the long run, area $(a + b)$ measures the increases in the input surpluses only. (Our convention is to call the long-run quasi-rents to fixed family labor family labor surplus.)

The efficiency gain (loss) needed to effect the preceding redistribution of market benefits through direct payments can be measured graphically. From the benefits to producers (including input suppliers) and consumers, we subtract the tax needed to finance the payments. Letting NB equal net benefits, we have

$$\begin{aligned} NB &= \text{area } (a + b) + \text{area } (c + d + h) - \text{area } (a + b + c + d + h + g) \\ &= -\text{area } (g) \end{aligned} \tag{9-1}$$

Area (g) is therefore a graphic representation of the efficiency loss.

A Closer Look at Farm Labor

If the welfare effects for a particular input are of interest, it becomes necessary to study the market for that input; the concept of producer surplus is too aggregative for such a purpose. Take agricultural labor as an important example. For this example, we assume that family and hired labor are identical and that the long-run supply function for farm labor is upward sloping. A direct payment program would in the long run generate benefits to labor, but what determines whether the benefits would amount to much? More generally, what conditions determine whether a direct payment program is an effective means for generating benefits to farm labor?

To answer this important question, we hark back to the aggregative farm sector model of Chapter 4. Abstracting from uncertainty and dynamics, we assume an aggregate production function as follows:

$$Q = \alpha_0 A^{\alpha_1} K^{\alpha_2} L^{\alpha_3}$$

where L is here defined as the sum of family and hired labor. The t subscripts that appeared in Chapter 4 are dropped because we abstract from lags in production. It is assumed that the supply functions for land A and producer goods K are perfectly inelastic and perfectly elastic, respectively. Let the farm labor supply be given by

$$L = aW^b \tag{9-2}$$

where W equals the return (wage) to labor, b is the elasticity of the labor supply function, and a is a multiplicative constant. Think of a graphic representation of this supply function with L on the vertical axis. If W increases from the competitive value W_c to the new level W_g under a direct payment program, the benefit to farm labor, that is, the change in worker surplus, is given by the area under the labor supply curve and between W_g and W_c. Applying integral calculus, this area equals

$$WS = \frac{1}{b+1}\left(W_g L_g - W_c L_c\right) \tag{9-3}$$

where WS equals the change in worker surplus. Recall that with a Cobb–Douglas production function and assuming profit maximization $WL = \alpha_3(PQ)$, where α_3 is labor's production elasticity. Thus

$$WS = \frac{\alpha_3}{b+1}\left(P_0^* Q_g - P_c Q_c\right) \tag{9-4}$$

where the product $P_c Q_c$ equals total receipts to farmers under competition and $P_0^* Q_g$ equals total receipts to farmers including total direct payments. Other things being equal, if the elasticity of labor supply is large (b is large) and/or if labor is a relatively minor farm input (α_3 is small), a direct payment program tends to be an ineffective means for elevating returns to farm labor.

To gain a rough idea of the quantitative magnitudes involved, we take as plausible estimates $b = 1.9$ and $\alpha_3 = 0.124$ (see Helmberger, 1991). If a direct payment program elevates the total receipts to agriculture by $4 billion, then the increase in worker surplus would equal $172 million, which amounts to 4.3 percent of the increase in receipts. Judging from past trends, farm labor is becoming less important with each passing decade relative to land and capital. In addition, the labor supply function appears to be elastic. These considerations lead us to believe that a direct payment program is likely an ineffective means for generating benefits to farm labor. Much the same applies to other farm programs as well.

9.2 DIRECT PAYMENTS AND MARKET PRICE SUPPORTS (TWO-PRICE PLAN) WITH INTERNATIONAL TRADE

Up to this point, our analysis has centered on a direct payment program assuming no international trade. We now relax this assumption, assuming instead that the country of interest is a food exporter. (As noted, the case of a food-importing country is taken up in Chapter 10.) In addition, we describe and analyze a market price support program and compare its effects with those generated under direct payments. Market price support programs are of great importance in the United States as well as in many other countries, including those comprising the European Community.

The domestic competitive demand and supply curves for the output of food in the United States are given by D and SS in panel b, Fig. 9.2; the demand and supply

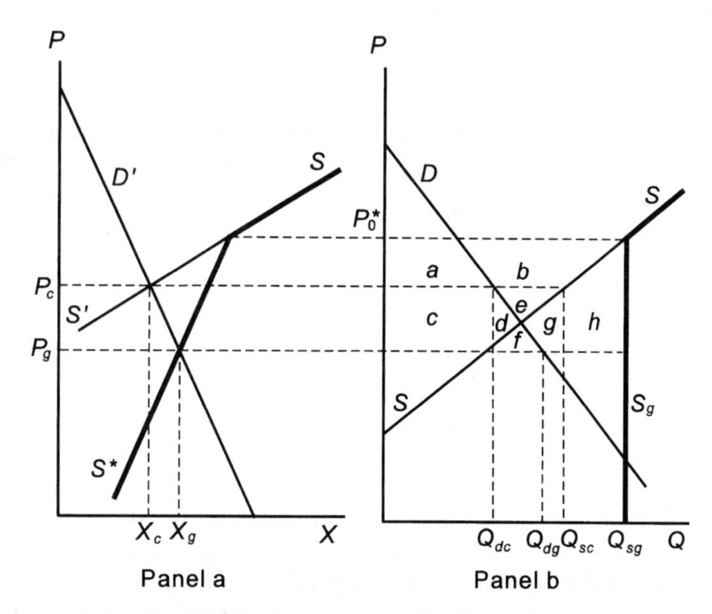

Figure 9.2 U.S. domestic demand and supply curves and the demand and supply curves for U.S. exports under perfect competition and a direct payment program, with program costs and benefits.

curves for U.S. exports X in the international market are given by D' and SS' in panel a. (At this stage, the student might like to review Chapter 5, which shows how the demand and supply for exports are derived.) Under competition and free trade, the equilibrium world price and the U.S. price both equal P_c, U.S. production and consumption equal Q_{sc} and Q_{dc} in panel b, and exports equal Q_{sc} minus Q_{dc}, which by construction equals X_c in panel a.

Direct Payments

A direct payment program is introduced in the United States with the target price P^* set equal to P_0^*. Direct payments are not, of course, given to foreign producers. The direct payment program shifts the U.S. supply curve from SS to the kinked curved SS_g. To find the new world equilibrium, we need first derive the new supply for U.S. exports. This can be done by simply subtracting laterally the demand curve D from the kinked supply curve SS_g. The resulting kinked supply for exports is given by SS^* in panel a. The intersection of the demand for U.S. exports D' and the new U.S. supply for exports SS^* determines the new world equilibrium price P_g, and U.S. exports equal $(Q_{sg} - Q_{dg})$, which, by construction equals X_g. The direct payment program lowers world price, increases U.S. output and world consumption, and decreases rest-of-world (ROW) production.

Turning to the welfare effects, we note that U.S. consumer surplus goes up by area $(c + d + f)$ in panel b. Producer surplus rises by area $(a + b)$. Since U.S. taxes rise by area $(a + b + c + d + e + f + g + h)$, the efficiency loss to the United States is given by area $(e + g + h)$.

By drawing a diagram showing how D' is derived for the ROW, the student can easily show that a direct payment program in the United States generates an increase in the ROW consumer surplus that exceeds the loss of ROW producer surplus. Therefore, the U.S. program results in a net efficiency gain to the ROW. Obviously, the ROW farmers will be displeased by U.S. policy, and those exporting nations that compete with the United States in the world market for food can be expected to complain, whether or not they themselves subsidize their farmers.

Market Price Support

We now consider a market price support program for food in the United States. Under this program, the government stands willing to purchase food in whatever quantity farmers wish to sell at the support price P^+. Exports from the rest of the world (ROW) to the United States are not allowed. What the U.S. government purchases in the domestic market is sold in the world market for whatever price can be obtained. Thus the market price support program under consideration is a two-price plan, involving a high support price in the domestic market and a low price in the international market. (Price support operations often result in large government stocks, but we abstract from storage for the moment.)

The competitive demand and supply curves are shown in Fig. 9.3. The government's demand for output becomes perfectly elastic at the support price P_0^+. The U.S. domestic demand DD, in panel b, shows how much private buyers would be willing to purchase at various prices. At the support price P_0^+, the quantity demanded equals Q_{dg}. The lower part of the demand, for prices less than P_0^+, never comes into play because the government does not allow the domestic price to fall below P_0^+. The total demand for farm output, for both private buyers and the government, is given by the darkened curve DD_g, kinked at $P = P_0^+$.

To derive the U.S. supply for exports, we horizontally subtract the supply SS from the demand DD_g. Then U.S. prices less than P_0^+ are no longer relevant. The export supply for the United States changes from SS' under competition (in panel a) to SS^+ under the price support program. The perfectly inelastic part of SS^+ can be explained as follows: As the world price falls below P_0^+, the U.S. price remains fixed at P_0^+. At the support price, exports equal Q_{sg} minus Q_{dg}, which equals X_g in panel a.

Equilibrium in the U.S. market is given by the intersection of DD_g and SS. The domestic price is elevated from P_c to P_0^+; domestic output grows from Q_{sc} to Q_{sg}. Domestic consumption falls from Q_{dc} to Q_{dg}. United States exports increase from $(Q_{sc} - Q_{dc}) = X_c$ to $(Q_{sg} - Q_{dg}) = X_g$. Letting P_0^* equal P_0^+, price supports increase U.S. exports more than do direct payments ceteris paribus. The reason for this is that U.S. output is the same under both programs if $P_0^* = P_0^+$. But price supports decrease do-

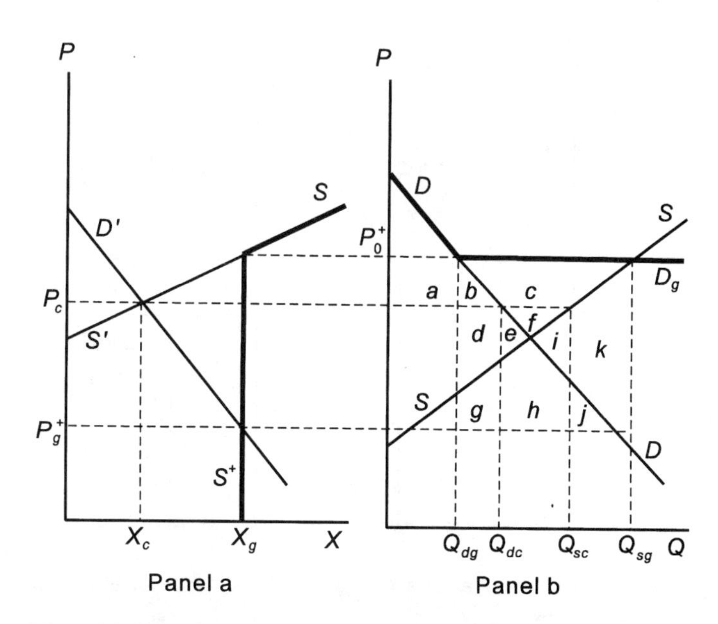

Figure 9.3 U.S. domestic demand and supply curves and the demand and supply curves for U.S. exports under perfect competition and a price support program, with program costs and benefits.

mestic consumption, whereas direct payments do just the opposite. The welfare impacts of a market price support program are left to the student as an important exercise.

Figure 9.4 allows a comparative analysis of direct payments and price supports on the assumption that the U.S. domestic demand is perfectly inelastic, an assumption that is probably realistic. The support level under price supports P_0^+ is set equal to the target price P_0^* under direct payments. Thus U.S. farmers produce the same level of output Q_{sg} under either program. Since demand is perfectly inelastic, U.S. exports are the same in either case. Foreign consumers and producers are affected in exactly the same way by the two programs.

The only differences in the economic effects of the two programs occur in the U.S. market. The student should understand that in either case the gain to the farm sector equals area $(a + b)$ and that the efficiency loss equals the area $(d + e + f)$. Consumer surplus falls by area (a) under price supports, but rises by area (c) under direct payments. Taxes under price supports are less than taxes under direct payments by the area $(a + c)$. Thus the choice between direct payments and price supports depends on how the government prefers to finance a given level of benefits to the farm sector. If the government prefers a higher food bill and a lower tax bill, it chooses price supports. If the government prefers a lower food bill and a higher tax bill, it chooses direct payments. Income taxes are painfully apparent; food taxes are hidden in higher food prices. Guess which approach politicians prefer.

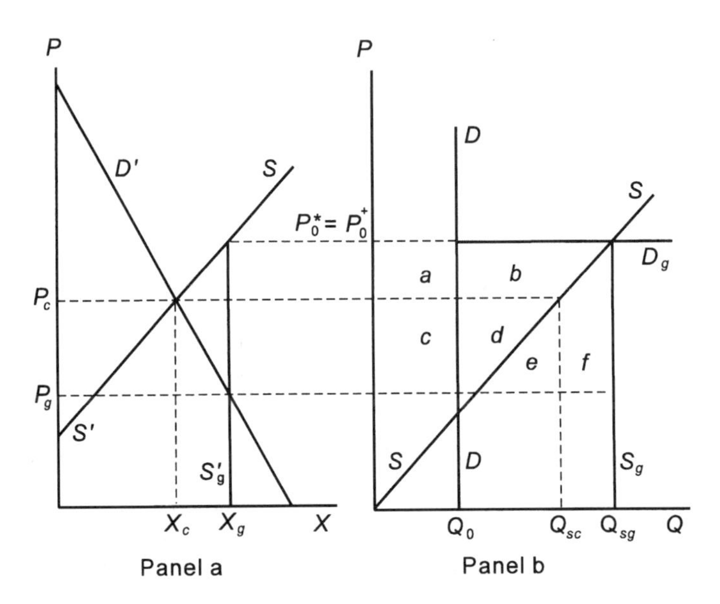

Figure 9.4 U.S. domestic demand and supply curves and the demand and supply curves for U.S. exports under perfect competition, direct payments, and price supports, with program costs and benefits.

9.3 DIRECT PAYMENT AND MARKET STABILIZATION PROGRAMS UNDER UNCERTAINTY

In Section 9.2 we viewed direct payment and price support programs as alternative ways for raising farm prices and incomes. We now analyze these programs from a different perspective and under different conditions. Direct payments and price supports are treated as alternative means for stabilizing farm markets under uncertainty, with the elevation of farm income being a secondary objective. Consideration is also given to a program in which arbitrageurs receive subsidies to store commodities.

The extreme volatility of farm prices is often seen as evidence of instability in farm markets. Price variation is claimed to be excessive; market instability, it is argued, calls for government intervention. Before traveling too far down this road, however, we must stop and ask why price volatility is objectionable and whether the government can do anything about it and whether the benefits—if benefits there be—are worth the cost.

Three main reasons have been given to support the need for farm price stabilization programs. First, although crop production is seasonal and subject to shocks such as the weather, the consumption needs of people for food are remarkably stable. Obviously, storing food in periods of relative abundance for consumption in periods of relative scarcity stabilizes both consumption and price and is vitally important for human welfare.

A second reason why stabilizing price is thought to be a worthy government objective is the belief that price variation leads to variation in farm income. Holding average income constant, farmers likely prefer stable to unstable incomes. As shown in Chapter 2, increasing the variance of price, holding the expected price constant, can increase the marginal cost of risk, which decreases output. Farmers may seek to lower risk through diversification and hedging and by stabilizing consumption through borrowing and saving, but there is no assurance that risk can be eliminated entirely. This raises the question of whether the cost of a government stabilization program might be more than offset by the decreased cost of risk, whether stabilization might generate net benefits.

It should not be assumed, however, that a program that stabilizes price will necessarily stabilize farm income. We will indeed show that price stabilization might be associated with farm income destabilization. This raises a difficult but important question. Under what market conditions is price stabilization associated with income stabilization or with destabilization?

A third possible benefit from a price stabilization program concerns marketing costs. Just as risk imposes costs on farmers, risk might also impose costs on marketing firms. Take an example. Marketing firms construct flexible food manufacturing plants that can reasonably accommodate unstable levels of food production. Flexibility might entail higher cost than could be achieved by a plant specialized for an even and continuous flow of raw material. Unstable production might, to take another example, require overtime and the higher wages that overtime ordinarily requires. Periods of overtime and multiple shifts alternating with periods of idle capacity and worker layoffs are inimical to efficient marketing.

In what follows we first center on direct payments as a means for stabilizing markets in an uncertain and volatile world, abstracting from commodity storage. It will be shown that, although a direct payment program can be used to stabilize the effective prices received by farmers, other dimensions of performance might not be stabilized. The analysis will also shed some light on the conditions under which programs that stabilize farm prices also stabilize farm income. After analyzing direct payments in the absence of storage, we then consider three alternative approaches for stabilizing farm markets taking storage into account. In addition to direct payments and price supports, we also consider an approach that involves payments to private storers that lessen the costs of carrying old crop into the new crop year.

Do Direct Payments Stabilize Farm Markets?

Throughout this section we will assume that, for whatever reasons, farmers seek to maximize expected profit. On this assumption, the competitive model given by Eqs. (2-27) can be modified as follows:

$$\begin{aligned} P_{t+1} &= \beta_0 - \beta_1 Q_{t+1} + e_{t+1} && \text{demand} \\ Q_{t+1} &= S[E_t(P_{t+1})] && \text{supply} \\ E_t(P_{t+1}) &= \beta_0 - \beta_1 Q_{t+1} && \text{expected price function} \\ V_t(P_{t+1}) &= \sigma_e^2 && \text{variance of price} \end{aligned} \qquad (9\text{-}5)$$

The reader should recall that Q_{t+1} is determined by producer decisions made in period t. The model can be solved graphically. In Fig. 9.5, the supply curve and the expected price (demand) curve are given by S_Θ and D_Θ, respectively. The intersection of these two curves gives the equilibrium expected price, $E_t'(P_{t+1})$, and output, Q'_{t+1}. Price in period $t + 1$ is determined by the intersection of the perfectly inelastic S' (output equals Q'_{t+1}) curve and actual demand in period $t + 1$, which might lie to either the right or left of D_Θ.

We now suppose that the government introduces a direct payment program, setting the target price at P_0^* in Fig. 9.5. We further suppose that in years of particularly high demand the actual market price exceeds the target, but that in years of low demand, the target exceeds market price. On these assumptions, it is clear that the effective price farmers expect to receive under the program, $E_t^*(P_{t+1})$, exceeds both P_0^* and $E_t'(P_{t+1})$. With the expected price given by $E_t^*(P_{t+1})$, output equals $Q^*_{t+1} > Q'_{t+1}$; actual price will be determined by the intersection of S^* and actual demand in period $t + 1$. In this simple world, production and consumption are perfectly stable with or without the program. The variance of market price is also unaffected.

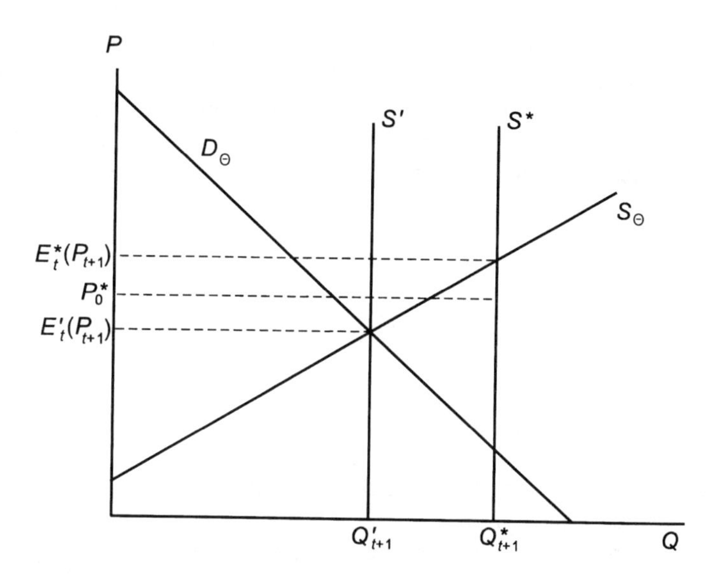

Figure 9.5 Expected demand and supply curves and output under perfect competition and a direct payment program.

That variance of the effective price to producers falls as a result of the program can easily be seen in a simple numerical example. As on previous occasions, the student should follow the example using paper and pencil. Consider a competitive model given by

$$P_{t+1} = 20 - Q_{t+1} + e_{t+1}$$

$$Q_{t+1} = 1.5E_t(P_{t+1}) \tag{9-6}$$

$$E_t(P_{t+1}) = 20 - Q_{t+1}$$

where it is supposed that e_{t+1} equals +2.0 half the time and –2.0 half the time. Using the last two equations, it is easy to show that $Q_{t+1} = 12$ and $E_t(P_{t+1}) = 8$. Market performance depends on the value of e_{t+1} as follows:

e_{t+1}	Q_{t+1}	P_{t+1}	$TR_{t+1} = Q_{t+1}P_{t+1}$
–2.0	12.0	6.0	72.0
+2.0	12.0	10.0	120.0

On the basis of these calculations, it can be shown that $V_t(P_{t+1}) = 4.0$ and that $E_t(TR_{t+1}) = 96.0$ and $V_t(TR_{t+1}) = 576.0$. Recalling that the coefficient of variation (CV) is defined as the square root of the variance divided by the expected value, all expressed as a percentage, we find that the CV of price equals 25 percent, as does the CV of total revenue.

We next suppose that a direct payment program is begun with the target price P^* equal to 8.00. Letting $\bar{P}_{t+1}$ equal the effective price received by the farmer, we make a guess, to be tested shortly, that half the time, when $e_{t+1} = +2.0$, the market price will exceed 8.0, in which case $\bar{P}_{t+1} = P_{t+1}$. Otherwise, when $e_{t+1} = -2.0$, $P_{t+1} < \bar{P}_{t+1} = 8.0$. Therefore,

$$\begin{aligned} E_t(\bar{P}_{t+1}) &= \frac{1}{2}(22 - Q_{t+1}) + \frac{1}{2}(8) \\ &= 15 - \frac{1}{2}Q_{t+1} \end{aligned} \tag{9-7}$$

Equation (9-7), along with the second equation from Eqs. (9-6), allows solving for Q_{t+1}, which equals 12.8571. Under the assumed conditions, market performance once again depends on the value of e_{t+1} as follows:

e_{t+1}	Q_{t+1}	P_{t+1}	$\bar{P}_{t+1}$	$\overline{TR} = Q_{t+1}\bar{P}_{t+1}$
−2.0	12.8571	5.1429	8.0	102.8568
+2.0	12.8571	9.1429	9.1429	117.5512

Notice that the market price exceeds the target price half the time, confirming our earlier guess. On the basis of these calculations it can be shown that $E_t(\bar{P}_{t+1}) = 8.5715$ and $V_t(\bar{P}_{t+1}) = 0.33$ and that $E_t(\overline{TR}) = 110.2040$ and $V_t(\overline{TR}) = 53.98$. The CV of the effective price received by the farmer falls from 25 percent under competition to 6.7 percent under the program. The CV of total farm revenue also falls from 25 percent under competition to 6.7 percent under the program. Relative to the competitive model, a direct payment program stabilizes both the effective farm price and total farm revenue. Since production cost in this simple model is determined in period t and is therefore not random, the variance of quasi-rent (farm income) depends solely on and in fact equals the variance of total revenue. *Hence in this numerical example a direct payment program stabilizes both price and quasi-rent, given that demand (but not production) is random.*

It is left to the student to show that the variance of market price P_{t+1} is the same with or without the program. Since the program causes the expected market price to fall from 8.0 to 7.14, the coefficient of variation for price actually rises from 25 percent under competition to 28 percent under the program. The program stabilizes the effective farm price, but destabilizes the market price. Consumption and production are, of course, nonrandom under both regimes, by assumption.

As it happens, the demand is inelastic in the relevant region in the above example. Would the qualitative results have been different if demand had been elastic? The answer is no. To see this, change the supply function such that $Q_{t+1} = 0.3E_t(P_{t+1})$. The demand is then elastic in the relevant region. Under this new specification, the coefficient of variation of total revenue equals 13 percent under perfect competition and 5 percent under a direct payments program with the target price set equal to 15.3846, the price expected under competition.[1]

We next consider an alternative model in which the source of instability is yield instead of stochastic demand as in the previous example. *It will be seen shortly that with stable demand and random yield a direct payment program that tends to stabilize price stabilizes quasi-rent if demand is inelastic; otherwise quasi-rent is destabilized.* The model is as follows:

$$\begin{aligned} P_{t+1} &= 20 - Q_{t+1} \\ A_t &= 0.5E_t(P_{t+1}) \\ Q_{t+1} &= A_t Y_{t+1} \\ E_t(P_{t+1}) &= 20 - E_t(Y_{t+1})A_t \end{aligned} \tag{9-8}$$

It is assumed that yield is a random variable equaling 0.5 half the time and 1.0 half the time. The expected value therefore equals $E_t(Y_{t+1}) = 0.75$. Under competition, the second and fourth equations can be used to show that $A_t = 7.2727$ and $E_t(P_{t+1}) = 14.5455$. Competitive market performance depends on yield as follows:

Y_{t+1}	Q_{t+1}	P_{t+1}	$TR_t = Q_{t+1}P_{t+1}$
0.5	3.6364	16.3636	59.5041
1.0	7.2727	12.7273	92.5620

On the basis of these calculations, it can be shown that $V_t(P_{t+1}) = 3.31$ and that $E_t(TR_{t+1}) = 76.03$ and $V_t(TR_{t+1}) = 273.21$. The CV of price equals 12.5 percent; the CV of total revenue equals 21.7 percent. The student should recognize the negative correlation between price and yield in this example.

Next suppose that a direct payment program is begun with the target price set equal to the expected competitive price of 14.5455. We make a guess, to be tested later, that the market price exceeds the target half the time, when yield equals 0.5, and is less than the target half the time, when yield equals 1.0. We have

$$E_t\left(\overline{P}_{t+1}\right) = \frac{1}{2}\left(20 - 0.5A_t\right) + \frac{1}{2}\left(14.5455\right) \tag{9-9}$$

[1]Notice that $V(TR) = V(PQ)$. (The $t + 1$ subscripts are omitted.) In this simple model, Q is not random. If P^* is high enough, then $P^* = \overline{P}_{t+1}$ and $V(TR) = 0$.

Using this result together with the second equation from Eqs. (9-8), we find that $E_t(\overline{P}_{t+1}) = 15.3536$ and $A_t = 7.6768$. Acreage rises under the program. Market performance depends on random yield as follows:

Y_{t+1}	Q_{t+1}	P_{t+1}	$\overline{P}_{t+1}$	$\overline{TR} = Q_{t+1}\overline{P}_{t+1}$
0.5	3.8384	16.1616	16.1616	62.0345
1.0	7.6768	12.3232	14.5455	111.6626

On the basis of these calculations, it can be shown that $V_t(\overline{P}_{t+1}) = 0.653$ and that $E_t(\overline{TR}_{t+1}) = 86.85$ and $V_t(\overline{TR}_{t+1}) = 615.73$. The CV of the effective farm price equals 5.3 percent. The CV of farm total revenues equals 28.6 percent. *With random yield being the only source of variation, a direct payment program stabilizes the effective price received by farmers, relative to the competitive outcome, but the variance of farm total revenue and quasi-rent rises.*

Turning to other dimensions of performance, we find that the variation of consumption Q_{t+1} (production), as measured by the coefficient of variation, is unaffected by the program. The program causes the coefficient of variation of price, on the other hand, to rise from 12.5 percent under competition to 13.5 percent. A direct payment program stabilizes the effective price received by the farmer in both of the numerical models considered previously; it stabilizes neither consumption (production) nor market price.

In the above numerical example, a direct payment program stabilized the effective price received by farmers, but it destabilized total revenue. As it happens, however, the demand was elastic in the relevant region. Would our result have been different if demand had been inelastic? The answer is yes. To see this, change the supply function such that $A_t = 3E_t(P_{t+1})$. Demand is then inelastic in the relevant region. Under competition, the coefficients of variation for price and total revenue equal, respectively, 75 percent and 56 percent. Under a direct payment program, with the target price set equal to 3, the coefficients of variations for the effective farm price and total revenue equal 54 percent and 26 percent, respectively.[2]

The main lessons to be learned from our analysis of direct payments under uncertainty may be summarized as follows: First, a direct payment program can certainly be used to elevate the expected effective price (price plus per unit direct payment) received by farmers. This expands production. Expected profit rises as well. These results are consistent with the comparative-static models considered earlier in this chapter.

[2]Notice that $V(TR) = V(PYA) = A^2V(PY)$. (The $t + 1$ subscripts are omitted.) The expression for $V(PY)$ is complex; generalizations as to the effects of a direct payment program with stochastic yield are difficult to develop. Notice, however, the interesting special case where demand is unitarily elastic everywhere. Then $V(TR) = 0$ under competition; a direct payment program can only destabilize TR even where the effective farm price is completely stabilized, i.e., where $\overline{P}_{t+1} = P^*$.

Second, direct payments can be used to stabilize, at least to some extent, the effective price received by farmers, but whether farm total revenue and quasi-rent are stabilized depends on a number of conditions. It appears that in linear models direct payments stabilize both the effective farm price and quasi-rent where the source of instability is stochastic demand. The same result holds if random yield is the only source of instability *and* if demand is inelastic; elastic demand suggests quasi-rent will be destabilized.

Third, it is probably inappropriate to think of a direct payment program as a means for stabilizing markets. Neither production, consumption, nor market price was stabilized in the above examples. This conclusion will receive further support in the analysis that follows, an analysis that takes commodity storage into account. It will be seen presently that meaningful market stabilization requires commodity storage.

Market Stabilization with Commodity Storage

A proper analysis of market stabilization programs on the part of the government must recognize price arbitrage and private storage with or without government intervention. For such an analysis, we turn to a study of the U.S. soybean market by Glauber, Helmberger, and Miranda (GHM) (1989). The GHM model, similar to the numerical model set forth in Section 6.2, was estimated using time series data and various statistical and numerical techniques and calibrated to fit 1977 economic conditions. The model is as follows:

$$\begin{aligned} Q_t &= 5.18P_t^{-0.61} \\ Q_t + I_t &= H_t + I_{t-1} \\ A_t &= 13.0[E_t(P_{t+1})]^{0.89} \\ H_t &= A_{t-1}Y_t \\ E_t(P_{t+1}) &\leq 1.09P_t + \$0.36 \\ E_t(P_{t+1}) &= f_t(A_t, I_t) \end{aligned} \tag{9-10}$$

where Q_t equals the annual quantity of U.S. soybeans demanded for export and the domestic crushing industry, measured in billions of bushels; P_t equals dollars per bushel; I_t equals commercial stocks; A_t equals acreage planted, in millions of acres; and Y_t equals yield per acre. Yield is random with an estimated expected value equal to 29.84 bushels per acre.[3] Numerical methods similar to those described previously

[3]The probability distribution for yield was assumed to be log normal. The estimated coefficient of variation equaled 17.43 percent.

were used to estimate the expected price function $E_t(P_{t+1}) = f_t(A_t, I_t)$. As you might guess, $E_t(P_{t+1})$ falls with ceteris paribus increases in either I_t or A_t. A spline function was used to approximate $f_t(A_t, I_t)$, which would require many pages to summarize and is therefore not reported here. (See Fig. 6.4 as an example of a spline function in which estimated points that lie on the surface of a function are connected with straight lines.) The student should recognize that, if the inequality in the price arbitrage condition holds, then $I_t = 0$ and we have five remaining equations to solve for five endogenous variables. If the equality holds, then $I_t \geq 0$ and we have six equations to solve for six endogenous variables.

Simulation analysis was used to estimate long-run (steady-state) equilibrium values for all endogenous variables under the economic conditions of 1977. The estimates for the variables of greatest interest are given in Table 9.1. [For the moment, ignore $E_t(\overline{P}_{t+1})$ and $\overline{CVP}_{t+1}$.] The values in the first column, where $P^* = 0$, are the steady-state values under competition. The steady-state equilibrium expected price for soybeans, $E_t(P_{t+1})$, was $5.71 per bushel. (The actual price received by U.S. farmers in 1977 was $5.88.) The steady-state coefficient of variation for price CVP_{t+1} and for aggregate producer quasi-rent $CVQR$ equaled 19.5 and 12.7 percent, respectively.

We may now ask how the values of these critical variables would change if the government introduced a market stabilization program of some kind. We consider, in turn, three such programs. The first is a direct payment program. The second is a market price support program. The third involves subsidies paid to arbitrageurs for the express purpose of lowering the cost of carrying old crop into the new crop year.

Consider a direct payment program that involves setting the target price equal to the alternative values given in Table 9.1. How would the resulting programs affect market performance? To answer this question, the system given by Eqs. (9-10) was

TABLE 9.1. Estimates of the Long-run (Steady-state) Values for Expected Market Price $E_t(P_{t+1})$, Expected Effective Farm Price $E_t(\overline{P}_{t+1})$, and Coefficients of Variation of Market Price (CVP_{t+1}), Farm Price ($\overline{CVP}_{t+1}$), and Producer Quasi-rent ($CVQR$) under Competition and a Direct Payment Program, U.S. Market for Soybeans, 1977 Economic Conditions.

Endogenous Variables	Target Price P^*							
	0.00	$4.50	$4.75	$5.00	$5.25	$5.50	$5.75	$6.00
	Price per Bushel							
$E_t(P_{t+1})$	5.71	5.71	5.70	5.65	5.55	5.44	5.30	5.13
$E_t(\overline{P}_{t+1})$	5.71	5.72	5.73	5.77	5.84	5.92	6.04	6.18
	Percentage							
CVP_{t+1}	19.5	19.5	19.6	19.7	19.9	20.2	20.5	20.8
$\overline{CVP}_{t+1}$	19.5	19.5	18.8	17.4	15.0	12.5	9.90	7.2
$CVQR$	12.7	12.8	14.0	15.5	17.0	19.5	22.3	25.0

Source: Glauber, Helmberger, and Miranda (1989).

modified in the GHM study as follows: Acreage planted was no longer written as a function of expected market price, but was written instead as a function of the expected effective price $E_t(\bar{P}_{t+1})$, where $\bar{P}_{t+1}$ equals the market price plus the direct payment per unit $(P^* - P_{t+1})$. The introduction of the new variable $\bar{P}_{t+1}$ necessitated the derivation of an additional expected price function, using numerical methods, that shows the values of $E_t(\bar{P}_{t+1})$ for alternative combinations of values for A_t and I_t. A new expected price function for market price P_{t+1} was also estimated. On the basis of the new quantified model, simulation was used to estimate the steady-state values for variables of interest for alternative price targets P^* (Table 9.1).

The student will notice, first, that the expected or average effective price received by the soybean producer, $E_t(\bar{P}_{t+1})$, rises with increases in the target from $5.71 per bushel under competition to $6.18 under direct payments, the latter price associated with a $6.00 target. The difference between the target and the expected effective price diminishes as the target rises. Presumably, if the target were raised high enough, the difference would disappear altogether. Relative to the effective farm price, the market price responds in an opposite manner, falling with increases in the target. These results are intuitively appealing and consistent with the numerical examples given earlier, where storage was ignored.

The coefficient of variation of the effective farm price, $\overline{CVP}_{t+1}$, falls from 19.5 percent under competition to 7.2 percent as the target is raised to $6.00. The coefficient of variation of quasi-rent, $CVQR$, rises dramatically from 12.7 percent under competition to 25.0 percent with a $6.00 target. Stabilization of the effective farm price destabilizes producer quasi-rent, but recall that the only source of instability in the GHM model is random yield.

According to additional GHM results (not reported in Table 9.1), the direct payment program increased steady-state expected values for production and consumption, decreased slightly the variation of production, increased slightly the variation of consumption, and decreased slightly the expected value of the year-end stocks carried over into the new crop year.

Drawing on the GHM study and the earlier analysis that abstracted from storage, we may tentatively conclude the following: A direct payment program can be used to elevate the average effective price farmers receive over time, increasing production and consumption and generating benefits to both consumers and producers (farm input suppliers in the long run). The program can be used to stabilize the effective farm price, but whether quasi-rent is stabilized is another matter, depending on the sources of instability and the elasticity of total demand. In any event, a direct payment program has little effect on the stability of market price, production, and consumption. It seems appropriate to think of direct payments primarily as a means for elevating farm price and farm income, but not as a means for stabilizing markets.

We now take up the effects of a market price support program, again taking advantage of the GHM study. The support program envisaged here involves both a support price and a release level, not unlike support operations used frequently in the United States. Under this program the government stands willing to commit to government storage of as much commodity as required to keep price from falling below the support level P_s^+. The government also stands willing to sell any part or all of its stock at the market price, providing the market price is not less than the release level P_r^+. The idea is to try, at least, to keep the price within an acceptable band and thus avoid large price extremes and price variation. It may be noted straightway that one effect of a price support program, if only the support level is set high enough, is to wipe out the private storage industry. Private storage is the offspring of price variability. In particular, a government program that diminishes price differences over time will tend to decrease private storage. Price support operations clearly affect the private demand for stocks.

In Fig. 9.6, we consider a price support program on the assumption that private storage does not occur, which could easily happen if the support price is sufficiently high. The aggregate demand for both consumption and exports is given by the curve labeled DD_c. The government demand for commodity to be held as government stocks is given by D_g, being perfectly elastic at the support price P_s^+. The aggregate demand for consumption, exports, and government stocks is given by the kinked curve DD_g. The program of interest here is not a two-price program. It does not allow dumping the U.S. surplus in the world market at a price less than the domestic price.

Price support operations also affect the supply function. In Fig. 9.6, we let harvested output equal Q_0; government-held stocks equal $(Q_2 - Q_0)$. The release price

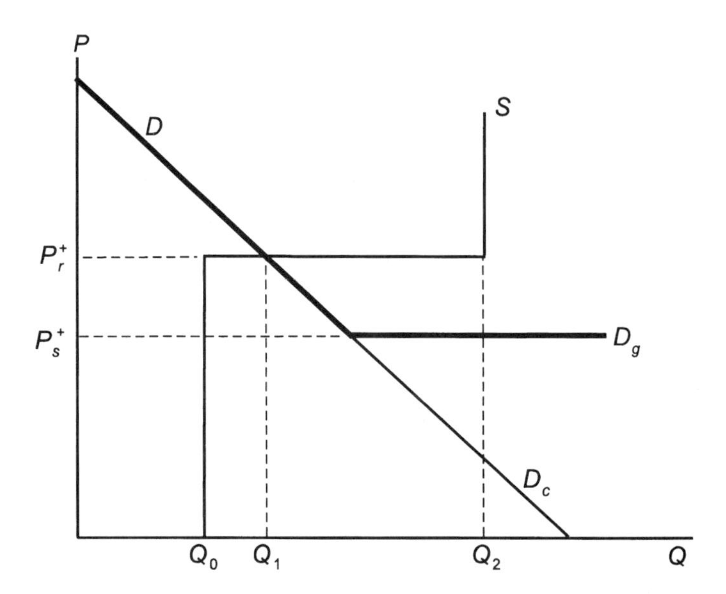

Figure 9.6 Demand and short-run supply curves under a price support program with government-held stocks.

is given by P_r^+. For any market price less than P_r^+, the private market is willing to supply Q_0, which, of course, equals the harvested output. At any $P \geq P_r^+$, the government is willing to sell as much as buyers want to buy up to the limit of its stock. The result is a stepped supply function, the nature of the step depending on the level of P_r^+ and the level of government stocks. In Fig. 9.6, the equilibrium market price equals P_r^+ and the government sells $(Q_1 - Q_0)$ from its stock, leaving $(Q_2 - Q_1)$ for the next period. At this juncture the student should experiment with various configurations involving a kinked demand and a step supply in order to understand the kinds of transactions that the government might need to undertake as a result of its program.

Introducing a price support program requires modification of the competitive market model given by Eqs. (9-10). Importantly, the price expectation $E_t(P_{t+1})$ becomes a function of A_t, I_t, and G_t, where the latter equals stocks held by the government. Both private and government stocks become relevant variables because the storage decisions of private storers and the government are not the same. The numerical methods used to estimate the expected price function must take account of the profit motive for holding private stocks and the rules used by the government in managing its stocks. The GHM study, to which we now return more explicitly, used the top five equations from Eqs. (9-10), but the last equation was altered by expressing the expected price $E_t(P_{t+1})$ as a function of A_t, I_t, and G_t. Various levels of support P_s^+ were considered, with the release price P_r^+ set equal to 120 percent of the support price. Estimates of long-run (steady-state) values for selected endogenous variables are given in Table 9.2. For convenience, the competitive values from Table 9.1 are repeated in the first column.

TABLE 9.2. Estimates of the Long-run (Steady-state) Values for Expected Price $E_t(P_{t+1})$ and Coefficients of Variation for Price CVP_{t+1} and for Producer Quasi-rent $CVQR$ under Competition and a Market Price Support Program, U.S. Market for Soybeans, 1977 Economic Conditions.[a]

Endogenous Variable	Support Price P_s^+				
	0.00	4.50	4.75	5.00	5.25
			Price per Bushel		
$E_t(P_{t+1})$	5.71	5.71	5.70	5.67	5.74*
			Percentage		
CVP_{t+1}	19.5	19.1	16.8	12.0	8.3*
$CVQR$	12.7	12.7	13.8	15.9	17.6*

[a]The release price P_r^+ equals 120 percent of the support level P_s^+. Cells marked with asterisks indicate a support program that gives rise to expected government stocks that tend toward infinity.

Source: Glauber, Helmberger, and Miranda (1989).

Table 9.2 supports several conclusions. First, a $5.25 price support level, well below the competitive expected price $5.71, causes expected government stocks to grow endlessly and to become infinitely large. Clearly, in the actual operation of a price support program, great care must be taken to avoid setting the support level too high. In what follows we will not be concerned with price support levels in excess of $5.00.

Second, in long-run equilibrium, a price support program has practically no effect on the expected market price. The decreases in the expected market price as the support price rises from zero to $5.00 are trivial. It may be noted, however, that this conclusion does not apply to the years immediately following the inception of a program. During the early years of operation, the government's stocks tend to be small; on average, the government adds more to stocks than it disgorges. This tends to drive prices up. Producers enjoy relatively high expected prices to start with; price support operations initially elevate expected farm income.

Third, although price support operations tend to stabilize market price (and consumption, also), they tend to destabilize producer quasi-rent. For example, setting the support level at $5.00 decreases the variation of market price by 38.5 percent of the variation under competition. The variation of quasi-rent rises by 25.2 percent. Again the student should recognize, however, that random yield is the only source of market instability by assumption.

Because of the decline in price variability, expected private storage falls quickly as the support price is increased. According to the GHM study, setting the support price at $4.75, well below the competitive expected value, lowers expected private year-end stocks by 77 percent.

The third form of government intervention of interest here entails government subsidies to private storers. Payments are made to cover at least part of the cost of bin space; government loans are made available to arbitrageurs at less than market rates of interest. To see the effects of this stabilization program, the reader should refer to Table 6.2, which reports the findings of the study by Lowry et al. To recall, the results given in that table show what happens to endogenous variables as the carrying charge, treated as an exogenous variable, is lowered toward zero. We now interpret the declines in the carrying charge appearing in Table 6.2 as the consequence of government subsidies to the private storers of the commodity. According to the results in Table 6.2, lowering the effective carrying charge from the competitive to a near-zero level decreases the steady-state coefficient of variation for the annual expected price from 15.41 to 7.5 percent. The association between price stability and quasi-rent stability is complex, doubtless reflecting, in part, two sources of market instability: random demand and random yield. As the carrying cost declines from high to near-zero levels, the coefficient of variation of quasi-rent at first declines (from 9.87 to 8.91 percent), but then rises. Because this study allows for both demand and production uncertainty, the results cast a serious doubt on the proposition that price stabilization can be relied on generally to stabilize income in a substantial way. It appears that commodity markets will need to be studied one by one.

Policy Implications

Intelligent policy choices involve fitting the right program to the policy objective. A tariff on tea will likely be an inappropriate means for providing shelter for the homeless, to take an extreme example. If the objective is to elevate farm prices and short-run incomes, direct payments (or a two-price program) would appear to be effective and in that sense appropriate. Market price supports and storage subsidies would appear to be inappropriate devices. If, on the other hand, market stabilization or, more particularly, price and consumption stabilization is the policy objective, then price supports and storage subsidies are appropriate devices, but direct payments are not.

If stabilization is the objective, much can be said in favor of relying on subsidies to the private storage industry. For one thing, the GHM study suggests that, for a given level of government expenditure, the storage subsidy program yields the greatest reduction in the variation of price and consumption. Relative to other programs, storage subsidies are probably cost effective. Equally important, a storage subsidy program is flexible in light of ever changing demand and supply factors that might soon make fixed support and release levels obsolete. The idea is to let the decentralized market mechanism decide when to add to stocks, because the current price is low relative to what is expected in the future, and when to disgorge, because the current price is high relative to what is expected in the future. The idea is to substitute private arbitrageurs for civil service employees and politicians. It should also be mentioned that, in the case of price supports, politicians might be tempted to raise support and release levels in order to elevate farm incomes in the short run and garner additional votes, all without regard to sensible management of commodity stocks.

PROBLEMS

9.1. Demand and supply are given by $P = 10 - 0.5Q$ and $P = 0.5Q$. Ignore international trade. Find the equilibrium market price, production, and consumption for each of the following regimes. Also, calculate change in consumer surplus, change in producer surplus, government expenditure, and the efficiency loss for the three farm programs described in parts b, c, and d.

- **a.** Perfect competition.
- **b.** A direct payment program with the target price P^* set equal to 6.
- **c.** A price support program with the support price P^+ set equal to 6. (Assume the government sells the surplus commodity acquired under the program for bird seed at a price equal to 2.)
- **d.** The farm program includes both the programs described under parts b and c above except that the support price equals 5.

9.2. The U.S. food demand and supply are $D_u = 10 - P_u$ and $S_u = 4\ P_u$. The ROW food demand and supply are $D_w = 15 - 0.5P_w$ and $S_w = (1.0)P_w$. Find price, production, consumption, and exports (imports) for both regions for the following regimes.

a. Perfect competition.

b. The United States starts a direct payment program with the target price P^* set equal to 4.

c. The United States starts a price support program with the support price P^+ set equal to 4. The U.S. surplus is dumped in the world market.

9.3. The aggregate production function is $Q = L^{0.5}A^{0.5}$. Suppose land is fixed at 16. The supply for labor is $L = 0.5W^3$. Food demand is $Q = P^{-0.2}$. The target price P^* equals 0.2 under a direct payment program. Calculate the increase in the total rent to land owners and the benefits to labor.

REFERENCES

Glauber, Joseph, Peter Helmberger, and Mario Miranda, "Four Approaches to Commodity Market Stabilization: A Comparative Analysis," *American Journal of Agricultural Economics*, 71, no. 2 (May 1989), 326–337.

Helmberger, Peter G., *Economic Analysis of Farm Programs*. New York: McGraw-Hill, Inc., 1991.

Houck, James P., *Elements of Agricultural Trade Policies*. New York: Macmillan Publishing Co., 1986.

Lowry, Mark, Joseph Glauber, Mario Miranda, and Peter Helmberger, "Pricing and Storage of Field Crops: A Quarterly Model Applied to Soybeans," *American Journal of Agricultural Economics*, 69, no. 4 (November 1987), 740–749.

10

Analysis of Farm Programs: Part II

Chapter 9 analyzed the use of direct payments and market price supports by a food exporting nation as alternative means for elevating farm prices and incomes (income redistribution) and as alternative means for stabilizing farm markets. It was shown that domestic prices can be supported through government removals of food from the domestic market and the dumping of surpluses in the international market. This, of course, results in high domestic prices relative to world prices. We begin this chapter by showing, in Section 10.1, that high domestic prices relative to world prices can also be achieved in food importing countries through import quotas and tariffs. Direct payments will again be seen to be of interest.

Following Section 10.1, this chapter returns to farm program analysis for food exporting nations. As we have seen, if price supports are used to stabilize markets, the support levels must be set below the market price expected under competitive conditions. Otherwise, governments stocks will grow endlessly. Even so, the governments of the United States and of many foreign nations have often used commodity acquisitions to elevate farm prices above competitive levels. One result has been vast commodity stocks of numerous commodities held in many parts of the world. Another has been the widespread dumping of surpluses in world markets. To avoid surpluses and/or dumping, governments have often turned to production control programs, and it is these programs that will occupy our attention in the latter parts of this chapter.

In Section 10.2, we analyze government-sponsored cartels as one such program. Section 10.3 centers on theoretical tools of use in Section 10.4. The latter section analyzes programs that offer farmers inducements to idle land as a means for decreasing production and increasing prices. Such programs have been the dominant approach to farm price and income problems in the United States since the early 1960s, and they are becoming increasingly popular in the European Community.

10.1 IMPORT QUOTAS, TARIFFS, AND DIRECT PAYMENTS: A COMPARATIVE ANALYSIS

Consider a country that ordinarily imports a substantial proportion of the food it consumes. As in the case of an exporting country, direct payments could be used to elevate farm prices and incomes, although, presumably, steps would need to be taken to ensure that payments are not made to foreign farmers. Unlike the case of an exporting country, however, import quotas and tariffs, often called *border measures*, could also be used to elevate domestic prices and incomes. In what follows we analyze the economic effects of quotas, tariffs, and direct payments under both the small- and large-country assumptions. According to the small-country assumption, the importing country's imports are small relative to world exports. Variation in its imports have no appreciable effect on world prices. Variation in a large country's imports, on the other hand, do affect world prices. The world price is exogenous in the case of a small country or importer, but endogenous in the case of a large importer. In what follows, we shall take the United States as an example of an importing country. The reader may think of sugar as an example of a commodity for which border measures are employed both in the United States and elsewhere.

Small-country Assumption

The demand and competitive supply for sugar in the United States are given by D and SS in panel b, Fig. 10.1. Abstracting from transportation costs and multiple currencies, the U.S. competitive demand for imports, given by DD' in panel a, is derived by subtracting laterally the supply curve SS from the demand curve D in panel b. The intersection of the demand for imports and the rest of the world supply of imports, the latter given by the flat curve labeled S' in panel a, determines the competitive level of U.S. imports M_c. By construction, M_c equals Q_{dc} minus Q_{sc}.

Suppose that the U.S. government seeks to elevate the domestic price to the target level given by P_0^*. One way to do this is to place a quota on imports equal to $(Q_{dg} - Q_{sg})$. This has the effect of raising the domestic price to P_0^*. At this price the total available supply, given by domestic production Q_{sg} plus imports $(Q_{dg} - Q_{sg})$, just equals consumption. A smaller quota would raise the U.S. price above P_0^*; a larger quota would leave the price below P_0^*. Exporting countries will be eager to ship as much as possible to the United States to take advantage of the high U.S. price. For the sake of fairness, the import quota might be apportioned among foreign suppliers on a proportionate basis. In other words, a foreign country's share of the quota might be set equal to that country's share of competitive imports. Notice that if the U.S. demand for imports is inelastic in the relevant range the total receipts to exporters will actually rise with the imposition of import quotas. Unfortunately for exporting nations, the demands for imports of many food products tend to be elastic.

As an alternative to an import quota, the United States could place a per unit tariff T on the commodity, with $T = (P_0^* - P_c)$. The importer must pay P_c to gain pos-

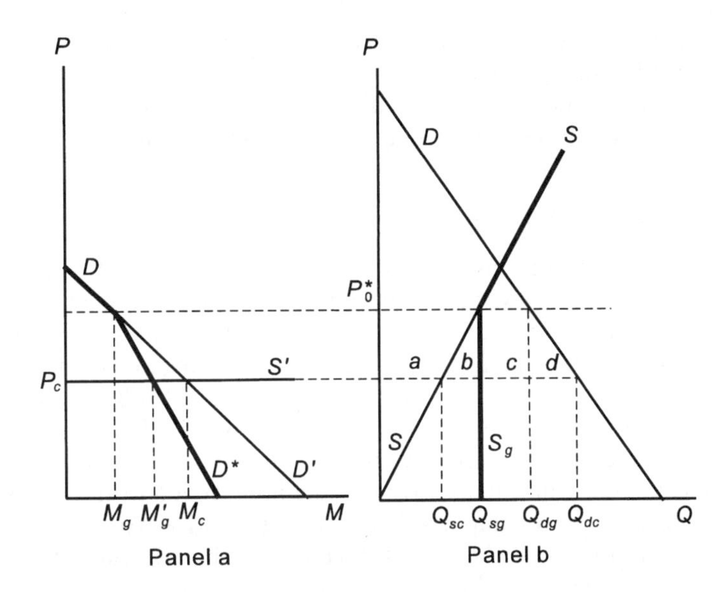

Figure 10.1 U.S. domestic demand and supply curves and the demand and supply curves for imports under the small-country assumption, with costs and benefits for import quotas, tariffs, and direct payments.

session of the commodity in the world market plus the per unit tariff $(P_0^* - P_c)$, which goes to the U.S. government, for the privilege of selling in the U.S. market. Under these circumstances, the importer can only earn normal profit if the domestic price rises to P_0^*. At this price, $(Q_{dg} - Q_{sg})$ will be imported, as in the case of the quota. Greater imports would drive price below P_0^*. Smaller imports would allow excess profit to be earned from importing sugar for U.S. consumption. Tariff revenue, which could be used to lower taxes, equals $(P_0^* - P_c)(Q_{dg} - Q_{sg})$.

Turning to direct payments, we note that if the target price is set equal to P_0^*, with a direct payment to U.S. sugar producers equaling $(P_0^* - P_c)$, then the U.S. supply curve shifts from SS to the kinked curve SS_g in panel b, Fig. 10.1. Subtracting horizontally the new supply curve from U.S. demand yields the kinked demand for imports DD^* in panel a. The intersection of the demand DD^* and the supply for imports S' determines the equilibrium level of imports, which equals M_g'. By construction, M_g' equals Q_{dc} minus Q_{sg}. Importantly, the U.S. domestic market price remains equal to the world price P_c. The tax required to finance the program is given by $(P_0^+ - P_c)Q_{sg}$.

Turning to an analysis of welfare in the United States, we note that under an import quota U.S. consumers lose area $(a + b + c + d)$, and producers gain area (a); the efficiency loss equals area $(b + c + d)$. The consumer loss and producer gain are the same under a tariff, but a tariff lowers taxes by the area (c). Hence the efficiency loss falls to the area $(b + d)$. From the point of view of U.S. interests, a tariff is much to be preferred to an import quota.

Under direct payments, producers again gain area (a), but consumers lose nothing. The tax required to finance the payments equals area $(a + b)$. The efficiency loss equals area (b). Obviously, direct payments is the most efficient means for effecting a given transfer of benefits to farmers under the present assumptions.

Large-country Assumption

We now analyze the case of a large importer. Here, variation in imports causes noticeable variation in world prices. Many of the results derived in the case of the small importer carry over to this one, but there is at least one surprising difference. Tariffs and direct payments might generate efficiency gains to the importing country!

The U.S. domestic demand and supply are given by D and SS in panel b, Fig. 10.2. The competitive demand for imports is given by DD' in panel b. For a large importer, the competitive supply curve for imports, given by SS', is upward sloping. Competitive equilibrium is given by the intersection of the supply for imports SS' and the demand for imports DD'. The competitive world price equals P_c. At this price, U.S. farmers produce Q_{sc}, U.S. consumers buy Q_{dc}, and $(Q_{dc} - Q_{sc})$ is imported. Also, $(Q_{dc} - Q_{sc}) = M_c$, by construction.

If the U.S. government sets an import quota equal to $(Q_{dg} - Q_{sg})$, the U.S. price rises from P_c to P_0^*. The world price drops to P_g. Exporters will be eager to sell as

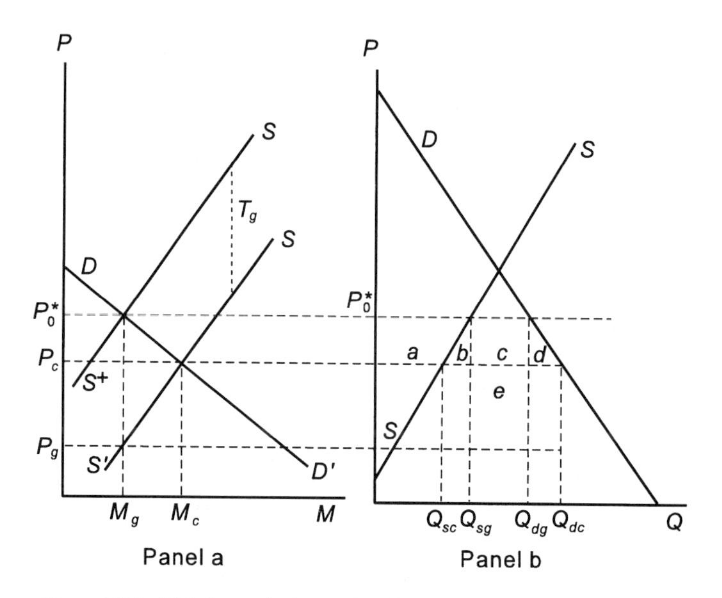

Figure 10.2 U.S. domestic demand and supply curves and the demand and supply curves for imports under the large-country assumption, with costs and benefits for import quotas and tariffs.

much as possible in the high-priced U.S. market, and we may again suppose that an exporter's share of the quota is, for reasons of equity, set equal to that country's share of competitive exports.

To analyze a tariff, we note that a per unit tariff shifts the supply for imports up in a parallel fashion. In panel a the per unit tariff is set at T_g such that the price of imports to U.S. consumers is driven up to the target price P_0^*. As in the case of the import quota analyzed previously, U.S. consumption falls from Q_{dc} to Q_{dg} and production increases from Q_{sc} to Q_{sg}. Imports drop to $(Q_{dg} - Q_{sg})$. Because the United States is a large buyer, the decreased imports drive the world price of sugar down from P_c to P_g.

Turning to benefit–cost analysis, we see that both the quota and the tariff generate a consumer loss in the United States equal to area $(a + b + c + d)$; U.S. farmers gain area (a). The efficiency loss to the United States caused by the quota equals area $(b + c + d)$.

To measure the efficiency loss in the case of the tariff, we need to calculate government revenue. Since the per unit tariff shifts the supply up linearly, we have in equilibrium $T_g = P_0^* - P_g$. Therefore, the tariff take equals $(P_0^* - P_g)(Q_{dg} - Q_{sg})$ or area $(c + e)$. To measure the efficiency loss, we have

$$\text{efficiency loss} = \text{area } (a + b + c + d) - \text{area } (a + c + e) = \text{area } (b + d - e) \qquad (10\text{-}1)$$

It is clear from the graph, however, that area (e) exceeds area $(b + d)$. Hence the efficiency loss indicated by Eq. (10-1) is negative, meaning that there is an efficiency gain in this particular instance. A gain is not the inevitable consequence of a tariff. Much depends on the shape of the import supply. The steeper it is, the more likely that a tariff results in an efficiency gain.

We expect that large importing nations would rush to impose tariffs were it not for the likelihood of retaliation. This explains why the nations of the world from time to time arrange international clambakes under GATT (General Agreement on Tariffs and Trade) to negotiate lower trade barriers. Life would be much simpler for the negotiators if countries relied on border measures only to protect their domestic industries, but many other approaches are possible, including direct payments. As we have already seen, a direct payment program is more efficient than tariffs as a means of elevating price to the target level where the importing nation is small. We now show this result might not hold if the importing nation is large.

Figure 10.3 has been drawn to facilitate a comparative analysis of tariffs and direct payments. Both programs raise the price to U.S. producers from the competitive price (not shown in the figure) to the target P_0^*. The analysis for tariffs is the same as before. Notice that the tariff T_g lowers imports to M_g and the world price falls to P_g. A direct payment program shifts the U.S. supply from SS to SS_g. Therefore, the demand for imports shifts from DD' to DD^*. The world and U.S. consumer price is lowered to P_g^*.

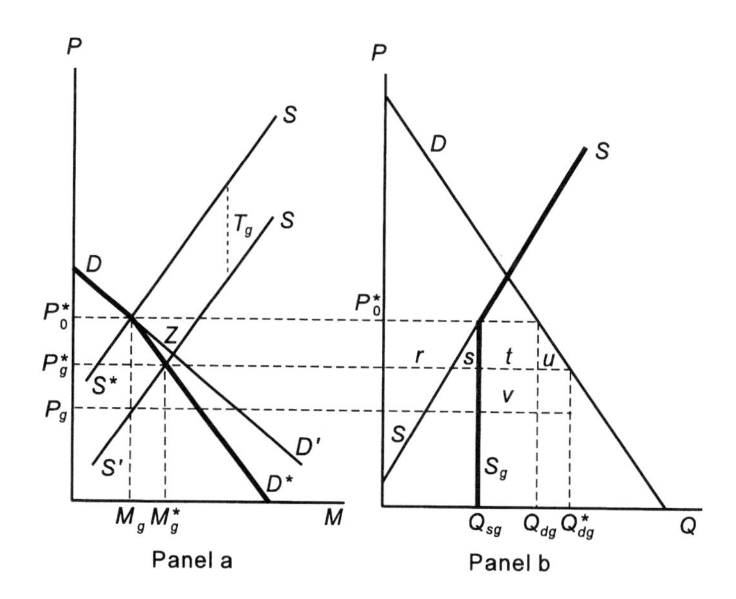

Figure 10.3 U.S. domestic demand and supply curves and the demand and supply curves for imports under the large-country assumption, with costs and benefits for tariffs and direct payments.

Turning to welfare analysis, the gain to U.S. producers is the same for either program. Under direct payments, the consumers gain area $(r + s + t + u)$ relative to the outcome under tariffs. Whereas the tariff generates government revenue equal to area $(t + v)$, direct payments cost the taxpayer area $(r + s)$. To measure the efficiency gain from direct payments *relative to the tariff*, we have

$$\text{efficiency gain} = \text{area } (r + s + t + u) - \text{area } (r + s + t + v) = \text{area } (u - v) \tag{10-2}$$

Figure 10.3 is drawn to assure that area (v) exceeds area (u). Therefore, a direct payment program is less efficient in this particular instance than is a tariff in generating the same level of benefits to U.S. producers. Since we already know that direct payments are more efficient than a tariff if SS' is perfectly flat, as in the case of the small-country assumption, we are led to believe that whether a tariff is more efficient than direct payments as a device for supporting U.S. producers turns on the steepness of the supply for imports. And this is exactly the case. To see this, consider a clockwise rotation of the SS' curve in panel a through the point Z, all the time keeping the target price at P_0^*. As SS' approaches perfect elasticity, area (u) shrinks somewhat, but area (v) vanishes. Thus, if SS' is sufficiently flat, a direct payment program is more efficient than is a tariff in generating benefits to the U.S. producers.

10.2 GOVERNMENT-SPONSORED CARTELS

Because price support programs often lead to government-owned surpluses, with all the problems that they entail, governments often link production controls to price supports. One approach to production control is a government-sponsored cartel. Important examples include the U.S. programs for tobacco and peanuts. Although considered in Chapter 8, a government cartel is now analyzed in more detail.

The government embarks on a program to control production through issuing certificates that give farmers the right to sell output. It is made illegal for the farmer to sell a unit of output, a bushel, say, without surrendering a certificate. Like money, certificates are issued in various denominations. They are distributed in an equitable manner as follows: Each farmer is assigned a production base q_c that equals his or her level of output in competitive equilibrium prior to the initiation of the program. The farmer's share of total competitive output, q_c/Q_c, determines his or her share of allowable production and his share of issued certificates. Take an example. In competitive equilibrium, a farmer produces 40,000 bushels of corn. This sets the farmer's base. The government decides to limit production to 80 percent of the competitive level. The farmer therefore receives 32,000 certificates. The base is an asset that entitles the farmer to a flow of certificates in future years. In the short run, we assume that the distribution of production bases and certificates among farmers is fixed. In the long run, we allow bases and certificates to be bought and sold. In the long run, buying certificates is akin to renting base; buying base amounts to the purchase of an asset.

Short-run Analysis

A production control program along the lines spelled out certainly can be used to raise quasi-rents of farmers in the short run. Suppose, for example, that demand is inelastic in the relevant range. Total revenue rises with decreased production at the same time that aggregate variable costs fall.

To model the program's effects more carefully, let λ be the proportion of base production each farmer is allowed to market, where $0 < \lambda \leq 1$. Letting q_i equal the allowable quantity of production (marketings) and q_{ci} equal base or competitive production, both for the ith farmer, we have $q_i = \lambda q_{ci}$, $i = 1, 2, \ldots, N$. Aggregate quasi-rent QR to the industry is given by

$$QR = P \sum_i q_i - \sum_i C_i(q_i) \tag{10-3}$$

where $C_i(q_i)$ is the ith farmer's total variable cost function. Letting the inverse output demand be $P = D(Q)$, we have

$$QR = D(Q)Q - \sum_i C_i(q_i)$$
$$= D(\lambda Q_c)\lambda Q_c - \sum_i C_i(\lambda q_{ci}) \quad (10\text{-}4)$$

where $Q = \sum_i q_i$ and $Q_c = \sum_i q_{ci}$.

Clearly, the government's decision or control variable in this program is λ. Of interest are the effects of a program in which the government's objective is to maximize QR through the optimal choice of λ. The first-order condition for a maximum of QR is

$$\frac{dQR}{d\lambda} = Q_c\left[\frac{dP}{dQ}Q + P\right] - \sum_i \frac{dc_i}{dq_i} q_{ci} = 0 \quad (10\text{-}5)$$

The first-order condition implies that

$$MR = \sum_i MC_i \frac{q_{ci}}{Q_c} \quad (10\text{-}6)$$

where MR equals industry marginal revenue and MC_i equals the marginal cost of the ith producer.[1] The optimum aggregate output can be found by equating the marginal revenue with the weighted sum of the marginal costs of farmers, where the ith weight equals the ith farmer's share of aggregate competitive output. Letting Q_g equal the optimum output, the government sets λ equal to Q_g/Q_c. This fixes the level of aggregate output together with its distribution among the N farmers.

If we assume that all marginal cost functions are identical, then $q_{ci} = q_c$ and $MC_i = MC$ for all i and, since $Q_c = Nq_c$, Eq. (10-6) implies that $MR = MC$. Here we need merely sum horizontally the MC curves of all N farmers and locate the point where the resulting aggregate MC curve intersects the industry marginal revenue curve. This solution is the same as that for a multiple-plant monopolist when the plants are identical. (We ignore the special case when some farms or plants must be shut down.) Under the assumption of identical costs, market performance under a government cartel dedicated to the maximization of aggregate quasi-rents is identical to that for a profit-maximizing monopolist.

The preceding short-run analysis may be modified in a significant way by supposing that certificates may be bought and sold. With negotiability, the production of whatever level of output is permitted will be distributed among farmers in a manner that equates the marginal costs of all farmers. The easiest way to see this is to consider an example. Suppose that Farmer Smith's and Farmer Big's marginal costs

[1]The second-order condition for a maximum is satisfied if the marginal revenue curve for the industry is downward sloping and the marginal cost curves of farmers are all upward sloping.

equal, respectively, 20 and 15 prior to the exchange of certificates. The value of Farmer Big's right to sell a unit of output exceeds that for Farmer Smith by 5. Hence Farmer Big buys the right to sell (a certificate) from Farmer Smith. As production shifts from Farmer Smith to Farmer Big, the former's marginal cost declines, while the latter's marginal cost increases. The trading of production rights (certificates) will continue until marginal costs are equated. This assumes that a perfect market develops for the buying and selling of certificates, an assumption that seems reasonable with a large number of relatively small farmers.

With negotiability of certificates, the government can engineer the perfect cartel solution even with nonidentical marginal cost curves of producers. Imagine a diagram in which the marginal cost curves of all farmers are summed horizontally. Let the level of output associated with the equating of marginal revenue and aggregate marginal cost be given by Q_m. The government sets λ equal to the ratio Q_m/Q_c. The buying and selling of certificates lead to the equation of the marginal costs of production among farmers. By setting λ equal to the ratio Q_m/Q_c and through allowing the buying and selling of certificates, the government is able to maximize the aggregate quasi-rents of farmers, but the benefits to farmers are distributed in a particular manner. The greater q_{ci}/Q_c is, for example, the greater will be the ith farmer's share of the production rights.

Clearly, the government need not operate its cartel in such a manner as to maximize aggregate short-run quasi-rents. Notice that quasi-rents can be elevated to whatever level is desired, subject, of course, to the maximum level by choosing λ such that $Q_m \le Q \le Q_c$. The extent to which the government desires to create short-run benefits to farmers at the expense of consumers is a policy decision involving value judgments.[2]

Long-run Analysis

In long-run analysis, we seek first to model the pricing of certificates on the assumption that all production bases are owned by nonfarmers. To farm, the family must buy certificates (i.e., rent base). (We will show later that this assumption is not as restrictive as it might at first seem.) We let r equal the level of certificates purchased by the representative farmer and P_r equal the price per certificate.

The farmer's profit function in the long run is

$$\pi = P_q q - C(q) - TE - P_r r \tag{10-7}$$

where $C(q)$ is the production cost function and, as before, TE equals the family's transfer earnings. Because $q = r$, we have

[2]In the present context, a value judgment is a statement that expresses a preference for one set of market results over another. Rational policy choices presumably involve choosing the policy alternative that delivers the most preferred market results.

$$\pi = (P - P_r)q - C(q) - TE \tag{10-8}$$

where $(P - P_r)$ equals the real or net price received per unit of allowable output. Profit maximization implies that $(P - P_r)$ equals the marginal cost of production, given by $C'(q)$, subject to the proviso that $(P - P_r)$ be not less than the average total cost of production AC. (To assure a maximum, $C''(q) > 0$.)

Entry and exit, on the other hand, imply that, for the marginal family farm, $\pi = 0$. This implies that

$$P - P_r = AC \tag{10-9}$$

Hence $P_r = P - AC$. The price per certificate equals the difference between output price and the average cost of production for the marginal farm.

Armed with these results, we turn to a graphic analysis with the aid of Fig. 10.4. The demand and long-run competitive supply for output are given by D and S. The marginal revenue curve is given by MR. Let $P = S(Q)$ be the supply function. Since long-run supply shows the minimum average cost of production for various levels of output, we have $AC = S(Q)$. The total industry cost of output TC is therefore given by

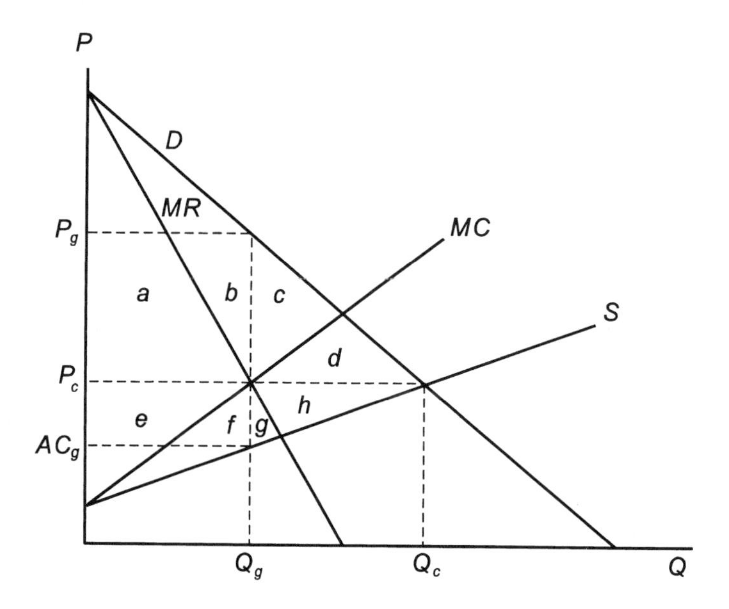

Figure 10.4 Demand and long-run supply curve and industry marginal revenue and marginal cost curves, with costs and benefits for a government cartel.

$$TC = S(Q)Q$$
$$= C(Q) \tag{10-10}$$

The industry marginal cost of production is given by $C'(Q)$. A graphic expression for industry marginal cost is given by MC in Fig. 10.4. Now assume that the government's long-run objective is to maximize what we will call the cartel residual, equaling the difference between industry total revenue and industry total cost of production, excluding the outlay on certificates. The cartel residual is maximized by equating marginal revenue and marginal cost in the spirit of a monopolist, setting output equal to Q_g in Fig. 10.4. With $Q = Q_g$, $P = P_g$, and $AC = AC_g$. The price of certificates equals $P_g - AC_g$.

A benefit–cost analysis points up the myopia behind this approach to farm policy. Relative to competitive equilibrium, consumers lose area $(a + b + c + d)$. Long-run producer surplus falls by area $(e + f + g + h)$. The owners of production bases (i.e., the long-run rights to produce) receive area $(a + b + e + f)$, but, of course, they must pay for this annual source of rent. In a perfect market for assets, the initial owners of the production rights, received gratis from the government, enjoy a capital gain CG as follows:

$$CG = \frac{\left(P_g - AC_g\right)Q_g}{i} \tag{10-11}$$

Subsequent owners get what they pay for and receive no surplus or benefit. Thus the consumers and farm input suppliers are all made worse off in the long run so that initial farmers may be given the ownership of a monopoly.

Our analysis is based on the assumption that the government wishes to maximize the cartel residual. This need not be the case, but no matter. Whatever feasible level of cartel residual the government seeks to generate can be analyzed using Fig. 10.4. The closer Q_g is to the competitive output Q_c, the lower the residual.

Finally, our analysis assumes that all farmers rent base (i.e., buy certificates). The impact of the program on the farmer who buys base, thus receiving the certificates directly from the government, can be analyzed in two stages. As a producer, the farmer pays $(P_g - AC_g)$ per certificate. As an asset owner, he receives from himself $(P_g - AC_g)$ per certificate. Through inheritance, of course, future generations of farmers could benefit from the capital gains that their ancestors received free of charge from the government in some earlier epoch.

10.3 THE INDIRECT PROFIT FUNCTION

A voluntary program that offers farmers inducements of various kinds to idle farmland has been the dominant approach to farm policy in the United States since the 1960s. An unusual challenge to the theorist grows out of the need to model the farmer's decision as to whether to participate in such programs. As it turns out, the concept of the indirect profit function is exceedingly useful in this endeavor and in showing that voluntary land diversion programs can have profound effects on agricultural pricing.

The firm's ordinary or direct profit function under conditions of certainty is

$$\pi = Pq - C(q) \tag{10-12}$$

where $C(q)$ is here defined as the total cost function. (We ignore transfer earnings for the present.) Assuming perfect competition, price P is a constant. Our objective is to explain (model) how the firm chooses values for output q and profit π. Viewed from this perspective, Eq. (10-12) has two unknowns, and another equation is required to complete the model. The standard way of proceeding is, of course, to assume that the entrepreneur desires to maximize profit. This implies that q is chosen in such a manner as to equate price and marginal cost subject to the constraint that marginal cost is not less than average cost. Mathematically, profit maximization implies that

$$P = \frac{dC}{dq} \tag{10-13}$$

Marginal cost is given by dC/dq, which is a function of q. Solving Eq. (10-13) for q as a function of P yields the supply

$$q = s(P) \tag{10-14}$$

Equations (10-12) and (10-14) constitute a complete structural model. Equation (10-14), as it happens, is the reduced form for q. Substituting $s(P)$ for q in Eq. (10-12) yields the reduced form for π:

$$\begin{aligned} \pi &= Ps(P) - C[s(P)] \\ &= \pi^*(P) \end{aligned} \tag{10-15}$$

Equation (10-15) is called the *indirect profit function*. The function $\pi^*(P)$ shows the maximum profit associated with various levels of price. Taking the first derivative of Eq. (10-15) with respect to price, we have

$$\begin{aligned} \frac{d\pi}{dP} &= s(P) + \frac{ds}{dP}P - \frac{dC}{dq}\frac{ds}{dP} \\ &= s(P) + \frac{ds}{dP}\left(P - \frac{dC}{dq}\right) \end{aligned} \tag{10-16}$$

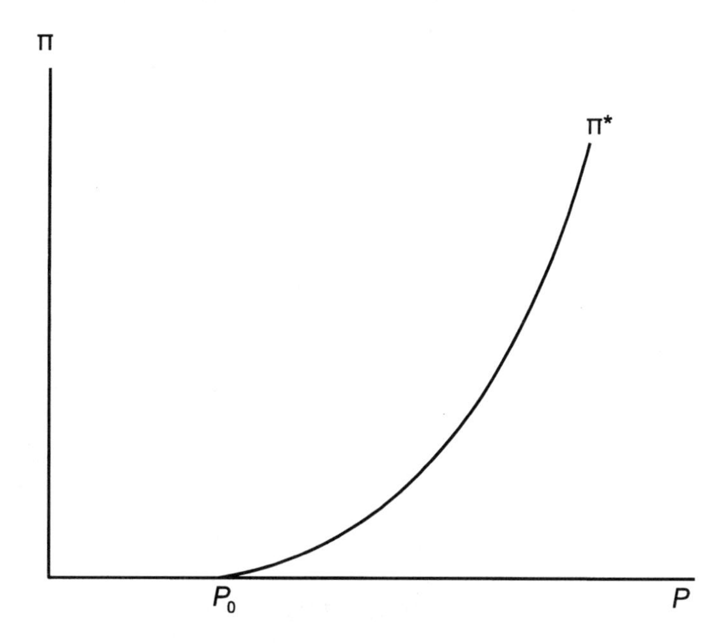

Figure 10.5 Indirect profit curve.

But $P = dC/dq$. Thus[3]

$$\frac{d\pi}{dP} = s(P) = q \tag{10-17}$$

The first derivative of $\pi^*(P)$ with respect to price yields the firm's supply function. Furthermore, since the second derivative of $\pi^*(P)$ with respect to price is positive, the firm's supply curve slopes upward.

The π^* curve in Fig. 10.5 is a graphic representation of the indirect profit function. This figure reflects the assumption that the minimum average cost of production equals price P_0. For prices less than P_0, the firm ceases all operations. For prices larger than or equal to P_0, production takes place, and profit rises at an increasing rate with increases in price.

Consider an example:

$$\pi = Pq - q^2 \tag{10-18}$$

[3]Equation (10-17) is called Hotelling's lemma in the economics literature. It is a special case obtained from applying the envelope theorem to the indirect profit function $\pi^*(P)$. See Appendix A.

The firm's supply function is $q = P/2$. Substituting $(P/2)$ for q in Eq. (10-18) yields

$$\begin{aligned} \pi &= \frac{P^2}{2} - \frac{P^2}{4} \\ &= \frac{P^2}{4} \end{aligned} \tag{10-19}$$

Equation (10-19) is the indirect profit function. Notice that $d\pi/dP = P/2$. The equation for supply is thus recovered. Also, $(d^2\pi/dP^2) > 0$. Profit rises with P at an increasing rate.

This analysis can be easily modified to take account of price uncertainty on the assumption of risk neutrality. Simply replace π with $E(\pi)$ and P with $E(P)$. A graphic representation of the indirect expected profit function would be the same as that given in Fig. 10.5. Only the definitions of the variables measured along the axes would need to be changed.

10.4 A VOLUNTARY LAND DIVERSION PROGRAM

Consider the following land diversion program designed by a government to elevate farm prices and incomes. The farmer is offered the option of idling or diverting a fixed proportion of his or her land in return for government payments. We assume that the farmer's land is part of the fixed plant. Let the proportion of land that must be idled be given by $(1 - \lambda)$ where $0 < \lambda \leq 1$. (Notice that λ is not defined here as it was in Section 10.3.) If the farmer joins or participates in the program, then λa_t is planted to a crop; $(1 - \lambda)a_t$ stands idle.

Obviously, no farmer would idle land voluntarily without an inducement of some kind. The inducement in the program at hand takes the form of a deficiency payment calculated as the product of four terms: $(P^* - P_{t+1})y^*\lambda a_t$. The payment per unit of output produced, bushel, say, is given by $(P^* - P_{t+1})$, where P^* is a target price set by the government in period t, as in the case of direct payments analyzed in Chapter 9, and where P_{t+1} equals the price in period $t + 1$ unknown in period t. In the event $P_{t+1} \geq P^*$, no payment is made. The term y^* is the program yield based on the farmer's historical yields. We may simply think of y^* as equaling the farmer's yield under perfect competition before the program is put into effect. As already explained, the product of the last two terms, λa_t, equals the acreage that the farmer is permitted to plant under the program.

A sufficiently low expected market price $E_t(P_{t+1}) = \Theta_p$ would encourage the farmer to join the program, idling land and limiting output. The opportunity cost of idling land is low when market prices are low. Contrariwise, a high expected price discourages program participation. The opportunity cost of idling land is high when market prices are high. What will become evident in the analysis that follows is that the indirect expected profit function is of great use in the derivation of a supply function that shows how much the farmer will produce at alternative expected market prices, given the option of participating in a land diversion program. The resulting supply function is the key concept in the analysis that follows.

Short-run Analysis for an Individual Farmer

If the farmer chooses not to join the program, the profit function is given by

$$\pi_{t+1,n} = P_{t+1}q_{t+1,n} - C_n(q_{t+1,n}) - TFC \qquad (10\text{-}20)$$

where we have added the subscript n to indicate that the farmer is *not* a program participant. Recall that this profit function (minus the n subscript) was analyzed at length in Chapter 2. Throughout this chapter we assume price uncertainty, but abstract from production uncertainty.

Letting Θ_n equal the expected quasi-rent for the nonparticipating farmer, we have

$$\Theta_n = \Theta_p q_n - C_n(q_n) \qquad (10\text{-}21)$$

The $t + 1$ subscripts are dropped to simplify notation. The total variable cost function for the nonparticipant, who is free to plant all of his or her land, is now written as $C_n = C_n(q_n)$.

If the farmer chooses to join the program, the profit function is given by

$$\pi_{t+1,j} = P_{t+1}q_{t+1,j} - C_j(q_{t+1,j}) - TFC + (P^* - P_{t+1})y^*\lambda a_{tj} \qquad (10\text{-}22)$$

where the subscript j is added to indicate program participation. Letting Θ_j equal the expected quasi-rent for the participating farmer, the expected quasi-rent function is given by

$$\Theta_j = \Theta_P q_j - C_j(q_j) + (P^* - \Theta_P)y^*\lambda a_j \qquad (10\text{-}23)$$

The $t + 1$ subscripts are again dropped to simplify notation. The total variable cost function for the participating farmer is given by $C_j = C_j(q_j)$. Because land is idled we assume that, for any given quantity of output, $C_j > C_n$. It costs more to produce a given level of output if some of the land must be idled.

The farmer is assumed to maximize expected profit. The maximization problem may be solved in a three-stage procedure. The farmer first maximizes Θ_n using Eq. (10-21). Let Θ_n^* be the resulting maximized value. Similarly, in the second stage, the farmer finds Θ_j^* by maximizing Θ_j using Eq. (10-23). In the third stage, the farmer decides whether to join the program. If $\Theta_n^* < \Theta_j^*$, the farmer joins. If $\Theta_n^* > \Theta_j^*$, the farmer does not join. If $\Theta_n^* = \Theta_j^*$, then the farmer is indifferent between being in or out of the program, and it is not possible to predict which choice will be made without further information.

Maximizing Θ_n in the first stage, using Eq. (10-21), implies that $\Theta_P = dC_n/dq_n$; the expected price is equated to the marginal cost of production. Solving this equation for q_n as a function of Θ_P yields the nonparticipating farmer's supply function:

$$q_n = s_n(\Theta_P) \tag{10-24}$$

The graphic representation of this supply function is given by the curve labeled ss_c in Fig. 10.6. The indirect expected quasi-rent function is given by

$$\begin{aligned}\Theta_n &= \Theta_P s_n(\Theta_P) - [C_n(s_n(\Theta_P)] \\ &= \Theta_n^*(\Theta_P)\end{aligned} \tag{10-25}$$

A graphic representation of the nonparticipant's indirect quasi-rent function is given by Θ_n^* in Fig. 10.7. From Section 10.3 we know that $d\Theta_n^*/d\Theta_P$ yields the optimum level of output q_n. Therefore, for any Θ_P, the slope of $\Theta_n^*(\Theta_P)$ in Fig. 10.7 gives the nonparticipant's optimum output. By implication, the minimum of the farmer's average variable cost curve equals Θ_{P0}; if $\Theta_P < \Theta_{P0}$, the farmer shuts down operations and does not equate marginal cost to Θ_P.

To maximize Θ_j in the second stage, we rewrite Eq. (10-23) thus:

$$\Theta_j = \Theta_m + \Theta_f \tag{10-26}$$

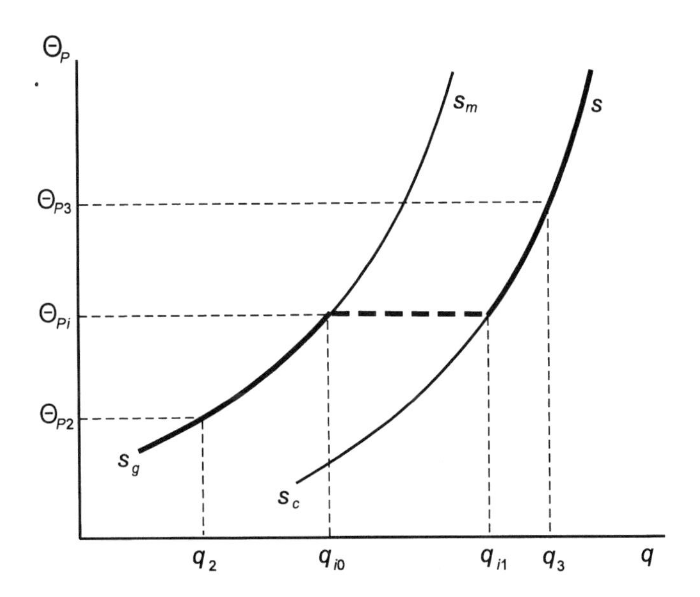

Figure 10.6 Farmer's supply curve, given the option of joining a land diversion program, with the indifference expected price.

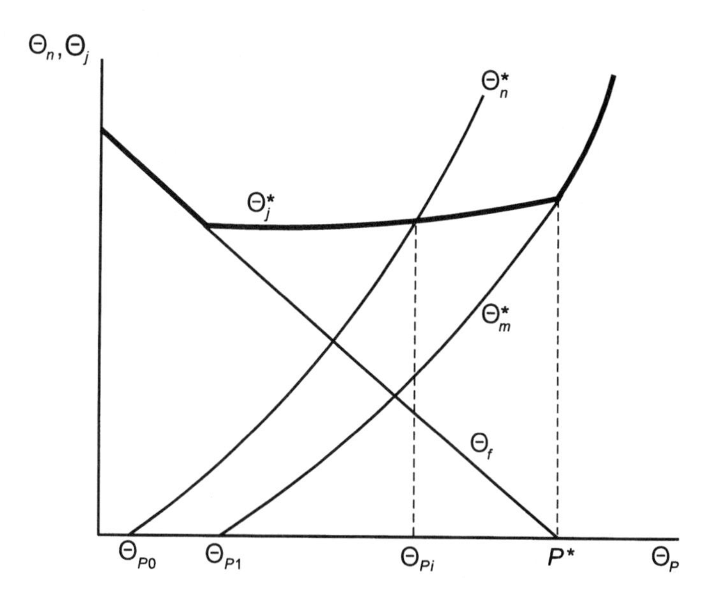

Figure 10.7 Farmer's expected indirect quasi-rent curves, given the option of joining a land diversion program, with the indifference expected price.

where $\Theta_m = \Theta_P q_j - C_j(q_j)$ and $\Theta_f = (P^* - \Theta_P)y^*\lambda a_j$. This decomposition reflects the two sources of quasi-rent for the participating farmer, the sale of q_j in the private market and the check from the government. Notice that Θ_f is a downward sloping linear function of Θ_P as in Fig. 10.7. With a_j fixed, the participating farmer's choice of q_j does not influence Θ_f. The same, of course, does not apply to Θ_m. Maximizing Θ_m implies that $\Theta_P = dC_j/dq_j$. The expected price is equated to the marginal cost. Solving this equation for q_j as a function of Θ_P yields the supply function for the participating farmer:

$$q_j = s_j(\Theta_P) \tag{10-27}$$

The graphic representation of this supply function is given by the curve labeled $s_m s_g$ in Fig. 10.6.

The indirect expected quasi-rent function for the participating farmer is given by

$$\begin{aligned} \Theta_j &= \Theta_P s_j(\Theta_P) - C_j[s_j(\Theta_P)] + (P^* - \Theta_P)y^*\lambda a_j \\ &= \Theta_m^*(\Theta_P) + \Theta_f \qquad (10\text{-}28) \\ &= \Theta_j^*(\Theta_P) \end{aligned}$$

The graphs of $\Theta_m^*(\Theta_P)$ and Θ_f are given in Fig. 10.7. Summing vertically the Θ_m^* curve and the Θ_f curve yields the darkened Θ_j^* curve, which is the graphic representation of $\Theta_j^*(\Theta_P)$. For any given expected price Θ_P, the participating farmer produces less output than does the nonparticipating farmer. For any Θ_P, at least beyond Θ_{P1}, the slope of the Θ_m^* curve (and of the Θ_j^* curve as well) is less than that for the Θ_n^* curve. The minimum of the participating farmer's average variable cost curve equals Θ_{P1}; for expected prices less than Θ_{P1}, the participating farmer produces nothing.

Armed with Figs. 10.6 and 10.7, we are prepared to take up the third and final stage of the maximization problem, deriving in the process the farmer's output supply function. For all expected prices less than Θ_{Pi}, the Θ_j^* curve lies above the Θ_n^* curve (Fig. 10.7). For any such expected price (for $\Theta_P < \Theta_{Pi}$), the farmer maximizes expected income by joining the program. For increases in expected prices in the range from Θ_{P1} to Θ_{Pi}, the farmer's output increases along the $s_m s_g$ curve in Fig. 10.6. If $\Theta_P = \Theta_{P2}$, for example, the optimum output equals q_2. For all expected prices above Θ_{Pi}, the Θ_n^* curve lies above the Θ_j^* curve (Fig. 10.7). The farmer maximizes expected profit by not joining the program. For increases in expected prices in the range of expected prices above Θ_{Pi}, the farmer's output increases along the ss_c curve in Fig. 10.6. If $\Theta_P = \Theta_{P3}$, for example, the optimum output equals q_3. At the expected price Θ_{Pi}, the Θ_j^* and Θ_n^* curves intersect. Expected profit is the same whether or not the farmer joins the program. At $\Theta_P = \Theta_{pi}$, program participation implies that output equals q_{i0}; nonparticipation implies that output equals q_{i1}. It is not possible to predict which output the farmer will choose without further information. The expected price Θ_{Pi} is called the *indifference expected price* because the farmer is indifferent between program participation and nonparticipation. More formally, the indifference expected price is defined implicitly by

$$\Theta_n^*(\Theta_{Pi}) = \Theta_j^*(\Theta_{Pi}) \tag{10-29}$$

As noted, the farmer's optimum output is indeterminate at the expected price Θ_{Pi}.

In conclusion, the farmer's supply curve is given by the discontinuous darkened curve labeled ss_g in Fig. 10.6, consisting of $s_m s_g$ for expected prices less than Θ_{Pi} and of ss_c for expected prices in excess of Θ_{Pi}. In the absence of the program, the farmer's supply curve would be the competitive supply given by ss_c. One effect of the land diversion program is now apparent. For all $\Theta_P < \Theta_{Pi}$, the program reduces the farmer's output.

Short-run Aggregative Analysis

Having derived the supply curve for a representative farmer, we may now derive the aggregate supply curve for all farmers on the assumption that the indifference expected prices vary among farmers. We might expect that the indifference expected prices for well-managed farms are less than for poorly managed farms. In any event, we sum horizontally all the individual farmer supply curves to obtain the aggregate supply curve SS_g in Fig. 10.8. The $S_m S_g$ curve is the aggregate supply assuming that

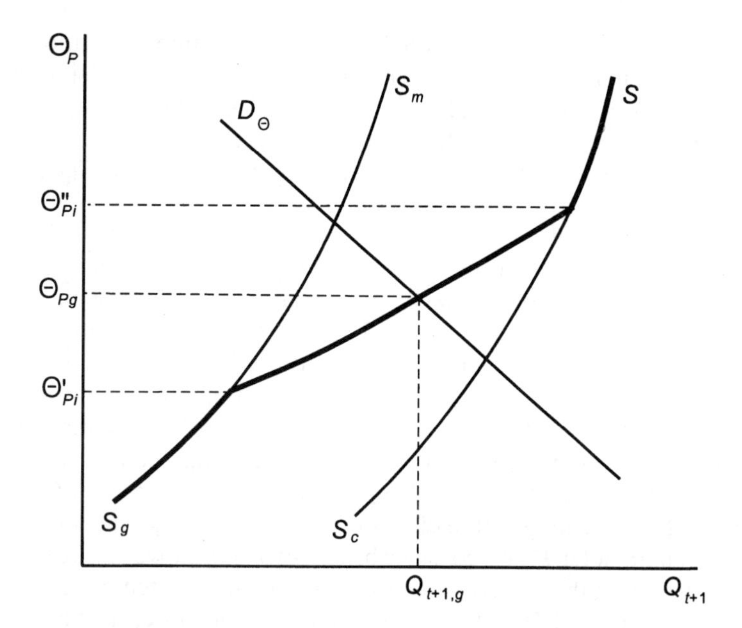

Figure 10.8 Aggregate expected demand curve and aggregate supply curve, given the option of joining a land diversion program, showing equilibrium output and expected price.

all farmers participate. The SS_c curve is the aggregate supply assuming that no farmer participates. (The latter curve is also the competitive supply.) The lowest indifference expected price, associated with the most efficient operation, is given by Θ'_{Pi}. The highest is given by Θ''_{Pi}, perhaps associated with the least efficient farmer. For all $\Theta_P < \Theta'_{Pi}$, the aggregate supply curve reflects 100 percent participation in the program. For all $\Theta_P > \Theta''_{Pi}$, no farmer participates. The flatness or relatively large elasticity of the scalloped line connecting S_mS_g with SS_c reflects two forces. As the expected price rises in the range from Θ'_{Pi} to Θ''_{Pi} every farmer intensifies production. In addition, some farmers exit the program, putting additional acres to work in the process. The aggregate supply curve, although kinked, is drawn as a continuous curve on the assumption of a large number of relatively small farm producers.

Up to this point the expected price has been treated as an exogenous variable. We may now write out the complete market model with the expected price made endogenous, as follows:

$$
\begin{aligned}
P_{t+1} &= \beta_0 - \beta_1 Q_{t+1} + e_{t+1} && \text{demand} \\
Q_{t+1} &= S(\Theta_P) && \text{supply} \qquad (10\text{-}30) \\
\Theta_P &= \beta_0 - \beta_1 Q_{t+1} && \text{expected price function}
\end{aligned}
$$

The demand is stochastic with $E_t(e_{t+1}) = 0$. The supply equation corresponds to the SS_g curve in Fig. 10.8. In period t, output for period $t + 1$ is determined by the last two equations of Eq. (10-30). In period $t + 1$, price is determined by the predetermined level of output and the value assumed by e_{t+1}.

Market equilibrium can be derived graphically using Fig. 10.9. The graph of the expected price function is given by D_Θ. Its intersection with SS_g gives the market equilibrium levels of output, $Q_{t+1,g}$, and expected price, Θ_{Pg}. Given the assumed configuration of the expected price and supply curves, all those farmers with indifference expected prices below Θ_{Pg} do not participate. Those with indifference expected prices above Θ_{Pg} do participate. Under perfectly competitive conditions, the equilibrium levels of output and expected price would be given by the intersection of the D_Θ and SS_c curves. Thus it is clearly the case that the land diversion program lowers output and raises the expected market price, which is, of course, what the program is intended to do. Equilibrium price $P_{t+1,g}$ is determined by the intersection of actual demand D in period $t + 1$ with a perfectly inelastic supply at $Q_{t+1,g}$.

Before taking up the effects of exogenous shocks, including changes in the program parameter P^*, we consider briefly two alternative configurations of the D_Θ and SS_g curves. If the D_Θ curve intersects SS_g at an expected price less than Θ'_{Pi}, all farmers participate. If D_Θ intersects SS_g at an expected price above Θ''_{Pi}, no farmer participates. In the latter case, the program has no effect. If the target price P^* is too low and the land idling requirement $(1 - \lambda)$ is too high, farmers will simply ignore the program.

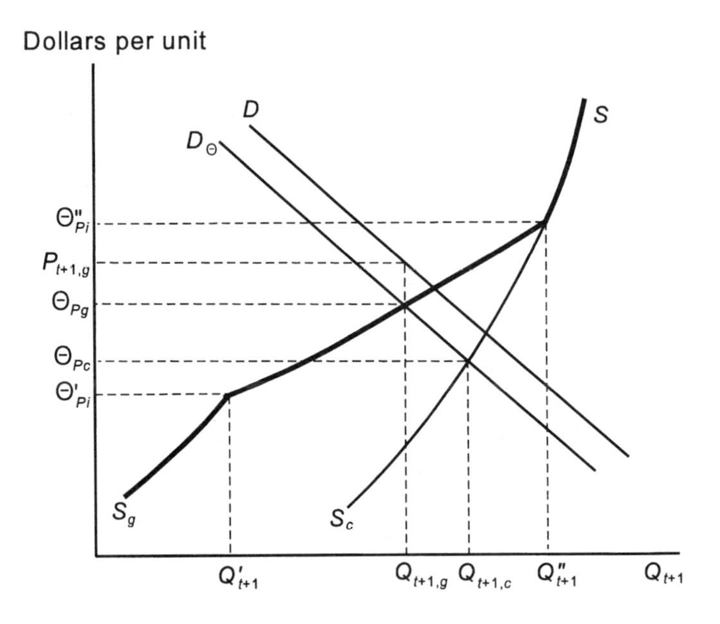

Figure 10.9 Aggregate expected and actual demand curves and aggregate supply curve, given the option of joining a land diversion program, showing equilibrium output, expected price, and actual price.

As always, the usefulness of models such as that given by Eq. (10-30) or Fig. 10.9 depends on what can be learned from them as regards the effects of exogenous shocks. Three examples are discussed. First, a ceteris paribus expansion of demand, holding the probability distribution of e_{t+1} constant, increases output and expected price. Program participation falls and program effects are diminished.

Second, a lowering of input prices shifts the Θ_m^* and Θ_n^* curves in Fig. 10.7 to the left and upward. The indifference expected price falls. Both the S_mS_g and SS_c curves in Fig. 10.8 shift to the right and downward. With lower indifference expected prices as well, the SS_g curve also shifts to the right and downward. Thus lower input prices mean increased output and a lower expected price.

Third, how does market performance respond to changes in program parameters? We limit ourselves to changes in the target price P^*. [The analysis of changes in the proportion of land that must be idled, $(1 - \lambda)$, is complex and inconclusive.] An increase in P^* shifts the Θ_f curve in Fig. 10.7 upward and to the right in a parallel fashion. This elevates the indifference price without affecting the Θ_n^* and Θ_m^* curves. This means that the scalloped line connecting the S_mS_g and SS_c curves in Fig. 10.8 rises. With the expected price function given by D_Θ, the effect is to increase the rate of program participation, lower output, and increase the expected price.

Numerical Example

We now consider an important numerical example of a special case when all farmers have identical operations and therefore the same indifference expected price. Under competitive conditions, the expected profit function for the representative farmer is given by

$$\Theta = \Theta_P q_{t+1} - q_{t+1}^2 - TFC \tag{10-31}$$

where $C = q_{t+1}^2$ is the total variable cost function. Equating expected price Θ_P and marginal cost $= 2q_{t+1}$ yields the farmer's supply function $q_{t+1} = \Theta_P/2$. The indirect expected profit function is $\Theta = \Theta_P^2/4 - TFC$. Assuming that there are 1000 identical farms, the aggregate supply is given by $Q = 500\ \Theta_P$. Let the expected price function (in this simple case, expected demand) be given by $Q_{t+1} = 10{,}000 - 500\Theta_P$. In competitive equilibrium, $Q_{t+1} = 5000$, $\Theta_P = 10$, and $q_{t+1} = 5$. Assume that each farmer has 5 acres of land planted in period t with yield $y_{t+1} = 1$.

Suppose that the government introduces a land diversion program with $\lambda = 0.8$ and $P^* = 18$. The established yield is one unit per acre for each farmer. The indirect expected quasi-rent function for the nonparticipating farmer is given by

$$\Theta_n = \Theta_P q_n - q_n^2 \tag{10-32}$$

Again we drop the $t + 1$ subscripts and use the n subscript to indicate nonparticipation. The nonparticipating farmer's total variable cost function is identical to that under competition. We have $C_n = q_n^2$. The nonparticipating farmer's supply function and indirect expected quasi-rent function are given by

$$q_n = \frac{\Theta_P}{2} \tag{10-33}$$

$$\Theta_n = \frac{\Theta_P^2}{4} \tag{10-34}$$

If all farmers refused to join the program, the aggregate supply function would be given by $Q_n = 500\Theta_P$. This, of course, is the same as the competitive supply function, and its graph corresponds to the SS_c curve in Fig. 10.8.

Now suppose that all farmers join the program. The expected quasi-rent function for the representative farmer is given by

$$\Theta_j = \Theta_P q_j - 2q_j^2 + (18 - \Theta_P)(1)(4) \tag{10-35}$$

where $C = 2q_j^2$ is the participant's total variable cost function. The participating farmer's supply function (which can be found by equating the expected price Θ_P to the marginal cost $= 4q_j$) and expected quasi-rent function are

$$q_j = \frac{\Theta_P}{4} \tag{10-36}$$

$$\Theta_j = \frac{\Theta_P^2 - 32\Theta_P + 576}{8} \tag{10-37}$$

It may be noted that $\Theta_m = \Theta_p^2/8$. The graph of $\Theta_p^2/8$ passes through the origin, but is otherwise similar to the Θ_m^* curve in Fig. 10.7. Also, $\Theta_f = 72 - 4\Theta_P$, which corresponds to the Θ_f curve in Fig. 10.7. If all farmers joined the program, the aggregate supply function would be given by $Q_j = 250\Theta_P$. This supply function corresponds to the supply curve $S_m S_g$ in Fig. 10.8.

The indifference expected price can be found by solving the following quadratic equation for Θ_{Pi}:

$$\frac{\Theta_{Pi}^2}{4} = \frac{\Theta_{Pi}^2 - 32\Theta_{Pi} + 576}{8} \tag{10-38}$$

Solving Eq. (10-38) for Θ_{Pi} yields $\Theta_{Pi} = 12.8444$. If the expected price equals 12.8444, then $\Theta_n = \Theta_j = 41.24$, and $q_n = 6.42$, and $q_j = 3.21$.

To find market equilibrium with the farm program in effect, we resort to trial and error. With zero participation, the expected price is the same as under competition, which equals 10. But $\Theta_{Pi} = 12.844$. At an expected price equal to 10, all farmers would want to participate. Therefore, in equilibrium, there must be at least some program participation.

Next suppose that participation is 100 percent. The expected price can be found by solving

$$Q_{t+1} = 10{,}000 - 500\Theta_P$$
$$Q_j = 250\Theta_P \tag{10-39}$$
$$Q_{t+1} = Q_j$$

We find that Θ_P equals 13.333, which, of course, exceeds Θ_{Pi}. At this high price, no farmer would choose to participate. It must be true, then, that if there is to be an equilibrium some farmers will join the program, but others will not. Since all farmers have identical operations, the equilibrium expected price *must* equal the indifference expected price.

Armed with this bit of information, we calculate that if Θ_P equals 12.8444 then $Q_{t+1} = 3{,}577.8$. (Use the demand function.) This is the level of output in equilibrium. To find the number of program participants N_j and nonparticipants N_n, we solve the following set of equations:

$$1000 = N_n + N_j$$
$$3577.8 = 6.42N_n + 3.22N_j \tag{10-40}$$

(Recall that at the indifference expected price $q_n = 6.42$ and $q_j = 3.21$.) We find that $N_j = 885.422$ and $N_n = 114.578$. The student should recognize that in the proposed solution, no farm can increase expected profit or quasi-rent by changing his or her output. This is why the proposed equilibrium is, in fact, the equilibrium. There is an element of indeterminacy in this problem since we cannot identify in advance who will and who will not participate. Fortunately, this element of indeterminacy is of no importance whatever, since all farming operations are the same. Also, the student should be aware that in this numerical example the scallop that connects the S_mS_g curve to the SS_c curve, as in Fig. 10.8, is perfectly flat. In addition, the expected demand passes through this flat segment of SS_g.

Benefit–Cost Implications of Land Diversion

We now take up long-run analysis, with special attention given to welfare effects. The main objective is to show that a land diversion program like that just described is to a considerable extent a rent or price support program for land. In other words, people who own land at the time when the program is announced or begun are the major beneficiaries.

Figure 10.9 gives the long-run supply curve SS_c under perfect competition and the kinked long-run supply curve SS_g under a voluntary land diversion program. The latter is drawn on the assumption that the government parameters P^* and λ are set such that at low levels of expected demand, with equilibrium output less than or equal to Q'_{t+1}, all farmers participate. For high levels of expected demand, with equilibrium output larger than or equal to Q''_{t+1}, no farmer participates. We analyze the case when expected demand encourages partial but not total participation. Absent the program, output and expected price equal $Q_{t+1,c}$ and Θ_{Pc}. With the program, output and expected price equal $Q_{t+1,g}$ and Θ_{Pg}.

Assuming that the production function is Cobb–Douglas, the production functions of nonparticipating and participating farmers are

$$q_n = \alpha_0 a_n^{\alpha_1} k_n^{\alpha_2} h_n^{\alpha_3} \tag{10-41a}$$

$$q_j = \alpha_0 a_p^{\alpha_1} k_j^{\alpha_2} h_j^{\alpha_3} \tag{10-41b}$$

where the t and $t + 1$ subscripts are suppressed to simplify notation and where $a_p = \lambda a_j$.[4] It is not the amount of land rented by the program participant that enters the production function; it is the rented land that actually gets farmed. The long-run expected quasi-rent functions for the nonparticipating farmer and the participating farmer are as follows:

$$\Theta_n = \Theta_P \alpha_0 a_n^{\alpha_1} k_n^{\alpha_2} h_n^{\alpha_3} - Ra_n - Gk_n \tag{10-42a}$$

$$\Theta_j = \Theta_P \alpha_0 a_p^{\alpha_1} k_j^{\alpha_2} h_j^{\alpha_3} - Ra_j - Gk_j + (P^* - \Theta_P)y^* \lambda a_j \tag{10-42b}$$

Equation (10-42b) can be rewritten as

$$\Theta_j = \Theta_P q_j - [R - (P^* - \Theta_P)y^* \lambda]a_j - Gk_j \tag{10-43}$$

Notice that the real rent paid by the participating farmer for land actually planted is a random variable given by $[R - (P^* - P_{t+1})y^* \lambda]$. Its expected value as of period t is the coefficient of a_j in Eq. (10-43).

Maximizing expected quasi-rent implies equating the expected marginal value product of each variable input to its price. Furthermore, on the basis of the analysis given in Chapter 4, we know that the production elasticity for each input equals that

[4]In Chapter 4, we assumed that

$$q_{t+1} = \alpha_0 a_t^{\alpha_1} k_t^{\alpha_2} h_{t0}^{\alpha_3},$$

where a_t equaled acreage planted, k_t equaled producers goods (capital), and h_{t0} equaled fixed family labor. The modifications in the text are required by the assumption of a government program.

input's share of expected total receipts to the industry. Therefore, in long-run competitive equilibrium, we have

$$R_c A_0 = \alpha_1(\Theta_{Pc} Q_c) \tag{10-44}$$

where R_c equals land rent in long-run competitive equilibrium and A_0 equals the fixed supply of land.

Under the land diversion program, we have

$$R_g A_n = \alpha_1(\Theta_{Pg} Q_n) \tag{10-45a}$$

$$[R_g - (P^* - \Theta_{Pg})y^*\lambda]A_j = \alpha_1(\Theta_{Pg} Q_j] \tag{10-45b}$$

where R_g equals land rent under a land diversion program; A_n and A_j equal, respectively, land rented by nonparticipators and participants; and Q_n and Q_j equal, respectively, the quantity of output produced by nonparticipators and participants. Note that $A_n + A_j = A_0$ and $Q_n + Q_j = Q_g$. Adding Eqs. (10-45a) and (10-45b), we have, after rearranging,

$$R_g A_0 = \alpha_1(\Theta_{Pg} Q_g) + (P^* - \Theta_{Pg})y^*\lambda A_j \tag{10-46}$$

The term on the far right of Eq. (10-46) is nothing more than the total revenue farmers expect to receive from the government (equals the total expected tax) in return for idling land. Thus, in long-run equilibrium under land diversion, the rent to land owners includes not only a share of expected total market receipts (equals α_1); the rent includes as well the amount taxpayers are expected to pay to idle land.

The increase in aggregate rent to land owners can be derived by subtracting $R_c A_0$ from $R_g A_0$. Using the left-hand sides of Eqs. (10-44) and (10-46), this yields

$$(R_g - R_c)A_0 = \alpha_1(\Theta_{Pg} Q_g - \Theta_{Pc} Q_c) + (P^* - \Theta_{Pg})y^*\lambda A_j \tag{10-47}$$

The increase in the total rent bill, assuming partial participation, depends in a crucial way on the elasticity of expected aggregate demand. If, for example, expected demand has unitary elasticity, then $\Theta_{Pg} Q_g = \Theta_{Pc} Q_c$, and the total rent bill rises by the amount of the total expected tax. The rent bill rises, on the assumption of at least some program participation, even if demand is elastic such that $\Theta_{Pg} Q_g < \Theta_{Pc} Q_c$. We can be sure of this because $P_g > P_c$, and an individual farmer who does not idle land can afford to pay a higher rent under the program than under perfect competition. Keen competition always puts the use of land in the hands of farmers who can pay the highest rent.

The analysis of the effects of land diversion on other inputs follows that for land. It can be shown, for example, that

$$\Theta_{wg} L_g - \Theta_{wc} L_c = \alpha_3(\Theta_{Pg} Q_g - \Theta_{Pc} Q_c) \tag{10-48}$$

where, as in Chapter 4, Θ_w and L equal, respectively, the expected implicit wage to family labor and total family labor, where the subscripts g and c indicate, respectively, long-run equilibrium values under land diversion and perfect competition. Whether a land diversion program benefits family labor in the long run again depends on the elasticity of expected aggregate demand. For example, unitary elasticity implies that $\Theta_{Pg}Q_g = \Theta_{Pc}Q_c$. Under this condition, family labor will receive no benefits whatever from a land diversion program.

10.5 CONCLUDING OBSERVATIONS

Scholars interested in agricultural pricing must be prepared to analyze government intervention for at least two reasons: farm programs are pervasive and farm programs have significant effects. Governments around the world seem unable to resist the temptation to intervene in farm markets. The forms of intervention are many and varied. In the United States, program mechanics vary considerably from one commodity to another. For some commodities, such as feedgrains, wheat, and cotton, program mechanics not only vary considerably over time, but they have also become exceedingly complex, involving, at the same time, land diversion, price support operations, and soil conservation.

In Chapters 9 and 10 we have analyzed a small but important menu of approaches to farm price and income problems, including direct payments, market price supports, border measures (for imported commodities), cartels, and land diversion. Theory suggests that these programs can have significant effects, and economic research indicates that they often do. Some examples from agricultural economics research will illustrate the point. In a 1984 study, Sumner and Alston (1984) estimated that the U.S. tobacco program, an example of a government-sponsored cartel, raised the U.S. farm price of tobacco by 20 to 30 percent of its free market value. Production was curtailed by between 50 and 100 percent. They estimated further that tobacco growers paid \$800 million per year to the owners of the production rights (output quotas) for the privilege of producing a crop. In a 1985 study, Johnson et al. (1985) estimated what would have happened in the crop years 1986–1987 through 1989–1990 under a free-market option (no price targets and diverted acres) and under a government program option according to the 1981 farm legislation (deficiency payments and land diversion). Relative to the government program option, free-market conditions would have lowered the U.S. corn price by 9 percent, both the wheat and cotton prices by 6 percent, and the rice price by 28 percent. According to a 1985 study by the Australian Bureau of Agricultural Economics, the European Community's farm program (involved tariffs and price support operations) elevated the EC's wheat price above the world price by 65 percent over the period 1967–1968 through 1982–1983. For barley and maize, the corresponding figure was 56 percent. In a 1986 study, Babcock and Schmitz (1986) found that the U.S. sugar program, which relies mainly on import quotas, maintained the U.S. domestic price of sugar at 21 cents per pound when the world price was only 5 cents. They estimated that the decrease in U.S. consumer surplus as a result of the program was $2.7 billion; the increase in U.S. producer surplus equaled $1.35 billion. Finally, Helmberger and Chen (1994) estimated that the U.S. price support program for milk used in manufacturing raised the price to producers, relative to estimated competitive values, by 14 percent in 1980, 19 percent in 1981, 20 percent in 1982, and 22 percent in 1983.

The main conclusion to be drawn from models of farm programs, together with the findings of research on many farm commodity markets, both in the United States and abroad, is that farm program analysis is indispensable to the explanation and understanding of agricultural product pricing.

PROBLEMS

10.1. You are given the following information: U.S. demand and supply are $Q_d = 10 - P$ and $Q_s = P/6$. The supply of imports to the United States is $M = 1 + P/8$. You may ignore multiple currencies and transportation cost. For each of the following regimes, find U.S. production, consumption, price, and imports. For parts b, c, and d, assume that the U.S. target price is 7, and calculate the producer gain, consumer gain or loss, tax increase or decrease (if any), and the efficiency gain or loss.

a. Perfect competition.

b. Import quota.

c. Per unit tariff.

d. A direct payment program.

10.2. You are given the following information. Demand is given by $P = 100 - Q$. The total cost functions for two firms are given by $C_1 = 2q_1 + 0.5q_1^2$ and $C_2 = q_2^2$. The two firms behave like perfectly competitive firms, equating price to marginal cost.

a. Find Q_c, P_c, q_{1c}, q_{2c}, and the profit to each firm.

b. The government forms a cartel. Firm 1 is told to produce q_{1g}, where $q_{1g} = \lambda q_{1c}, 0 < \lambda \leq 1$. Firm 2 is told to produce $q_{2g} = \lambda q_{2c}$. The firms are not allowed to exchange production rights. The government chooses λ to maximize the cartel residual. Find $\lambda, Q_g, P_g, q_{1g}, q_{2g}$, and the cartel residual.

10.3. For the following total variable cost functions, find the firm's supply function and the indirect quasi-rent function. Confirm that the derivative of the indirect quasi-rent function equals the firm's supply function.

a. $C = 0.8q^2$

b. $C = q^3/3$

10.4. In a voluntary farm program, farmers are guaranteed a price no less than P^* if they do not use pesticides. The farmer's total variable cost function is $C = q^2$ with pesticides and $C = 4q^2$ without pesticides. Derive the farmer's supply function for output.

10.5. In a voluntary whole-farm removal program, the government stands willing to pay the farmer B per acre idled if he or she idles all the farm acreage and produces nothing.

a. Using the indirect quasi-rent function, derive the farmer's short-run supply function.

b. Derive the aggregate short-run supply function on the assumption that all farmers are identical and explain how equilibrium market prices are determined.

REFERENCES

Babcock, Bruce, and Andrew Schmitz, "Look for Hidden Costs," *Choices,* Fourth Quarter, 1986, pp. 18–21.

Bureau of Agricultural Economics, *Agricultural Policies in the European Community, Their Origins, Nature and Effects on Production and Trade.* Canberra, Australia: Australian Government Publishing Service, 1985.

Gardner, Bruce L., *The Economics of Agricultural Policies.* New York: Macmillan Publishing Co., 1987.

Helmberger, Peter G., *Economic Analysis of Farm Programs*. New York: McGraw-Hill, Inc., 1991.

Helmberger, Peter G., and Yu-Hui Chen, "Economic Effects of U.S. Dairy Programs," *Journal of Agricultural and Resource Economics,* 19, no. 2 (December 1994), 225–238.

Houck, James P., *Elements of Agricultural Trade Policies.* New York: Macmillan Publishing Co., 1986.

Ives, Ralph, and John Hurley, *United States Sugar Policy: An Analysis*. Washington, D.C.: U.S. Department of Commerce, National Technical Information Service, PB88–204201, April 1988.

Johnson, Stanley R., Abner W. Womack, William H. Meyers, Robert E. Young II, and Jon Brandt, "Options for the 1985 Farm Bill: An Analysis and Evaluation," in *U.S. Agricultural Policy: The 1985 Farm Legislation,* ed. Bruce L. Gardner. Washington, D.C.: American Enterprise Institute for Public Policy Research, 1985.

Sumner, Daniel A., and Julian M. Alston, *Consequences of Elimination of the Tobacco Program,* North Carolina Agricultural Research Service, North Carolina State University, Bulletin 469, March 1984.

Appendices

Appendix A

Comparative Static Analysis

Consider a variable x defined in the domain $X = \{x: a \leq x \leq b\}$, where a and b are, respectively, the lower bound and the upper bound of x. The notation $x \in X$ simply means $a \leq x \leq b$. Consider a function of the variable x denoted by $f(x), x \in X$. Assume that $f(x)$ is differentiable for any $x \in X$. Let $f'(x) = \partial f(x)/\partial x$ denote the first derivative of $f(x)$ with respect to x. Similarly, let $f''(x) = \partial^2 f(x)/\partial x^2$ be the second derivative of $f(x)$ with respect to x.

Definition 1: The function $f(x)$ is (strictly) increasing if $f'(x)$ $(>) \geq 0$ for all $x \in X$. The function $f(x)$ is (strictly) decreasing if $-f(x)$ is (strictly) increasing for all $x \in X$.

Definition 2: The function $f(x)$ is (strictly) concave if $f''(x)$ $(<) \leq 0$ for all $x \in X$. The function $f(x)$ is (strictly) convex if $-f(x)$ is (strictly) concave for all $x \in X$.

Consider the following optimization problem. Choose x so as to

$$\text{Max}[f(x, \alpha), x \in X] \tag{A-1}$$

where $f(x, \alpha)$ is called the *direct objective function*, x being the decision variable and α denoting a parameter. Assuming that this problem has a solution, denote this solution by $x^*(\alpha)$. By definition, $x^*(\alpha)$ satisfies

$$f^*(\alpha) = f(x^*(\alpha), \alpha) \geq f(x, \alpha), \qquad \text{for all } x \in X \tag{A-2}$$

where $f^*(\alpha)$ is called the *indirect objective function*.

Assume that the function $f(x, \alpha)$ is differentiable with respect to x and α. Then, for an interior solution satisfying $a < x^*(\alpha) < b$, $x^*(\alpha)$ necessarily satisfies

$$\frac{\partial f(x, \alpha)}{\partial x} = 0 \tag{A-3}$$

when evaluated at $x = x^*(\alpha)$. Expression (A-3) is called the necessary *first-order condition* to the optimization problem (A-1). Assume in addition that

$$\frac{\partial^2 f(x, \alpha)}{\partial x^2} < 0, \quad \text{for all } x \in X \tag{A-4}$$

That is, the function $f(x, \alpha)$ is a strictly concave function of x. If the condition (A-4) is satisfied, then (A-3) becomes a necessary and sufficient condition for $x^*(\alpha)$ to be a *unique* interior solution of the optimization problem (A-1). For that reason, expression (A-4) is called the *sufficient second-order condition.*

Under condition (A-4), the implicit function theorem can be used to investigate the properties of the optimal decision function $x^*(\alpha)$. To see that, differentiate (A-3) to obtain

$$\begin{aligned} \frac{\partial^2 f}{\partial x^2}\frac{\partial x^*(\alpha)}{\partial \alpha} + \frac{\partial^2 f(x, \alpha)}{\partial x\, \partial \alpha} &= 0 \\ \frac{\partial x^*(\alpha)}{\partial \alpha} &= -\left[\frac{\partial^2 f}{\partial x^2}\right]^{-1} \frac{\partial^2 f(x, \alpha)}{\partial x\, \partial \alpha} \end{aligned} \tag{A-5}$$

Expression (A-5) gives the *comparative statics* of the optimal decision $x^*(\alpha)$: it evaluates the influence of a change in the parameter α on the optimal decision $x^*(\alpha)$. Given $\partial^2 f/\partial x^2 < 0$, it follows from (A-5) that $\partial x^*(\alpha)/\partial\alpha$ has always the same sign as $\partial^2 f(x, \alpha)/\partial x\, \partial\alpha$.

Finally, consider differentiating the indirect objective function (A-2) with respect to α. This gives

$$\frac{\partial f^*(\alpha)}{\partial \alpha} = \frac{\partial f}{\partial \alpha} + \frac{\partial f}{\partial x}\frac{\partial x^*(\alpha)}{\partial \alpha}$$

or, using the first-order condition (A-3),

$$\frac{\partial f^*(\alpha)}{\partial \alpha} = \frac{\partial f(x, \alpha)}{\partial \alpha} \tag{A-6}$$

evaluated at $x = x^*(\alpha)$.

Expression (A-6) gives the *envelope theorem*, which states that the derivative of the indirect objective function with respect to a parameter is equal to the derivative of the direct objective function with respect to this parameter, evaluated at the optimum.

Appendix B

Production Decisions in an Uncertain World

This appendix covers many of the issues analyzed in Chapter 2, but the treatment is more general. Section I centers on basic definitions of use in the study of risk behavior. Sections II and III analyze short- and long-run production decisions, respectively, under conditions of price uncertainty.

I. SOME BASIC CONCEPTS

Consider a decision maker facing an uncertain situation. Let his or her preference function be represented by the utility function $u = u(\Theta, \sigma)$, where Θ denotes expected income and σ is the standard deviation of income. We assume that $\partial u/\partial\Theta > 0$; that is, an increase in expected income always increases welfare.

As in Chapter 2, define the *risk premium* X as the maximum *sure* amount of money that the decision maker is willing to pay in order to eliminate income risk. This can be written as the amount of money r that implicitly satisfies

$$u(\Theta, \sigma) = u(\Theta - X, 0) \qquad \text{(B-1)}$$

Let $X(\Theta, \sigma)$ denote the solution of (B-1) for X. The risk premium $X(\Theta, \sigma)$ can be interpreted as a measure of the decision maker's willingness to insure against risk. Also, the expression $[\Theta - X(\Theta, \sigma)]$ is the *certainty equivalent* in the sense that it involves no risk and yet [from (B-1)] generates the same utility as the risky prospect (Θ, σ).

Definition 1: A decision maker exhibits risk aversion (risk neutrality or risk inclination) if $X(\Theta, \sigma) > 0$ ($= 0$ or < 0) for all Θ and σ.

This simply means that a risk-averse decision maker must be willing to pay a positive amount of money to eliminate the risk (i.e., to reduce the standard deviation

σ to zero). Similarly, a decision maker is said to be risk neutral if he or she is indifferent between any change in the standard deviation σ. It is commonly believed that when large sums of money are involved, large relative to initial wealth, most decision makers are risk averse; that is, they are willing to pay a positive amount of money to insure against risk.

After substituting $X(\Theta, \sigma)$ into (B-1), differentiating (B-1) with respect to σ yields

$$\frac{\partial u}{\partial \sigma} = -\frac{\partial u}{\partial \Theta}\frac{\partial X}{\partial \sigma} \quad \text{(B-2)}$$

or

$$\frac{\partial X}{\partial \sigma} = -\frac{\dfrac{\partial u}{\partial \sigma}}{\dfrac{\partial u}{\partial \Theta}} = MRS(\Theta, \sigma) \quad \text{(B-3)}$$

where $MRS(\Theta, \sigma)$ is the marginal rate of substitution between Θ and σ.

Definition 2: As an alternative to Definition 1, we may say that a decision maker exhibits risk aversion (risk neutrality or risk inclination) if $\partial u/\partial \sigma < 0$ ($= 0$ or > 0) for all Θ and σ.

In other words, a positive (zero or negative) willingness to pay for eliminating risk can be expressed equivalently in terms of a negative (zero or positive) derivative of $u(\Theta, \sigma)$ with respect to σ. That our two sets of definitions are equivalent is apparent from Fig. 2.4.

Definition 3: A decision maker exhibits decreasing (constant or increasing) absolute risk aversion if $\partial X(\Theta, \sigma)/\partial \Theta < 0$ ($= 0$ or > 0) for all Θ and σ.

Absolute risk aversion relates changes in expected income to the willingness-to-insure [as measured by the risk premium $X(\Theta, \sigma)$]. For example, under constant absolute risk aversion (CARA), a change in expected income has no effect on the willingness-to-pay for insurance. Alternatively, under decreasing absolute risk aversion (DARA), a higher expected income would tend to dampen the willingness-to-pay for insurance. It is commonly believed that most decision makers exhibit DARA preferences; that is, private wealth accumulation tends to be a substitute for insurance.

Differentiating (B-3) with respect to Θ, Definition 3 suggests the following result:

Result 1: A decision maker exhibits decreasing (constant or increasing) absolute risk aversion if $\partial MRS(\Theta, \sigma)/\partial \Theta = -(\partial^2 u/\partial \sigma\, \partial \Theta)/(\partial u/\partial \Theta) + (\partial^2 u/\partial \Theta^2)(\partial u/\partial \sigma)/(\partial u/\partial \Theta)^2 < 0$ ($= 0$ or > 0) for all Θ and σ.

II. SHORT-RUN PRODUCTION DECISION

Consider a firm facing the profit function

$$\pi = w + pq - C(q) \tag{B-4}$$

where p is output price, q is output, and $C(q)$ is the cost function satisfying $C'(q) > 0$ and $C''(q) > 0$. (The t subscripts used in Chapter 2 are here suppressed for easy notation.) The term w can be interpreted either as exogenous income or as the negative value of fixed cost. Assume that the output price p is not known for sure at the time of the production decisions: It has a subjective probability distribution with mean Θ_p and standard deviation σ_p. It follows that the expected profit is $\Theta = \Theta_p q - C(q)$, while the standard deviation of profit is $\sigma = \sigma_p q$.

The decision maker has a preference function represented by

$$u(\Theta, \sigma) = u[w + \Theta_p q - C(q), \sigma_p q] \tag{B-5}$$

or, using the certainty equivalent formulation of (B-1),

$$u(\Theta, \sigma) = u[w + \Theta_p q - C(q) - X(q, \cdot), 0] \tag{B-6}$$

where $X(q, \cdot)$ is the risk premium. Assume that the output q is chosen so as to maximize the utility function (B-5) or (B-6).

Consider first a short-run situation where the competitive firm makes its production decision taking the probability distribution of output price as exogenously given. The first-order condition associated with this optimization problem is

$$\frac{\partial u}{\partial \Theta}\left[\Theta_p - C'(q)\right] + \frac{\partial u}{\partial \sigma}\sigma_p = 0 \tag{B-7}$$

or

$$\Theta_p - C'(q) - X'(q, \cdot) = 0 \tag{B-8}$$

where $C'(q) = \partial C(q)/\partial q$ is the marginal cost and $X'(q, \cdot) = -\sigma_p(\partial u/\partial \sigma)/(\partial u/\partial \Theta)$ is the marginal risk premium. Denote by $q^*(w, \Theta_p, \sigma_p)$ the solution of the optimization problem, that is, the value of q that satisfies (B-7) or (B-8). It follows from (B-7) or (B-8) that the optimal output decision q^* corresponds to the point where the expected output price Θ_p is equal to the marginal cost C' plus the marginal risk premium X'.

On the basis of Definition 2, we can sign the marginal risk premium $X'(q, \cdot) = -\sigma_p(\partial u/\partial \sigma)/(\partial u/\partial \Theta)$, as follows:

Result 2: The optimal supply decision corresponds to the condition

$$\Theta_p - C'(q) = X'(q,\cdot)\begin{cases} = 0 & \text{under risk neutrality} \\ > 0 & \text{under risk aversion} \end{cases}$$

In other words, under risk neutrality, $X' = 0$ and the optimal supply condition reduces to equating expected output price with marginal cost. Alternatively, under risk aversion, expected output price *exceeds* the marginal cost of production at the optimum because of a positive marginal risk premium, $X' > 0$.

Assume that the second-order condition for a maximum is satisfied: $H = \partial^2 u/\partial q^2 < 0$. Then, using the results in Appendix A, a comparative static analysis of the optimal output decision $q^*(w, \Theta_p, \sigma_p)$ gives

$$\frac{\partial q^*}{\partial w} = -H^{-1}\left[\frac{\partial^2 u}{\partial \Theta^2}(\Theta_p - C') + \frac{\partial^2 u}{\partial \sigma\, \partial \Theta}\sigma_p\right] \tag{B-9}$$

$$\frac{\partial q^*}{\partial \Theta_p} = -H^{-1}\left[\frac{\partial^2 u}{\partial \Theta^2}(\Theta_p - C')q + \frac{\partial u}{\partial \Theta} + \frac{\partial^2 u}{\partial \sigma\, \partial \Theta}q\sigma_p\right] \tag{B-10}$$

and

$$\frac{\partial q^*}{\partial \sigma_p} = -H^{-1}\left[\frac{\partial^2 u}{\partial \Theta\, \partial \sigma}(\Theta_p - C')q + \frac{\partial u}{\partial \sigma} + \frac{\partial^2 u}{\partial \sigma^2}q\sigma_p\right] \tag{B-11}$$

Expression (B-9) along with Result 1 imply the following:

Result 3: $\dfrac{\partial q^*}{\partial w} = 0$ in the absence of risk, or under constant absolute risk aversion (CARA).

$\dfrac{\partial q^*}{\partial w} > 0$ under decreasing absolute risk aversion (DARA) if $u(\Theta, \sigma)$ is strictly concave in $\Theta(\partial^2 u/\partial \Theta^2 < 0)$.

Proof: In the absence of risk, $\sigma_p = 0$ and $X' = 0$, implying from (B-7) or (B-8) that $(\Theta_p - C') = 0$. It follows from (B-9) that $\partial q^*/\partial w = 0$.

Under CARA, $(\partial^2 u/\partial \Theta^2)(\partial u/\partial \sigma) = (\partial^2 u/\partial \sigma\, \partial \Theta)(\partial u/\partial \Theta)$ from Result 1. Substituting this expression into (B-9) and using (B-7) yields $\partial q^*/\partial w = 0$.

Finally, under DARA, $(\partial^2 u/\partial \Theta^2)(\partial u/\partial \sigma) < (\partial^2 u/\partial \sigma\, \partial \Theta)(\partial u/\partial \Theta)$ from Result 1. Substituting this expression into (B-9) and using (B-7) yields $\partial q^*/\partial w > 0$ if $\partial^2 u/\partial \Theta^2 < 0$. Q.E.D.

Thus, in the absence of risk, exogenous income (or fixed cost) has *no influence* on production decisions. This result still holds under CARA risk preferences, which include risk neutral preferences as a special case. However, it does not hold in general under risk aversion and DARA preferences. For example, under decreasing absolute risk aversion (DARA) and a strictly concave utility function, Result 3 shows that $\partial q^*/\partial w > 0$; that is, an increase in exogenous income (or a decrease in fixed cost) has a positive influence on supply. Thus the term $\partial q^*/\partial w$, which characterizes the effect of income on production decisions, reflects departures from CARA risk preferences.

Combining Eqs. (B-9) and (B-10) gives

$$\frac{\partial q^*}{\partial \Theta_p} = -H^{-1}\frac{\partial u}{\partial \Theta} + \frac{\partial q^*}{\partial w}q^* \tag{B-12}$$

The first term on the right-hand side of (B-12) is always positive. It has been called a *substitution effect* associated with an expected change in output price. The second term $[(\partial q^*/\partial w)q^*]$ is an *income effect*. Thus Eq. (B-12) decomposes an expected price effect into two components: a substitution effect (always positive) plus an income effect. This generates the following result.

Result 4: A sufficient condition to have an upward sloping supply function, $\partial q^*/\partial \Theta_p > 0$, is that the income effect be nonnegative, $\partial q^*/\partial w \geq 0$.

Combining Results 3 and 4, CARA is a sufficient condition to generate an upward sloping supply function. Similarly, DARA preferences, together with a strictly concave utility function, imply an upward sloping supply function.

Finally, Eq. (B-11) gives the marginal impact of a change in price risk (as measured by σ_p) on the supply function q^*. Under risk neutrality, $\partial u/\partial \sigma = 0$ and $\Theta_p = C'$, implying that risk has no impact on the production decision: $\partial q^*/\partial \sigma_p = 0$. Under risk aversion, this result no longer holds. In general, risk aversion implies that risk does influence supply: $\partial q^*/\partial \sigma_p \neq 0$. For example, consider the case when the utility function is linear: $u(\Theta, \sigma) = \Theta - \alpha\sigma$, where α is a positive parameter ($\alpha > 0$). Then, from Eq. (B-11), we have $\partial q^*/\partial \sigma_p < 0$. In other words, output price risk would have a negative effect on production.

III. LONG-RUN PRODUCTION DECISIONS

Under free entry and exit, industry equilibrium can exist only if there is no incentive for entry or exit. Assume that the industry is composed of N identical firms and that $C(0) = 0$; that is, any firm that decides to exit (or enter) the industry can do so at no cost. Then a long-run industry equilibrium is characterized by

$$u[w + \Theta_p q^* - C(q^*), \sigma_p q^*] = u(w, 0) \tag{B-13}$$

where w denotes exogenous income. The left-hand side of (B-13) is the utility of an active firm producing $q^* > 0$, while the right-hand side is the utility of an inactive firm (that produces $q = 0$). Clearly, replacing the = sign in (B-13) by the sign > implies that inactive firms can increase their utility by becoming active. This would provide incentives for firms to enter the industry. Similarly, replacing the = sign in (B-13) by the sign < implies that active firms can increase their utility by becoming inactive. This would provide incentives for firms to exit the industry. In either case, entries or exits imply that the industry is not in long-run equilibrium. Thus long-run industry equilibrium must necessarily satisfy (B-13).

Using (B-1), consider the certainty equivalent of the left-hand side of (B-13), $[w + \Theta_p q^* - C(q^*) - X(q^*)]$, where $X(q)$ is the risk premium. Expression (B-13) then implies the following result:

Result 5: A long-run equilibrium satisfies

$$\Theta_p = \frac{C(q^*)}{q^*} + \frac{X(q^*)}{q^*}$$

This states that, at equilibrium, the expected output price Θ_p is equal to the sum of two terms: the average cost $C(q^*)/q^*$, plus the average risk premium $X(q^*)/q^*$. This determines the equilibrium expected output price Θ_p^e.

It is of interest now to compare (B-8) with Result 5. While the short-run analysis equates expected price with *marginal cost plus a marginal risk premium* (B-8), the long-run analysis equates expected price with *average cost plus an average risk premium* (Result 5). In the case when the average cost plus the average risk premium is a U-shaped function of output q, these two results can be combined into one, as follows:

$$\Theta_p^e(w, \sigma_p) = \text{Min}_q \left[\frac{C(q)}{q} + \frac{X(q)}{q} \right] \quad \text{(B-14)}$$

which has for solution $q^e(w, \sigma_p) = q^*[w, \Theta_p^e(w, \sigma_p), \sigma_p]$. This illustrates that the short-run results discussed are fully consistent with long-run industry equilibrium. Expression (B-14) indicates that the optimal firm size is given by the size q that minimizes the sum of the average cost plus the average risk premium. Also, in the absence of risk or under risk neutrality, $X = 0$ and expression (B-14) reduces to choosing output so as to minimize the average cost of production. Finally, expression (B-14) shows that risk necessarily increases expected output price Θ_p^e under risk aversion ($X > 0$). In other words, the average risk premium $[X(q)/q]$ is added to the average cost $[C(q)/q]$ in the determination of the industry equilibrium expected output price.

Given the determination of Θ_p^e and q^e from (B-14), the only remaining variable to be examined is N, the equilibrium number of firms in the industry. Suppose that the industry faces the aggregate-price-dependent demand function $p = D(Q) + \sigma e$, where $\partial D/\partial Q < 0$, Q is aggregate consumption and e is a random variable with mean 0 and variance 1. Since the aggregate supply is (Nq), the market equilibrium condition is $Q = Nq^e$, which equates aggregate supply with aggregate demand. It follows that the industry equilibrium must satisfy

$$p = D(Nq^e) + \sigma_p e$$

which has for expected value

$$\Theta_p^e = D(Nq^e) \tag{B-15}$$

This defines the equilibrium number of firms in the industry: the value N^e that implicitly satisfies (B-15). It follows that expressions (B-14) and (B-15) provide a complete characterization of the long-run firm and industry equilibrium under risk and free entry and exit.

IV. THE MULTIPRODUCT FIRM

So far we have focused our attention on the single-product firm, where output q is a scalar. In this section, we extend the analysis to the multiproduct firm, where y is a $(m \times 1)$ vector of outputs, $q = (q_1, q_2, \ldots, q_m)$ with corresponding prices $p = (p_1, p_2, \ldots, p_m)$. Assuming that the output prices p are not known for sure at the time of the production decisions, let their mean be denoted by $\theta_p = (\theta_1, \theta_2, \ldots, \theta_m)$. And let their variance be denoted by the $(m \times m)$ matrix Σ, where

$$\Sigma = \begin{bmatrix} \Sigma_{11} & \Sigma_{12} & \Sigma_{13} & \cdots \\ \Sigma_{12} & \Sigma_{22} & \Sigma_{23} & \cdots \\ \Sigma_{13} & \Sigma_{23} & \Sigma_{33} & \cdots \\ . & . & . & . \end{bmatrix}$$

Σ_{ii} is the variance of p_i and Σ_{ij} is the covariance between p_i and p_j.

Then the objective function of the firm is to maximize

$$u(\Theta, \sigma) = u[w + \Theta_p' q - C(q), (q' \Sigma q)^{1/2}] \tag{B-16}$$

where $\Theta = w + \Theta_p' q - C(q)$ is the firm's expected profit, $(q' \Sigma q) \geq 0$ is the variance of profit, and $\sigma = (q' \Sigma q)^{1/2}$ is the standard deviation of profit.

But why would a firm want to produce more than one output? First, there may be technological reasons why a multiproduct firm is more efficient than a single-product firm. This corresponds to the existence of *economies of scope*. To define economies of scope, consider splitting the original firm producing outputs q into K

smaller firms ($K \leq m$), where each output q_i is now produced entirely by one of the new firms. Let q^k be the vector of outputs produced by the kth new firm, and denote its cost of production by $C_k(q^k)$.

Definition 4: The technology exhibits economies of scope if

$$C(q_1, q_2, \ldots, q_m) < C_1(q^1) + C_2(q^2) + \cdots + C_K(q^K) \quad \text{(B-17)}$$

where $(q^1, q^2, \ldots, q^K)$ is a partition of $q = (q_1, q_2, \ldots, q_m)$ as defined previously.

Expression (B-17) simply means that it is less costly to produce all the outputs $q = (q_1, q_2, \ldots, q_m)$ from a single multiproduct firm than from separate firms, each producing different output(s) independently. Thus the existence of economies of scope is equivalent to the existence of a cost reduction associated with a multiproduct production technology. The source of this cost reduction is often associated with the existence of *public inputs*. Public inputs are inputs that, once acquired for the production of one output, are costlessly available for the use in the production of other outputs. More generally, economies of scope are expected to be present whenever a production decision associated with a given output has positive spill-over effects on the production of other outputs. Thus economies of scope can provide a technological motivation for the existence of a multiproduct firm. Alternatively, the absence of economies of scope would imply that multiproduct firms cannot be justified on the basis of the underlying technology.

Another possible motivation for the existence of a multiproduct firm is *risk diversification*. To see that, consider the maximization of the objective function (B-16). This maximization problem can be written as the following two-stage decomposition:

$$\text{Max}_{\Theta}[\ \text{Max}_{q,\sigma}[u(\Theta, \sigma): \Theta = w + \Theta_p' q - C(q); \sigma^2 = q' \Sigma q]] \quad \text{(B-18)}$$

where the first stage involves choosing *q given* Θ, while the second stage involves choosing the expected profit Θ.

Assuming risk averse behavior ($\partial u / \partial \sigma < 0$), the first stage in (B-18) can be expressed as

$$\sigma^*(\Theta)^2 = \text{Min}_q[q' \Sigma q: \Theta \leq w + \Theta_p' q - C(q)] \quad \text{(B-19a)}$$

while the second stage is

$$\text{Max}_{\Theta}\ u[\Theta, \sigma^*(\Theta)] \quad \text{(B-19b)}$$

The first-order conditions for an interior solution in the first-stage problem (B-19a) are

$$2\Sigma_q - \lambda\left[\Theta_p - \frac{\partial C}{\partial q}\right] = 0$$

and

$$\Theta = w + \Theta_p' q - C(q)$$

where $\lambda \geq 0$ denotes the Lagrange multiplier associated with the expected profit constraint. Assuming that the second-order conditions are satisfied, solving these first-order conditions gives $q^+(\Theta)$, the solution of (B-19a) for q.

The solution to the first-stage problem (B-19a) also gives $\sigma^*(\Theta)^2 = q^+(\Theta)'\Sigma\, q^+(\Theta)$. The function $\sigma^*(\Theta)$ is called the *mean-standard deviation frontier*: it expresses the trade-offs between mean profit Θ and risk (as measured by σ^*) for *any risk-averse decision maker*. Indeed, the first stage (B-19a) depends only on market prices, on the technology [as reflected by $C(q)$], and on the variance Σ. It does *not* involve the risk preferences of the decision maker [as reflected by $u(\Theta, \sigma)$]. Since measuring the utility function of the decision maker is often difficult, focusing on the first stage (B-19a) is often seen as a convenient way of analyzing decision making under risk and risk aversion.

Note that the function $\sigma^*(\Theta) \geq 0$ is necessarily nondecreasing. This follows from the fact that increasing Θ in (B-19a) tends to reduce the feasible set and thus increase the optimal value of the objective function in a minimization problem. Given $\partial\sigma^*/\partial\Theta \geq 0$, it follows that a risk-averse decision maker would always prefer the lowest possible standard deviation σ given some level of expected profit Θ. Alternatively, he or she would always prefer the highest possible expected profit Θ for a given level of risk (as measured by σ). The feasible trade-offs between Θ and σ are conveniently summarized by the mean-standard deviation frontier $\sigma^*(\Theta)$ defined in (B-19a). This frontier always depends on the variance matrix Σ. In general, the more positive correlations that exist between the p's, the more difficult it will be to reduce σ for a given Θ. Alternatively, the more negative correlations that exist between the p's, the easier it will be to reduce σ for a given Θ. This risk reduction is associated with choosing the q's that have prices with a strong negative correlation. It illustrates the fact that risk aversion can provide an incentive to diversify. By appropriately choosing the activities that are more negatively correlated (or less positively correlated), the formulation (B-19a) shows how a diversification across outputs can reduce risk σ while maintaining a given level of expected profit Θ.

Now consider the second-stage problem (B-19b). The first-order condition associated with the second-stage problem is

$$-\frac{\partial u/\partial\Theta}{\partial u/\partial\sigma} = \frac{\partial\sigma^*}{\partial\Theta}$$

where $-(\partial u/\partial\Theta)/(\partial u/\partial\sigma)$ is the marginal rate of substitution between the expected profit Θ and risk σ, and $\partial\sigma^*/\partial\Theta$ is the slope of the mean standard deviation frontier $\sigma^*(\Theta)$. The marginal rate of substitution between Θ and σ reflects the risk preferences of the decision maker. Thus the second-stage decision involves risk preferences: it consists of choosing a point on the mean-standard deviation frontier $\sigma^*(\Theta)$ where the slope of this frontier is equal to the marginal rate of substitution between Θ and σ. Let this point be given by Θ^*, the solution of (B-19b) for Θ. Then, combining the two stages of the analysis, the optimal firm decisions associated with problem (B-18) are $q^* = q^+(\Theta^*)$.

Appendix C

Farm Output and Input Pricing

Consider a farm choosing land a and other inputs k in the production of output. For simplicity, we will consider only the single-output case. However, the reader should note that the analysis presented here could be easily extended to multiple outputs. Let p denote the expected output price, J be the price of inputs k, and R be the rental price of land. Assume that output and output price are independently distributed random variables and that the farmer is risk neutral. Then the farmer makes decisions by maximizing expected profit:

$$\pi(p, J, R, TFC) = \text{Max}_{k,a}[pf(k, a) - J'k - Ra - TFC] \tag{C-1}$$

where $f(k, a)$ denotes expected output, pf is expected revenue, Ra is the rental cost of land, $J'k$ is the cost of inputs other than land, and TFC denotes total fixed cost. Assuming that the farm chooses to produce and to use positive amounts of inputs (i.e., quasi-rent is not negative), the first-order conditions associated with the maximization problem (C-1) are

$$p\frac{\partial f}{\partial k} = J \tag{C-2a}$$

and

$$p\frac{\partial f}{\partial a} = R \tag{C-2b}$$

Expressions (C-2) simply state that, at the optimum for a producing farmer, the expected marginal value product [the left-hand side in (C-2)] equals the input cost [the right-hand side in (C-2)]. Let $k^*(p, J, R)$ and $a^*(p, J, R)$ denote the optimal input choices for k and a corresponding to (C-1) and (C-2). The optimal expected supply function is then given by $q^*(p, J, R) = f(k^*, a^*)$.

Note that Eq. (C-1) could alternatively be written as

$$\pi(p, J, R, TFC) = \text{Max}_q[pq - C(J, R, q) - TFC] \tag{C-3a}$$

where q denotes expected output and

$$C(J, R, q) = \text{Min}_{q,a}[J'k + Ra: q = f(k, a)] \tag{C-3b}$$

is the cost function characterizing the input choices that minimize the cost of producing a given expected output level q. Denote by $k^c(J, R, q)$ and $a^c(J, R, q)$ the corresponding cost-minimizing input demand functions. The equivalence of (C-1) and (C-3) implies the following results:

$$k^*(p, J, R) = k^c[J, R, q^*(p, J, R)] \tag{C-4a}$$

and

$$a^*(p, J, R) = a^c[J, R, q^*(p, J, R)] \tag{C-4b}$$

Expressions (C-4) simply indicate that the cost minimization problem (C-3b) is fully consistent with the profit maximization problem (C-1).

Our discussion has assumed that the farmer decides to produce. Under what conditions would he or she decide not to produce? Assume that land is necessary input such that $f(k, 0) = 0$ for any $k \geq 0$. Define quasi-rent ϕ as total revenue pq minus total variable cost $C(J, R, q)$. Then the farmer would decide to produce only if $\phi(p, J, R) \geq 0$. Indeed, if $\phi(p, J, R) < 0$, then the farmer would be better off producing nothing. In this case, $\phi = 0$ and $\pi = -TFC$. An additional condition for production decisions therefore is

$$p \geq \frac{J'k^* + Ra^*}{q^*} = \frac{C\left(J, R, q^*\right)}{q^*} \tag{C-5a}$$

or, equivalently,

$$R \leq \frac{pq^* - J'k^*}{a^*} \tag{C-5b}$$

Expression (C-5b) indicates that the relevant range for optimal land allocation corresponds to the region where the rental price of land R is less than the *expected average revenue product* $(pq^* - J'k^*)/a^*$.

I. PARTIAL EQUILIBRIUM ANALYSIS

Expressions (C-2) and (C-5) characterize a partial equilibrium analysis of the supply function q^* and the input demand functions k^* and a^*. Such functions are partial equilibrium because they take all prices p, J, and R as exogenously given. Although appropriate in the context of a single competitive firm, this analysis stands in sharp contrast to the market equilibrium analysis presented later which will address the issue of price determination at the market level.

Propositions 1A and 1B summarize the properties of the profit function π and the cost function C. The derivation of these properties is beyond this textbook and can be found in advanced microeconomic textbooks.

Proposition 1A. The profit function $\pi(p, J, R, TFC)$ in (C-1) has the following properties:

- It is linear homogeneous in (p, J, R, TFC).
- It is convex in (p, J, R).
- It satisfies $\partial\pi/\partial p = q^* \geq 0$
  $\partial\pi/\partial J = -k^* \leq 0$
  $\partial\pi/\partial R = -a^* \leq 0$

Proposition 1B. The cost function $C(q, J, R)$ in (C-3b) has the following properties:

- It is linear homogeneous in (J, R).
- It is concave in (J, R).
- It satisfies $\partial C/\partial J = k^c \geq 0$
  $\partial C/\partial R = a^c \geq 0$

Let $\alpha = (p, J, R)$ denote the vector of prices facing the firm. Then, from the convexity of π in Proposition 1A, we have the following results:

$$\frac{\partial^2 \pi}{\partial \alpha^2} = \begin{bmatrix} \dfrac{\partial q^*}{\partial p} & \dfrac{\partial q^*}{\partial J} & \dfrac{\partial q^*}{\partial R} \\ -\dfrac{\partial k^*}{\partial p} & -\dfrac{\partial k^*}{\partial J} & -\dfrac{\partial k^*}{\partial R} \\ -\dfrac{\partial a^*}{\partial p} & -\dfrac{\partial a^*}{\partial J} & -\dfrac{\partial a^*}{\partial R} \end{bmatrix} = \text{a symmetric, positive, semidefinite matrix} \quad \text{(C-6)}$$

The positive semidefiniteness implies that the diagonal elements of the matrix in (C-6) are nonnegative. It follows that the expected supply function is necessarily *upward sloping* ($\partial q^*/\partial p \geq 0$) and that the input demand functions are always *downward sloping* ($\partial k_i^*/\partial J_i \leq 0$ and $\partial a^*/\partial R \leq 0$).

The symmetry of the matrix in (C-6) generates useful information on cross-price effects. It implies that the marginal effect of a rise in output price p on input demand is equal in absolute magnitude, but of opposite sign, compared to the marginal effect of a rise in the corresponding input price on output supply. For example, it implies that $\partial q^*/\partial R = -\partial a^*/\partial p$.

Additional results on cross-price effects can be obtained from differentiating (C-4):

$$\frac{\partial(k^*, a^*)}{\partial p} = \frac{\partial(k^c, a^c)}{\partial q} \cdot \frac{\partial q^*}{\partial p} \tag{C-7a}$$

and

$$\frac{\partial(k^*, a^*)}{\partial(J, R)} = \frac{\partial(k^c, a^c)}{\partial(J, R)} + \frac{\partial(k^c, a^c)}{\partial q} \cdot \frac{\partial q^*}{\partial(J, R)} \tag{C-7b}$$

Given $\partial q^*/\partial p > 0$, it follows from (C-7a) that $\partial(k^*, a^*)/\partial p$ has the same sign as $\partial(k^c, a^c)/\partial q$. In general, the marginal effect of output on cost-minimizing input demands $\partial(k^c, a^c)/\partial q$ can be either positive or negative. This marginal effect has been used in economics to characterize the nature of inputs.

Definition 1. *Normal inputs* are defined to be inputs for which the cost-minimizing input demand function is positively affected by an increase in output q.

Most inputs typically found in economic analysis seem to be *normal* inputs. If the inputs (k, a) are normal, $\partial(k^c, a^c)/\partial q \geq 0$ by Definition 1. Then equation (C-7a) implies that $\partial(k^*, a^*)/\partial p \geq 0$. Thus, for normal inputs, an increase in output price tends to stimulate input demand. And from the symmetry of (C-6), an increase in the cost of a normal input tends to decrease output supply.

Equation (C-7b) relates the slopes of the profit-maximizing input demand functions (k^*, a^*) to the slopes of the cost-minimizing demand functions (k^c, a^c). It yields the following result:

Proposition 2

- $\partial(k^*, a^*)/\partial(J, R)$ = a symmetric, negative semidefinite matrix.
- $\partial(k^c, a^c)/\partial(J, R)$ = a symmetric, negative semidefinite matrix.
- $\partial(k^*, a^*)/\partial(J, R) - \partial(k^c, a^c)/\partial(J, R)$ = a symmetric, negative semidefinite matrix.

Proof: The symmetric, negative semidefiniteness properties of $\partial(k^*, a^*)/\partial(J, R)$ and $\partial(k^c, a^c)/\partial(J, R)$ follow from the convexity properties in Propositions 1A and 1B.

Next, consider the following relationships:

$$\frac{\partial(k^c, a^c)}{\partial q} \cdot \frac{\partial q^*}{\partial(J, R)} = \frac{\partial(k^c, a^c)}{\partial q} \cdot \left[\frac{-\partial(k^*, a^*)}{\partial p}\right]', \quad \text{from the symmetry of (C-6)}$$

$$= -\left[\frac{-\partial(k^c, a^c)}{\partial q}\right] \cdot \frac{\partial q^*}{\partial p} \cdot \left[\frac{\partial(k^c, a^c)}{\partial q}\right]', \quad \text{from (C-7a)}$$

But $\partial q^*/\partial p \geq 0$ from Proposition 1, implying that this expression is symmetric, negative semidefinite. This result along with equation (C-7b) concludes the proof.

Q.E.D.

Proposition 2 implies that both profit-maximizing input demand functions and cost-minimizing input demand functions are *downward sloping*. It also implies that the profit-maximizing input demand functions tend to be *more price responsive* than the cost-minimizing input demand functions. The reason is that profit maximization provides more flexibility for the firm than cost minimization: output level is a choice variable in the former case, but not in the latter. In other words, constraining the firm's production choices (here by forcing it to choose a particular level of output) tends to reduce its ability to react to changing prices. This general result is called the *Le Chatelier principle*. It applies also to a comparison of short-run versus long-run behavior. In the short run, some inputs are fixed, while in the long run all inputs are variable. The preceding result then implies that *long-run* input demand and output supply functions are *more* price responsible than their short-run counterparts.

II. MARKET EQUILIBRIUM ANALYSIS

Now consider that farming is a competitive industry consisting of a large number of relatively small farms. The number of farms will be assumed constant in the short run. Under competition, it is assumed that all farms face the same market prices (p, J, R). It is not required that all farms face the same technology. For simplicity, we will abstract from price or production uncertainty in this section. Then the aggregate supply and demand functions from this industry are simply the sum across all farms of the individual supply and demand functions. More specifically, let $Q(p, J, R) = \Sigma_i\, q_i^*(p, J, R)$ denote the aggregate supply, $K(p, J, R) = \Sigma_i\, k_i^*(p, J, R)$ be the aggregate input demands for k, and $A(p, J, R) = \Sigma_i\, a_i^*(p, J, R)$ denote the aggregate demand for land.

First, note that addition across firms preserves the properties of the input/output choice functions. Thus the properties of the firm-level input demand–output supply functions discussed in the previous section carry over to the aggregate functions. For example, the symmetric, positive semidefiniteness of equation (C-6) applies as well to the aggregate choice functions (Q, K, A):

$$B = \begin{bmatrix} \frac{\partial Q}{\partial p} & \frac{\partial Q}{\partial J} & \frac{\partial Q}{\partial R} \\ -\frac{\partial K}{\partial p} & -\frac{\partial K}{\partial J} & -\frac{\partial K}{\partial R} \\ -\frac{\partial A}{\partial p} & -\frac{\partial A}{\partial J} & -\frac{\partial A}{\partial R} \end{bmatrix} = \text{a symmetric, positive semidefinite matrix} \tag{C-8}$$

At the market level, competitive prices are determined by the interactions of aggregate supply and aggregate demand. With respect to output, the aggregate supply function is $Q(p, J, R)$. Assume that the industry faces the aggregate demand function for output $D(p)$, which is downward sloping ($\partial D/\partial p < 0$). Then the market equilibrium condition for output is

$$Q(p, J, R) - D(p) = 0 \tag{C-9}$$

Let $p^*(J, R)$ denote the market equilibrium price that implicitly satisfies (C-9). From the implicit function theorem applied to (C-9), we have

$$\frac{\partial p^*}{\partial(J, R)} = -\left[\frac{\partial Q}{\partial p} - \frac{\partial D}{\partial p}\right]^{-1}\left[\frac{\partial Q}{\partial(J, R)}\right] = \text{sign}\left[-\frac{\partial Q}{\partial(J, R)}\right] \tag{C-10}$$

since $\partial Q/\partial p \geq 0$ from Proposition 1A and $\partial D/\partial p < 0$ by assumption. We know that $\partial Q/\partial(J, R) \leq 0$ for normal inputs. It follows that, for normal inputs, $\partial p^*/\partial(J, R) \geq 0$. Then, in a market equilibrium context, any exogenous rise in input cost tends to induce an increase in output price.

Next consider the inputs k, the corresponding aggregate demand function being $K(p, J, R)$. Assume that the industry faces the aggregate supply function for those inputs $S(J)$, where $\partial S/\partial J$ is a symmetric, positive definite matrix. Then the market equilibrium condition for the inputs k is

$$S(J) - K[p, J, R] = 0 \tag{C-11}$$

Let $J^e(R)$ and $p^e(R) = p^*[J^e(R), R]$, the market prices that satisfy (C-9) and (C-11) simultaneously. They are the market equilibrium prices in the markets for inputs k and in the market for output q, respectively. They depend on the rental value of land R. Given market equilibrium conditions in the markets for q and k, consider next the aggregate demand for land:

$$A^e(R) = A[p^e(R), J^e(R), R] \tag{C-12}$$

Differentiating (C-12) with respect to R gives

$$\frac{\partial A^e}{\partial R} = \frac{\partial A}{\partial R} + \frac{\partial A}{\partial p} \cdot \frac{\partial p^e}{\partial R} + \frac{\partial A}{\partial J} \cdot \frac{\partial J^e}{\partial R} \qquad \text{(C-13)}$$

Expression (C-13) decomposes the net effect of a change in R on A into three parts: the direct impact of R on A (the first term), the indirect impact of R through p^e (the second term), and the indirect impact through J^e (the third term). The first term is the partial equilibrium effect (at the aggregate) discussed in the previous section. The last two terms reflect indirect price effects associated with the market equilibrium conditions for q and k. They indicate that expression (C-12) is a market equilibrium demand function for land. They also make it clear that the properties of $A^e(R)$ are in general different from the properties of the partial equilibrium demand function $A(p, J, R)$ discussed previously [e.g., see (C-8)].

The next proposition presents the general properties of the market equilibrium demand function $A^e(R)$.

Proposition 3. The market equilibrium function $A^e(R)$ satisfies

- $\partial A^e/\partial R \leq 0$.
- $\partial A/\partial R \leq \partial A^e/\partial R$.

Proof: First note from (C-8) that the matrix B is symmetric, positive semidefinite. Define the matrix

$$W = \begin{bmatrix} -\frac{\partial D}{\partial p} & 0 & 0 \\ 0 & \frac{\partial S}{\partial J} & 0 \\ 0 & 0 & 0 \end{bmatrix} \qquad \text{(C-14)}$$

By definition, the matrix W is symmetric, positive semidefinite. It follows that the matrix $[B + W]$ is also symmetric, positive semidefinite. Define the matrix M to be a submatrix of $[B + W]$, including all rows and columns of $[B + W]$ except the last row and the last column:

$$[B + W] = \begin{bmatrix} M & N' \\ N & Z \end{bmatrix} \qquad \text{(C-15)}$$

where $Z = -\partial A/\partial R$ and $N = [-\partial A/\partial p - \partial A/\partial J]$. The matrix M is symmetric and positive definite. Then consider the quadratic form:

$$\begin{bmatrix} -NM^{-1} & I \end{bmatrix} \begin{bmatrix} M & N' \\ N & Z \end{bmatrix} \begin{bmatrix} -M^{-1}N' \\ I \end{bmatrix} = -NM^{-1}N' + Z \geq 0 \qquad \text{(C-16)}$$

which is nonnegative from the positive semidefiniteness of $[B + W]$.

From (C-13), note that $\partial A^e/\partial R$ can be written as

$$\frac{\partial A^e}{\partial R} = \frac{\partial A}{\partial R} + \begin{bmatrix} \frac{\partial A}{\partial p} & \frac{\partial A}{\partial J} \end{bmatrix} \begin{bmatrix} \frac{\partial p^e}{\partial R} \\ \frac{\partial J^e}{\partial R} \end{bmatrix} \qquad \text{(C-17)}$$

Using the implicit function theorem on (C-9) and (C-11), as well as the symmetry condition (C-8), this can be written as

$$\frac{\partial A^e}{\partial R} = -Z + NM^{-1}N' \qquad \text{(C-18)}$$

where $Z = -\partial A/\partial R$ and $N = [-\partial A/\partial p - \partial A/\partial J]$. Since we have shown that $-NM^{-1}N' + Z \geq 0$, it follows that $\partial A^e/\partial R \leq 0$. Also, M being a symmetric positive definite matrix implies that $[NM^{-1}N'] = \partial A^e/\partial R - \partial A/\partial R \geq 0$. Q.E.D.

Proposition 3 shows that the market equilibrium demand for land $A^e(R)$ is still downward sloping, $\partial A^e/\partial R \leq 0$, like its partial equilibrium counterpart $A(R)$. It also shows that the presence of price adjustments in related markets tends to reduce the quantity adjustments in the demand for land. Note that these results are very general and do not require any of the inputs to be normal.

The difference between the partial equilibrium effects $\partial A/\partial R$ and the market equilibrium effects $\partial A^e/\partial R$ will depend on the extent of the adjustments for p^e and J^e. Proposition 3 states that *$\partial A^e/\partial R$ is always bounded between 0 and $\partial A/\partial R$*. Note that the bound $\partial A/\partial R$ will be attained if the aggregate output demand function $D(p)$ and the input supply functions $S(J)$ are perfectly elastic. In this case, prices are effectively exogenous and are not affected by changes in the rental value of land, implying that $\partial A/\partial R = \partial A^e/\partial R$. However, as the aggregate output demand and/or the aggregate input supplies become more inelastic, the prices p and J will adjust to R and we can expect to find $\partial A^e/\partial R > \partial A/\partial R$.

Appendix D

Spatial Equilibrium Analysis

Consider the spatial allocation of a given commodity in J regions. Let $S_i(x_i)$ represent the price-dependent supply function for the commodity in region i, x_i denoting the quantity produced in region i, $i = 1, 2, \ldots, J$. Let $D_i(y_i)$ represent the price-dependent demand function for the commodity in region i, y_i being the quantity consumed in region i, $i = 1, 2, \ldots, J$.

The competitive spatial equilibrium over n regions can be written as the solution of the following optimization problem:

$$\text{Max}_{y,x,t}\left\{\sum_{i=1}^{J}\left[\int_0^{y_i} D_i(q_i)\,dq_i - \int_0^{x_i} S_i(q_i)\,dq_i\right] - \sum_{i=1}^{J}\sum_{j=1}^{J} t_{ij}c_{ij}\right.$$

$$: x_i \geq \sum_{j=1}^{J} t_{ij}, \quad \forall i \tag{D-1}$$

$$y_i \leq \sum_{j=1}^{J} t_{ji}, \quad \forall i$$

$$y_i \geq 0, \quad x_i \geq 0, \quad t_{ij} \geq 0$$

where i (or j) denotes the ith (jth) region, $i = 1, \ldots, J$
J = number of regions
c_{ij} = unit transportation cost from region i to region j
$c_{ii} = 0$, for all i
$c_{ij} > 0$, for $i \neq j$
t_{ij} = quantity transported from region i to region j, for all i and j

The optimization problem (D-1) can be interpreted as the maximization of producer plus consumer surpluses across all regions, net of transportation cost. This optimization is subject to two feasibility constraints. The first constraint states that the total shipments out of region i cannot be larger than the quantity produced in region i. The second constraint indicates that the quantity consumed in region i cannot be larger than the total shipments into that region. Given $c_{ii} = 0$, note that t_{ii} (the shipment from region i to region i) measures the quantity produced in region i that is also consumed in region i. Finally, production, consumption, and trade are restricted to be nonnegative.

Consider the Lagrangian associated with the maximization problem (D-1):

$$L = \left\{ \sum_{i=1}^{J} \left[\int_0^{y_i} D_i(q_i)\,dq_i - \int_0^{x_i} S_i(q_i)\,dq_i \right] - \sum_{i=1}^{J} \sum_{j=1}^{J} t_{ij} c_{ij} \right.$$

$$+ \sum_{j=1}^{J} \lambda_i \left[x_i - \sum_{j=1}^{J} t_{ij} \right] + \sum_{i=1}^{J} \delta_i \left[\sum_{j=1}^{J} t_{ji} - y_i \right], \tag{D-2}$$

$$\left. y \geq 0, \quad x \geq 0, \quad t \geq 0, \quad \lambda \geq 0, \quad \delta \geq 0 \right\}$$

where the λ's and δ's are the Lagrange multipliers associated with the corresponding constraints.

The Kuhn–Tucker conditions associated with this optimization problem give

$$\frac{\partial L}{\partial y_i} = D_i - \delta_i \leq 0 \qquad = 0, \quad \text{if } y_i > 0 \tag{D-3}$$

$$\frac{\partial L}{\partial x_i} = -S_i + \lambda_i \leq 0 \qquad = 0, \quad \text{if } x_i > 0 \tag{D-4}$$

$$\frac{\partial L}{\partial t_{ij}} = -c_{ij} - \lambda_i + \delta_j \leq 0 \qquad = 0, \quad \text{if } t_{ij} > 0 \tag{D-5}$$

Equation (D-3) states that the consumer price in region i, D_i, cannot be larger than the Lagrange multiplier δ_i. And in the case where consumption is positive, the Lagrange multiplier δ_i equals D_i, the actual consumer price in region i. Equation (D-4) indicates that the producer price in region i, S_i, cannot be smaller than the Lagrange multiplier λ_i. And in the case when production is positive, the Lagrange multiplier λ_i equals S_i and is the actual producer price for region i. Finally, Eq. (D-5) shows the relationship between transportation cost and the Lagrange multipliers δ and λ. If $t_{ij} = 0$, it suggests that the difference between δ_i (the consumer price in region i) and λ_j (the producer price in region j) will always be no greater (and in general less) than the unit transportation cost between the two regions. Alternatively, if $t_{ij} > 0$, then, given the feasibility constraints and positive transportation cost across regions ($i \neq j$), we necessarily have $x_i > 0$ and $y_i > 0$ at the optimum. It follows from (D-3) and (D-4) that $\delta_j = D_j$ and $\lambda_i = S_i$. Then, from (D-5), the difference between the producer price in region i and the consumer price in region j will be exactly equal to the unit transportation cost between the two regions.

To see that this problem does characterize a spatial equilibrium competitive market, note that Eq. (D-5) implies that $t_{ij}[-c_{ij} - \lambda_i + \delta_j] = 0$ for all i and j. If $t_{ij} = 0$, there is no incentive for transportation between regions i and j because the price difference does not cover transportation cost. Alternatively, if $t_{ij} > 0$, then $\delta_j = D_j$ and $\lambda_i = S_i$, yielding

$$t_{ij}D_i - t_{ij}S_i - t_{ij}c_{ij} = 0$$

which simply states that the profit from transportation activities between regions i and j is zero. In this context, the solution obtained corresponds to a competitive equilibrium: there is no economic incentive for further spatial redistribution of the commodity.

Appendix E

Welfare Analysis

I. THE CONSUMER CASE

Demand Behavior

Consider a consumer choosing an $(n \times 1)$ vector $x = (x_1, x_2, \ldots, x_n)'$ of consumer goods. His or her preferences are represented by the (direct) *utility function* $u(x)$. We assume that the utility function is differentiable and increasing in x, with $\partial u/\partial x_i > 0$, for $i = 1, 2, \ldots, n$. The consumer is purchasing the commodity x_i at a market price p_i, $i = 1, 2, \ldots, n$. Letting $I > 0$ denote consumer income, the budget constraint takes the form $\sum_i p_i x_i \leq I$. It simply states that consumer expenditures cannot exceed money income I. Assuming that the consumer wants to maximize his or her welfare level [as represented by $u(x)$], his or her behavior can be represented by the following problem:

$$V(p, I) = \max_x\{u(x), \text{subject to } \textstyle\sum_i p_i x_i \leq I\} \tag{E-1}$$

where $V(p, I)$ is called the *indirect utility function* giving the utility level of an optimizing consumer facing prices $p = (p_1, p_2, \ldots, p_n)$ and income I. If utility were observable, the indirect utility level $V(p, I)$ would provide all the information necessary to evaluate the level of consumer well-being.

The utility maximization problem (E-1) is a constrained optimization problem that can be alternatively written as

$$\max_x\{u(x) + \lambda(I - \textstyle\sum_i p_i x_i\} \tag{E-1'}$$

where λ is a Lagrange multiplier measuring the marginal utility of income. Note that, given $\partial u/\partial x_i > 0$, the consumer will always choose to spend all his or her income, implying that the budget constraint is binding ($\sum_i p_i x_i = I$), and that the marginal utility of income is always positive ($\lambda > 0$). The first-order conditions associated with the utility maximization problem (E-1) or (E-1′) are

$$\frac{\partial u}{\partial x_i} - \lambda p_i = 0, i = 1, 2, \ldots, n \tag{E-2a}$$

and

$$I = \Sigma_i \, p_i x_i \tag{E-2b}$$

Equation (E-2a) can be rewritten as

$$\frac{\partial u / \partial x_i}{\partial u / \partial x_j} = \frac{p_i}{p_j}, \qquad \text{for } i \neq j = 1, 2, \ldots, n \tag{E-2a′}$$

which states that, at the optimum, the marginal rate of substitution between any two commodities i and j, $(\partial u/\partial x_i)/(\partial u/\partial x_j)$, equals their respective price ratio, p_i/p_j.

For a given price vector $p = (p_1, p_2, \ldots, p_n)$ and a given income level I, let the optimum solution to problem (E-1) or (E-1′) be denoted by $x_i^*(p, I), i = 1, 2, \ldots, n$. The function $x_i^*(p, I)$ is the *Marshallian demand function* for the ith commodity, $i = 1, 2, \ldots, n$. These demand functions represent consumer behavior. They can be interpreted as the consumption levels that satisfy the first-order conditions (E-2a) and (E-2b). Since consumption levels are observable, it follows that the demand functions $x_i^*(p, I), i = 1, 2, \ldots, n$, are in general empirically tractable. They have been the subject of much scrutiny in the investigation of consumer behavior.

Expenditure Function

As noted previously, if the indirect utility function $V(p, I)$ were observable, the analysis of consumer welfare would be easy. Unfortunately, this is not the case. Measuring consumer well-being is always difficult. However, if it is not possible to investigate consumer welfare directly, it is often possible to do it indirectly. To see this, consider the hypothetical situation where the consumer would try to find the smallest expenditure, $\Sigma_i \, p_i x_i$, that can keep her or him at a given utility level U. This corresponds to the following minimization problem:

$$E(p, U) = \min_x \{\Sigma_i \, p_i \, x_i, \text{ subject to } U = u(x)\} \tag{E-3}$$

where $E(p, U)$ is called the *expenditure function*. The expenditure function $E(p, U)$ gives the smallest expenditure that would keep the consumer at a given welfare level U under prices $p = (p_1, p_2, \ldots, p_n)$.

Note that problem (E-3) is a constrained optimization problem that can be alternatively written as

$$\min_x\{\Sigma_i p_i x_i + \gamma (U - u(x))\} \tag{E-3'}$$

where γ is a Lagrange multiplier that measures the marginal cost of utility. Note that, given $\partial u/\partial x_i > 0$, the marginal cost of utility is always positive ($\gamma > 0$). The first-order conditions associated with the utility maximization problem (E-3) or (E-3′) are

$$p_i - \gamma \frac{\partial u}{\partial x_i} = 0, \qquad i = 1, 2, \ldots, n \tag{E-4a}$$

and

$$U = u(x) \tag{E-4b}$$

Equation (E-4a) can be rewritten as

$$\frac{\partial u/\partial x_i}{\partial u/\partial x_j} = \frac{p_i}{p_j}, \qquad \text{for } i \neq j = 1, 2, \ldots, n \tag{E-4a'}$$

which states that, at the optimum, the marginal rate of substitution between any two commodities i and j, $(\partial u/\partial x_i)/(\partial u/\partial x_j)$, equals their respective price ratio, p_i/p_j.

For a given price vector $p = (p_1, p_2, \ldots, p_n)$ and a given welfare level U, let the optimum solution to problem (E-3) or (E-3′) be denoted by $x_i^c(p, U), i = 1, 2, \ldots, n$. The function $x_i^c(p, U)$ is the *Hicksian demand function* for the ith commodity, $i = 1, 2, \ldots, n$. These demand functions are sometimes called *compensated* demand functions since they correspond to problem (E-3), where $E(p, U)$ is the smallest income needed to keep the consumer at the welfare level U under prices p. The Hicksian demand functions $x_i^c(p, U), i = 1, 2, \ldots, n$, can be interpreted as the consumption levels that satisfy the first-order conditions (E-4a) and (E-4b).

Note that applying the envelope theorem (see Appendix A) to expression (E-3) yields the following result:

$$\frac{\partial E(p, U)}{\partial p_i} = x_i^c(p, U), \qquad i = 1, 2, \ldots, n \tag{E-5}$$

Expression (E-5) states that the derivative of the expenditure function with respect to the ith price is equal to the Hicksian demand function for the ith commodity, $x_i^c(p, U), i = 1, 2, \ldots, n$. We will make use of this convenient result next.

The expenditure minimization problem (E-3) is typically purely hypothetical. In the real world, consumers are rarely being kept at a constant welfare level. As a result, the Hicksian demand functions $x_i^c(p, U), i = 1, 2, \ldots, n$, are in general not observable. However, they are of interest here for two reasons: they have a convenient welfare interpretation (see later), and they have a close relationship with the Marshallian demand functions just discussed, $x_i^*(p, I), i = 1, 2, \ldots, n$ (which are themselves empirically tractable).

To see the close relationship between x^* and x^c, note that equation (E-2a′) and (E-4a′) are exactly equivalent. In other words, the first-order conditions (E-2a) and (E-4a) are fully consistent with each other. It follows that the Marshallian and Hicksian demand functions are consistent with each other if the remaining first-order conditions [i.e., (E-2b) and (E-4b)] are also consistent with each other. This will be the case if the expenditure function $E(p, U)$ in (E-3) is equal to income I in the utility maximization problem (E-1): $I = E(p, U)$. This generates the following identity:

$$x_i^c(p, U) = x_i^*[p, E(p, U)], i = 1, \ldots, n \tag{E-6}$$

Differentiating Eq. (E-6) with respect to p_i and using Eqs. (E-5) and (E-6) yields

$$\begin{aligned} \frac{\partial x_i^c}{\partial p_i} &= \frac{\partial x_i^*}{\partial p_i} + \left(\frac{\partial x_i^*}{\partial I}\right)\left(\frac{\partial E}{\partial p_i}\right) \\ &= \frac{\partial x_i^*}{\partial p_i} + \frac{\partial x_i^*}{\partial I} x_i^* \end{aligned} \tag{E-7}$$

Equation (E-7) is called the Slutsky decomposition. The left-hand side of (E-7), $\partial x_i^c/\partial p_i$, is the slope of Hicksian (or compensated) demand with respect to a price change. The right-hand side of (E-7) is the sum of two terms: the first term, $\partial x_i^*/\partial p_i$, is the slope of the Marshallian demand with respect to a price change; and the second term, $(\partial x_i^*/\partial I)\, x_i^*$, is an *income effect* (since it involves the effect of income I on demand, $\partial x_i^*/\partial I$). As a result, Eq. (E-7) states that the Hicksian price slope is equal to the Marshallian price slope, plus an income effect. In other words, the income effect, $(\partial x_i^*/\partial I)\, x_i^*$, measures exactly the difference between Marshallian and Hicksian price slopes. Note that Eq. (E-7) can be alternatively expressed in elasticity form as follows:

$$\frac{\partial x_i^c}{\partial p_i}\frac{p_i}{x_i} = \frac{\partial x_i^*}{\partial p_i}\frac{p_i}{x_i} + \frac{p_i x_i}{I}\frac{\partial x_i^*}{\partial I}\frac{I}{x_i} \tag{E-7′}$$

Equation (E-7′) establishes the relationship between the price elasticity of the Hicksian demand, $(\partial x_i^c/\partial p_i)(p_i/x_i)$, and the price elasticity of the Marshallian demand,

$(\partial x_i^*/\partial p_i)(p_i/x_i)$, for the ith commodity. From Eq. (E-7′), the difference between these two elasticities is given by $[(p_i x_i)/I]\ [(\partial x_i^*/\partial I)(I/x_i)]$. This income effect is equal to zero whenever the income elasticity of demand for the ith commodity, $[(\partial x_i^*/\partial I)(I/x_i)]$, is equal to zero. In this case, Marshallian and Hicksian demand functions have the same price elasticity. This implies that, as far as price effects are concerned, the (observable) Marshallian demand function can be used instead of the (unobservable) Hicksian demand function.

The problem is that, for many commodities, income effects are often nonzero. In other words, a zero income elasticity of demand is not very likely for most commodities. In this case, Eq. (E-7′) indicates that the Hicksian price elasticity of demand, $(\partial x_i^c/\partial p_i)(p_i/x_i)$, in general differs from the Marshallian price elasticity of demand, $(\partial x_i^*/\partial p_i)(p_i/x_i)$. Equation (E-7′) also shows that these two elasticities will be close to each other whenever $[(p_i x_i)/I]\ [(\partial x_i^*/\partial I)(I/x_i)]$ is small. In other words, the Hicksian price elasticity of demand, $(\partial x_i^c/\partial pi)(p_i/x_i)$ will be *approximately equal* to the corresponding Marshallian price elasticity of demand, $(\partial x_i^*/\partial p_i)(p_i/x_i)$, under the following conditions:

1. The budget share of the ith commodity, $(p_i x_i)/I$, is small.

and/or

2. The income elasticity of demand, $(\partial x_i^*/\partial I)(I/x_i)$, is small.

When either of these two conditions is satisfied, it follows that, as far as price effects are concerned, the (observable) Marshallian demand function can be used *approximately* in place of the (unobservable) Hicksian demand function. In general, the quality of this approximation can be expected to be good when the budget share, $(p_i x_i)/I$, and the income elasticity, $(\partial x_i^*/\partial I)(I/x_i)$, are small. This approximation plays a crucial role in the empirical evaluation of consumer welfare, as discussed next.

Welfare Analysis

Consider the case of an exogenous change in p_i, the price of the ith commodity, from p_i^0 to p_i^1. We are interested in measuring the welfare effect of this price change on the consumer. A good candidate for this measure is given by

$$\Delta E = E(p^0, U) - E(p^1, U) \tag{E-8}$$

where $p^0 = (p_1, \ldots, p_i^0, \ldots, p_n)$ is the price vector before the change, and $p^1 = (p_1, \ldots, p_i^1, \ldots, p_n)$ is the price vector after the change. The term $\Delta E = E(p^0, U) - E(p^1, U)$ in (E-8) can be interpreted as the change in income that is needed to keep the consumer at some given utility level U while the price vector changes from p^0 to p^1. If $\Delta E < 0$, then the consumer needs a higher income to remain at the same welfare level U. Then, for a given U, $(-\Delta E)$ can be interpreted as the *income compensation that the consumer must* ***receive*** *to be indifferent to the price change*. Alternatively, if $\Delta E > 0$, the consumer needs a lower income to remain at the same welfare level U. Then, for a given U, ΔE can be interpreted as the *income compensation that the consumer is willing to* ***pay*** *to be indifferent to the price change*. Based on these convenient interpretations, expression (E-8) can provide a basis for the evaluation of consumer welfare. Its greatest advantage is the fact that it measures consumer well-being using monetary units [in contrast with the utility function $u(x)$, which measures well-being in unobservable "utils"].

Two possible choices for U in (E-8) appear particularly attractive. On the one hand, U can be chosen to be the welfare level reached by the consumer before the price change: $U^0 = V(p^0, I)$. In this case, $\Delta E^0 = E(p^0, U^0) - E(p^1, U^0)$ in (E-8) is an income compensation measure reflecting *the willingness-to-pay (or willingness-to-receive if negative) of the consumer to accept a price change from* p^0 *to* p^1. This measure has been called *compensating variation* in welfare economics. On the other hand, U can be chosen to be the welfare level reached by the consumer after the price change: $U^1 = V(p^1, I)$. In this case, $\Delta E^1 = E(p^0, U^1) - E(p^1, U^1)$ in (E-8) is an income compensation measure reflecting *the willingness-to-pay (or willingness-to-receive if negative) of the consumer to give up the price* p^1 *in favor of* p^0. This measure has been called *equivalent variation* in welfare economics. Unfortunately, *compensating* and *equivalent* variations can in general differ from each other. As a result, in welfare economics, a lively debate has been generated by the issue of choosing U in (E-8). We will not attempt to address this issue here. Rather, we want to investigate how we might empirically measure expression (E-8).

From calculus, Eq. (E-8) can be written as

$$\Delta E = E(p^0, U) - E(p^1, U) = \int_{p_i^1}^{p_i^0} \frac{\partial E(p, U)}{\partial p_i} dp_i$$

Using (E-5), this takes the form

$$\Delta E = E(p^0, U) - E(p^1, U) = \int_{p_i^1}^{p_i^0} x_i^c(p, U) dp_i \tag{E-9}$$

Equation (E-9) indicates that the income compensation ΔE associated with a price change from p^0 to p^1 can be measured *exactly* as the area under the *Hicksian* (compensated) demand $x_i^c(p, U)$ and between the two prices p_i^0 and p_i^1. Equation (E-9) provides a possible basis for evaluating the welfare impact of a price change on consumers. The problem is that the Hicksian demand function is typically not directly observable. Thus Eq. (E-9) cannot be estimated directly.

However, as seen previously, there is a close relationship between the Hicksian and Marshallian price slopes. This suggests that expression (E-9) can be approximated as follows:

$$\Delta E = E(p^0, U) - E(p^1, U) \cong \int_{p_i^1}^{p_i^0} x_i^*(p, I) dp_i \tag{E-10}$$

The right-hand side of Eq. (E-10) is called *change in consumer surplus*. It is the area under the *Marshallian* demand $x_i^*(p, I)$ and between the two prices p_i^0 and p_i^1. Since the Marshallian demand is observable, the estimation of $x^*(p, I)$ and thus of the change in consumer surplus in (E-10) is empirically tractable.

The only question left is the issue of the approximation involved when Eq. (E-10) is used instead of Eq. (E-9) in consumer welfare evaluation. We have shown that Hicksian and Marshallian price slopes are identical in the absence of income effects. It follows that Eq. (E-10) provides an *exact measure* of (E-9) when the income elasticity of demand for the ith commodity, $(\partial x_i^*/\partial I)(I/x_i)$, is zero. And, using Eq. (E-7) or (E-7′), it follows that Eq. (E-10) can be expected to provide a *good approximation* to (E-9) whenever the income elasticity of the ith commodity, $(\partial x_i^*/\partial I)(I/x_i)$, is small, and/or the budget share, $(p_i x_i)/I$, is small. Under these conditions, the change in consumer surplus [as measured by the right-hand side in (E-10)] provides an empirically tractable way of approximating the income compensation $\Delta E = E(p^0, U) - E(p^1, U)$ in (E-8). (For a more precise discussion of the quality of this approximation, see Willig, 1976, cited at the end of Chapter 7.) In many situations, the approximation error involved in Eq. (E-10) is likely fairly small. As a result, throughout this book, we make extensive use of "change in consumer surplus" as a proxy measure of the effect of price changes on consumer welfare.

II. THE PRODUCER CASE

Consider a competitive industry facing an aggregate production function $y = f(x)$, where y denotes output and $x = (x_1, \ldots, x_n)'$ is a $(n \times 1)$ vector of variable inputs. (For simplicity, we limit our discussion to a single-output technology. However, it should be noted that the analysis presented here could be easily extended to a multiple-output technology.) Under competition, industry behavior is consistent with the maximization of quasi-rent:

$$\pi(p, r) = \text{Max}_{x,y}\{py - r'x: y = f(x)\} \tag{E-11}$$

where p denotes output price, $r = (r_1, \ldots, r_n)'$ is the $(n \times 1)$ vector of variable input prices, and all prices are assumed to be exogenously determined. Let $x^*(p, r)$ and $y^*(p, r)$ denote the profit-maximizing input demand and output supply functions of the industry solving the optimization problem (E-11). Equation (E-11) defines $\pi(p, r) = py^*(p, r) - r'x^*(p, r)$ as the quasi-rent generated by the industry.

Applying the envelope theorem (see Appendix A) to expression (E-11) yields the following results:

$$\frac{\partial \pi(p, r)}{\partial p} = y^*(p, r) \tag{E-12a}$$

and

$$\frac{\partial \pi(p, r)}{\partial r_i} = -x_i^*(p, r), \qquad i = 1, \ldots, n \tag{E-12b}$$

Expression (E-12a) states that the derivative of the quasi-rent $\pi(p, r)$ with respect to output price p is equal to the output supply function $y^*(p, r)$. Similarly, (E-12b) shows that the derivative of the quasi-rent $\pi(p, r)$ with respect to the ith input price r_i is equal to the negative of input demand function $x_i^*(p, r)$. We will make use of these convenient results later.

Assume that we are interested in measuring the welfare effects of a change in market prices (p, r) for the industry. A good candidate for measuring these effects is the quasi-rent $\pi(p, r)$ defined in (E-11). Thus, given a price change from (p^0, r^0) to (p^1, r^1), we would like to evaluate the associated change in the industry quasi-rent:

$$\Delta\pi = \pi(p^1, r^1) - \pi(p^0, r^0) \tag{E-13}$$

From (E-13), finding $\Delta\pi > 0$ (< 0) would imply that the industry gains (loses) as a result of the price change. And (E-13) provides a monetary measurement of the gains (or losses) by the industry.

How can we evaluate expression (E-13) empirically? One obvious way is to estimate directly the quasi-rent of the industry both before and after the price change. However, a simple alternative method is also available. To see this, first, consider a change in the price p from p^0 to p^1 (assuming that other prices remain constant). From calculus, Eq. (E-13) can be written as

$$\Delta\pi = \pi(p^1, r) - \pi(p^0, r) = \int_{p^0}^{p^1} \frac{\partial \pi(p, r)}{\partial p} dp$$

Using (E-12a), this becomes

$$\Delta\pi = \pi(p^1, r) - \pi(p^0, r) = \int_{p^0}^{p^1} y^*(p, r) dp \tag{E-14}$$

The expression on the right-hand side of (E-14) is called *change in producer surplus*: it is the area to the left of the supply curve and between the two prices p^0 and p^1.

Thus expression (E-14) shows that, in the case of a change in the price p, the induced *change in industry quasi-rent can be measured* ***exactly*** *by the change in producer surplus*. The supply function $y^*(p, r)$ is empirically tractable and can be used through (E-14) to estimate the associated change in industry quasi-rent.

Second, consider a change in the ith input price r_i from r_i^0 to r_i^1 (assuming that other prices remain constant). Let $r^0 = (r_1, \ldots, r_i^0, \ldots, r_n)$ and $r^1 = (r_1, \ldots, r_i^1, \ldots, r_n)$. Again, from calculus, Eq. (E-13) can be written as

$$\Delta\pi = \pi\left(p, r^1\right) - \pi\left(p, r^0\right) = \int_{r_i^0}^{r_i^1} \frac{\partial \pi\left(p, r\right)}{\partial r_i} dr_i$$

Using (E-12b), this becomes

$$\Delta\pi = \pi\left(p, r^1\right) - \pi\left(p, r^0\right) = -\int_{r_i^0}^{r_i^1} x_i^*\left(p, r\right) dr_i \qquad \text{(E-15)}$$

Expression (E-15) shows that, in the case of a change in the input price r_i, the induced *change in industry quasi-rent can be measured* ***exactly*** *by the negative of the area under the derived demand function $x_i^*(p, r)$ and between the two prices r_i^0 and r_i^1.* The demand function $x^*(p, r)$, being empirically tractable, can be used through (E-15) to estimate the corresponding change in industry quasi-rent.

III. MARKET EQUILIBRIUM ANALYSIS

In the two previous sections, we assumed that competitive market prices were given exogenously. We considered the welfare impacts of an *exogenous change in some prices* on consumers (Section I) and producers (Section II). We showed how such changes can affect economic behavior as well as the welfare of competitive agents. Except for the exogenous price change, we did not allow for any price adjustments. Such analyses are called *partial equilibrium analysis*, allowing *only for endogenous quantity adjustments*.

Prices are determined in markets, however, and it is often relevant to consider price and quantity adjustments simultaneously. This is particularly true in markets that are closely linked with each other. An example is given by a marketing channel composed of multiple stages (e.g., fertilizer producers, farmers, food processors, food retailers, and finally consumers). In this situation, any economic change at one stage of the marketing channel is expected to have some impact on the other stages. In general, these impacts will affect market prices throughout the marketing channel. For instance, an increase in the demand for corn can be expected to stimulate the derived demand for fertilizers, which will put upward pressure on fertilizer prices. Also, an increase in the cost of food production can be expected to influence food prices at the wholesale and retail levels. Analyses that allow for price adjustments as well as quantity adjustments across markets are called *market equilibrium analyses*. They go beyond the partial equilibrium analyses presented previously by *incorporating possible price adjustments through the functioning of linked markets*.

To illustrate these arguments, consider the simple case of a vertical sector constituted of two successive stages. In stage 1, a competitive industry (e.g., the nitrogen fertilizer industry) uses input x (e.g., natural gas) in the production of output y (e.g., nitrogen fertilizer), given the production function $y = f(x)$. In a way similar to (E-11), the behavior of the stage 1 industry is given by

$$\pi_1(p, r) = \text{Max}_{x,y}\{py - rx: y = f(x)\} \tag{E-16}$$

where p and r denote output and input prices, respectively, and π_1 is the quasi-rent of industry 1. Denote by $y_s^*(p, r)$ and $x^*(p, r)$ the partial equilibrium industry supply and demand functions for y and x, respectively.

In stage 2, a competitive industry (e.g., farming) uses the output of stage 1, y (e.g., nitrogen fertilizer), as an input in the production of a new output z (e.g., food) according to the production function $z = g(y)$. Following (E-11), the behavior of the stage 2 industry is given by

$$\pi_2(q, p) = \text{Max}_{y,z}\{qz - py: z = g(y)\} \tag{E-17}$$

where q and p denote output and input prices, respectively, and π_2 is the quasi-rent of industry 2. Denote by $z^*(q, p)$ and $y_d^*(q, p)$ the partial equilibrium industry supply and demand functions for z and y, respectively. For simplicity, we limit our discussion to two industries and to price adjustments in a single market. However, it should be noted that the arguments presented here can be extended to any number of related industries and to any number of price adjustments.

The industry quasi-rents $\pi_1(p, r)$ and $\pi_2(q, p)$ are *partial equilibrium* quasi-rents in the sense that they take all prices (q, p, r) as exogenously given. However, we now have all the information we need to investigate the competitive determination of the market price p for the intermediate good y: we have the aggregate supply function $y_s^*(p, r)$ and the aggregate demand function $y_d^*(q, p)$ for the corresponding commodity. We are interested in two issues: (1) What can we learn from considering the determination of the price p? (2) What can be said about the total aggregate quasi-rent $(\pi_1 + \pi_2)$ under competitive market equilibrium?

First, consider the determination of the equilibrium competitive market price p. Under market equilibrium, aggregate supply must equal aggregate demand, implying that

$$y_s^*(p^e, r) = y_d^*(q, p^e) \tag{E-18}$$

where p^e is the market equilibrium price for p. Denote by $p^e(q, r)$ the implicit solution of (E-18) for p, given some value of (q, r). Thus *$p^e(q, r)$ is the market equilibrium price equation for p*. Substituting this equation into the partial equilibrium functions $x^*(p, r)$, $y_s^*(p, r)$, $y_d^*(q, p)$, and $z^*(q, p)$ gives the *market equilibrium supply–demand functions*:

$$x^e(q, r) = x^*[p^e(q, r), r] \tag{E-19a}$$

$$y^e(q, r) = y_s^*[p^e(q, r), r] = y_d^*[q, p^e(q, r)] \tag{E-19b}$$

and

$$z^e(q, r) = z^*[q, p^e(q, r)] \tag{E-19c}$$

Equations (E-19) are market equilibrium supply–demand functions in the sense that they measure the effects of prices (q, r) on aggregate quantities, *allowing for induced price adjustments of the market price p*.

Second, consider the total quasi-rent $(\pi_1 + \pi_2)$, given market equilibrium for y [as given in (E-18)]. This *market equilibrium total quasi-rent* can be written as

$$\pi^e(q, r) = \pi_1[p^e(q, r), r] + \pi_2[q, p^e(q, r)] \tag{E-20}$$

We are interested here in measuring the total welfare impact of a change in market prices (q, r) on the two industries under market equilibrium conditions (i.e., allowing for induced price adjustments for p). This effect can be measured in terms of the market equilibrium total quasi-rent $\pi^e(q, r)$ given in (E-20). Thus, given a price change from (q^0, r^0) to (q^1, r^1), we would like to evaluate the associated change in π^e:

$$\Delta\pi^e = \pi^e(q^1, r^1) - \pi^e(q^0, r^0) \tag{E-21}$$

From (E-21), finding $\Delta\pi^e > 0$ (< 0) would imply that the two industries together gain (lose) as a result of the price change. And (E-21) provides a monetary measurement of these gains (or losses).

How can we evaluate expression (E-21) empirically? One obvious way is to estimate directly the profit or quasi-rent of the two industries both before and after the price change. However, a simple alternative method is also available. To see this, first consider a change in the price q from q^0 to q^1 (assuming that the price r remains constant). From calculus, Eq. (E-21) can be written as

$$\Delta\pi^e = \pi^e(q^1, r) - \pi^e(q^0, r) = \int_{q^0}^{q^1} \frac{\partial \pi^e(q, r)}{\partial q} dq$$

Using Eq. (E-20), (E-12), the market equilibrium condition (E-18), and expression (E-19), this can be alternatively written as

$$\Delta\pi^e = \pi^e(q^1, r) - \pi^e(q^0, r) = \int_{q^0}^{q^1} \left[\left(\frac{\partial \pi_1}{\partial p} + \frac{\partial \pi_2}{\partial p} \right) \frac{\partial p^e}{\partial q} + \frac{\partial \pi_2}{\partial q} \right] dq$$

$$= \int_{q^0}^{q^1} \left[\left[y_s^*(p^e, r) - y_d^*(q, p^e) \right] \frac{\partial p^e}{\partial q} + z^*(q, p^e) \right] dq \qquad \text{(E-22)}$$

$$= \int_{q^0}^{q^1} z^e(q, r) dq$$

Expression (E-22) shows that, in the case of a change in the price q, the induced *change in market equilibrium total quasi-rent for* ***both*** *industries can be measured* ***exactly*** *by the area under the market equilibrium supply function* $z^e(q, r)$ *and between the two prices* q^0 *and* q^1. The market equilibrium supply function $z^e(q, p)$, being empirically tractable, can be used through (E-22) to estimate the associated change in total quasi-rent under market equilibrium conditions for p.

Second, consider a change in the price r from r^0 to r^1 (assuming that the price q remains constant). From calculus, Eq. (E-21) can be written as

$$\Delta\pi^e = \pi^e(q, r^1) - \pi^e(q, r^0) = \int_{r^0}^{r^1} \frac{\partial \pi^e(q, r)}{\partial r} dr$$

Using Eq. (E-20), (E-12), the market equilibrium condition (E-18), and expression (E-19), this can be alternatively written as

$$\Delta\pi^e = \pi^e(q, r^1) - \pi^e(q, r^0) = \int_{r^0}^{r^1} \left[\left(\frac{\partial \pi_1}{\partial p} + \frac{\partial \pi_2}{\partial p} \right) \frac{\partial p^e}{\partial r} + \frac{\partial \pi_2}{\partial r} \right] dr$$

$$= \int_{r^0}^{r^1} \left[\left[y_s^*(p^e, r) - y_d^*(q, p^e) \right] \frac{\partial p^e}{\partial r} - x^*(p^e, r) \right] dr \qquad \text{(E-23)}$$

$$= \int_{r^0}^{r^1} -x^e(q, r) dr$$

Expression (E-23) shows that, in the case of an exogenous change in the price r, the induced *change in market equilibrium total quasi-rent for* ***both*** *industries can be measured* ***exactly*** *by the negative of the area under the market equilibrium demand function* $x^e(q, r)$ *and between the two prices* r^0 *and* r^1. The market equilibrium demand function $x^e(q, p)$, being empirically tractable, can be used through (E-23) to estimate the associated change in total quasi-rent under market equilibrium conditions for p.

These results show that market equilibrium functions can provide a basis for measuring total welfare effects in a market equilibrium context, allowing for both quantity and induced price adjustments. While our results were derived in the context of two industries, they can be extended for any number of industries. Then Eq. (E-22) or (E-23) provides a convenient measure of *total welfare change in all the industries affected by a change in prices q or r*, either directly or indirectly through the induced price adjustments in related markets.

Answers to Problems

Chapter 1

1.1: (a) $Q = 8.4 + 0.32Z_1 + 0.36Z_2$; $P = 3.2 + 0.16Z_1 - 0.72Z_2$; $Q_c = 10.16$; $P_c = 0.48$.
(b) $\epsilon_d = -0.024$; $\epsilon_s = 0.094$.
(c) $\epsilon_r(Q|Z_1) = 0.04$; $\epsilon_s(Q|Z_2) = 0.71$; $\epsilon_r(Q|Z_1) = 0.03$; $\epsilon_r(Q|Z_2) = 0.14$; $\epsilon_r(P|Z_1) = 0.33$; $\epsilon_r(P|Z_2) = 6$.
(d) arc $\epsilon_d = 1.0$; arc $\epsilon_s = 0.0945$.
1.2: (a) $(\partial Q/\partial P)(P/Q) = -\beta P^{-\beta-1}(P/P^{-\beta}) = -\beta$.
(b) Dollars.
1.3: $E(X) = 4$; $V(X) = 2$; $CX = 35.4$ percent; $E(Z) = 18$; $V(Z) = 132$; $CZ = 63.8$ percent.
1.4: $\mathrm{Cov}(Y, X) = 0$.
1.5: $\mathrm{Cov}[(a + bX), X] = E[X(a + bX)] - E(X)E(a + bX) = bV(X)$, but
$\rho(Y, X) = \mathrm{Cov}(Y, X)/\sigma_X\sigma_Y$ and $V(Y) = b^2V(X)$. Hence $\rho(Y, X) = 1$.
1.6: (a) $E(e) = 0$; $V(e) = 2$; $E(P) = 5$; $V(P) = 2$; $E(TR) = 150$; $V(TR) = 1800$.
(b) $\mathrm{Cov}(TR, P) = 60$ and $\sigma_{TR}\sigma_P = 60$.

Chapter 2

2.1: (a) For farmer A, $u'_\Theta = 10$, $u'_\sigma = -10\sigma$, and $(d\Theta/d\sigma) = \sigma$; for farmer B, $u'_\Theta = \phi_1$, $u'_\sigma = -\phi_2$, and $(d\Theta/d\sigma) = \phi_2/\phi_1$.
(b) Certainty equivalent equals 2; $(10)(10) - (5)(16) = 10(10 - X)$ implies that $X = 8$.
(c) For farmer A, $X = \sigma^2/2$; for farmer B, $X = \phi_2\sigma/\phi_1$.
2.2: $q_{t+1} = (\Theta_P - 2)/\sigma_P^2$; since $X = q_{t+1}^2\sigma_P^2/2$, $(\partial X/\partial q_{t+1}) = \sigma_P^2 q_{t+1}$.
2.3: $q_{t+1} = (\phi_1\Theta_P - \phi_2\sigma_P)/2a\phi_1$; since $X = \phi_2 q_{t+1}\sigma_P/\phi_1$, $(\partial X/\partial q_{t+1}) = \phi_2\sigma_P/\phi_1$.
2.4: $Q_{t+1} = 390.196$, $\Theta_P = 9.804$, $\sigma_P^2 = 16$; $e_{t+1} = 4$ implies that $P_{t+1} = 13.804$.
2.5: $Q_{t+1} = 447.073$, $\Theta_P = 12.547$, $\sigma_P^2 = 52$; $e_{t+1} = -8.0$ together with $u_{t+1} = 5$ implies that $P_{t+1} = 14.547$.
2.6: (a) $\Theta_Q = (200\Theta_G)/(4 + \sigma_G^2)$.
(b) $\Theta_G = 300 - 104\Theta_q$; $\sigma_G = (3704 - 4800\Theta_q + 1600\Theta_q^2)^{1/2}$.
(c) In addition to the equations given in (a) and (b), we have $P_{t+1} = 300 - Q_{t+1} + e_{t+1}$; $\Theta_Q = 100\Theta_q$; and $Q_{t+1} = 100\Theta_q v_{t+1}$.

Chapter 3

3.1: For $\psi = 0$, $q_{t+1} = 3.33$, and $u_{t+1} = 68.67$. For $\psi = 0.5$, $q_{t+1} = 6.67$, $u_{t+1} = 125.33$. For $\psi = 1$, $q_{t+1} = 10$ and $u_{t+1} = 182$.

3.2: (a) $E(X) = 1.5$; $V(X) = 1.25$.
(b) $E(X) = 1.5$; $V(X) = 0.25$.

3.3: (a) If $\rho = +1$, then $a_c = 275$, $a_s = 125$, $\Theta = 88{,}125$, $\sigma^2 = 3{,}062{,}500$, and $u_{t+1} = 23{,}125$. If $\rho = 0$, then $a_c = 208.8$, $a_s = 191.2$, $\Theta = 76{,}544$, $\sigma^2 = 1{,}419{,}117$, and $u_{t+1} = 82{,}132$. If $P = -1$, then $a_c = 204.7$, $a_s = 195.3$, $\Theta = 75{,}820$, $\sigma^2 = 191{,}406$, and $u_{t+1} = 142{,}070$.
(b) If $\rho = +1$, then $a_c = 400$, $a_s = 0$, $\Theta = 110{,}000$, $\sigma^2 = 1{,}440{,}000$, and $u_{t+1} = 148{,}000$. If $\rho = 0$, then $a_c = 397.1$, $a_s = 2.9$, $\Theta = 109{,}486$, $\sigma^2 = 1{,}419{,}126$, and $u_{t+1} = 148{,}015$. If $\rho = -1$, then $a_c = 304.7$, $a_s = 95.3$, $\Theta = 93{,}320$, $\sigma^2 = 191{,}406$, and $u_{t+1} = 177{,}070$.
(c) If $\rho = +1$, then $a_c = 0$, $a_s = 400$, $\Theta = 40{,}000$, $\sigma^2 = 1{,}440{,}000$, and $u_{t+1} = 8000$. If $\rho = 0$, then $a_c = 157.8$, $a_s = 242.2$, $\Theta = 67{,}611$, $\sigma^2 = 1{,}424{,}222$, and $u_{t+1} = 64{,}011$. If $\rho = -1$, then $a_c = 176.5$, $a_s = 223.5$, $\Theta = 70{,}895$, $\sigma^2 = 151{,}234$, and $u_{t+1} = 134{,}228$.

3.4: (a) $g_s = [4F_t(P_{t+1}) - 4\Theta_P]/0.2\,\sigma_P^2$.
(b) $g_d = [4\Theta_P - 4F_t(P_{t+1})]/0.2\,\sigma_P^2$.
(c) $F_t(P_{t+1}) = 2.8$.

3.5: (a) $q_{t+1} = 0.5F_t(P_{t+1})$; $g_s = [6F_t(P_{t+1}) + 0.1\sigma_P^2 F_t(P_{t+1}) - 6\,\Theta_P]/0.2\sigma_P^2$.
(b) If $F_t(P_{t+1}) = 10$ and $\sigma_P^2 = 1$, then $q_{t+1} = 5$. $\Theta_P = 5$, 10, and 15 implies, respectively, that $g_s = 155$, 5, and -145.
(c) If $\sigma_P^2 = 60$, then $q_{t+1} = 5$; for $\Theta_P = 5$, 10, and 15, $g_s = 7.5$, 5, and 2.5, respectively.

Chapter 4

4.1: (a) $a = 1.5k$.
(b) $k = 271.6$; $a = 407.4$.
(c) $k = 58.8$; $a = 88.2$.

4.2: (a) $TRP = 0.45\Theta_G^{1.25} a_t^{0.625}$; $MRP = 0.281\Theta_G^{1.25} a_t^{-0.375}$.
(b) $ARP = 0.45\Theta_G^{1.25} a_t^{-0.375}$; $NARP = ARP - 0.5a_t^{-1}$.
(c) $a_t = 0.862$; $k_t = 0.689$; $\Theta_q = 0.862$; $\Theta = 1.568$; $TRP = 5.515$; $MRP = 4.0$; $ARP = 6.4$; $NARP = 5.82$.

4.3: (a) $(\partial A_t/\partial\Theta_Z) > 0$; for inelastic f_1 function, we have $(\partial A_t/\partial Y_t) > 0$, $(\partial A_t/\partial J_t) < 0$, $(\partial A_t/\partial\alpha_0) < 0$, $(\partial A_t/\partial\alpha_2) < 0$; for elastic f_1 function, we have $(\partial A_t/\partial Y_t) < 0$, $(\partial A_t/\partial J_t) > 0$, $(\partial A_t/\partial\alpha_0) > 0$, $(\partial A_t/\partial\alpha_2) > 0$.
(b) $(\partial\Theta_G/\partial B_t) > 0$; $(\partial\Theta_Q/\partial B_t) < 0$; $(\partial R_t/\partial B_t) > 0$; $(\partial A_t/\partial B_t) < 0$; for inelastic demand, we have $(\partial\Theta_w/\partial B_t) > 0$, $(\partial L_t/\partial B_t) > 0$, $(\partial K_t/\partial B_t) > 0$; for elastic demand, we have $(\partial\Theta_w/\partial B_t) < 0$, $(\partial L_t/\partial B_t) < 0$, $(\partial K_t/\partial B_t) < 0$.

4.4: $k_t = (4\phi_1^2\Theta_G^2)/(\phi_1 J_t + 16\phi_2\sigma_G^2)^2$.

4.5: $P_a = (0.5)/(iA^{1.333})$.

Chapter 5

5.1: (a) $(0.8P_r - P) = 0.2 + 0.4q$; $(0.8P_r - P) = 0.2 + 0.004Q$; $AVC = 0.2 + 0.2q$; $AVC' = 0.2 + 0.002Q$.
(b) $P = 799.8 - 0.324Q$.
(c) $P = 313.8$; $P_r = 400$; $MM = 6.2$; $\pi = 25$.
5.2: (c) $\partial P/\partial \delta = 0$ for $\delta = 0.752$.
5.3: $P' = 9.678$; $P'' = 8.068$; $Q' = 17.355$, $Q'' = 30.342$; $P_0 = 13.277$; $P_m = 10.058$; $Q_0 = 13.447$; $Q_m = 13.447$.
5.4: (a) $\delta_o = (0.5 - \delta_m^2)^{0.5}$.
(b) $\delta_o = 0.3162, \delta_m = 0.6325, q = 2.16$; for $P_o = 4, \delta_o = 0.5, \delta_m = 0.5, q = 3$.
5.5: (a) $P_u = 22.9653 - 1.0333Q_{xd}$.
(b) $P_w = 5.74$, $P_u = 8.1$, $Q_d = 13.95$, $Q_s = 28.34$, $Q_{wd} = 22.13$, $Q_{ws} = 7.74$, and $Q_x = 14.39$.
(c) $P_u = 23.7333 - 1.0667Q_{xd}$; the new export demand lies everywhere above the old one.

Chapter 6

6.1: $P_1 = 7.45$; $Q_1 = 3.14$; $I_1 = 26.86$; and $E_1(P_2) = 8.45$.
6.2: (a) $E_2(P_3) = 16/I_2^2$; $V_2(P_3) = 42.67/I_2^4$.
(b) $P_2 = (16 - I_2^2)/I_2^2$.
6.3: (a) 5.38; 5.33; and 5.25.
(b) 5.29.
(c) 5.28.
(d) Increases in the carrying cost lowers the demand for stocks in period 2. With given values for A_1 and I_1, this tends to decrease prices in period 2.

Chapter 7

7.1: (a) There is neither loss nor gain.
(b) Capital loss equals 62,500.
(c) No. Neither demand nor supply for land input is affected.
7.2: $CS = 14$; quasi-rent rises by 14; the area under derived demand and between the two price lines equals 28.
7.3: (a) $Q = 25.17P^{0.429}$.
(b) For $P = 1, Q = 25.17, L = 3.178, W = 3.168, R = 0.151$. For $P = 2, Q = 33.876, L = 6.679, W = 4.057, R = 0.407$.
(c) $PS = 29.523$; increase in total rent = 25.549; $WS = 4.385$.
7.4: $CS = 22.5$; $PS = 4.5$; efficiency gain = 27.

Chapter 8

8.1: (a) $Q_M = 4$; $P_M = 8$; value of market = 32.
(b) $Q = 6$; maximum value of market = 36; deadweight loss = 4.

8.2: $Q = 12.771$; $MRP = 14.771$; $ARP = 50.161$; $P = 8.386$; $\pi = 533.513$.
8.3: (a) $TCS = 32.4$; $TPS = 162$; tax = 32.4; value of market = 162.
(b) $TCS = 27.225$; $TPS = 136.125$; tax = 0; value of market = 163.35.
8.4: (a) $P = 16$; $q_1 = 8$; $q_2 = 12$; $\pi_1 = 62$; $\pi_2 = 119$.
(b) $P = 16.615$; $q_1 = 9.231$; $q_2 = 9.231$; $\pi_1 = 66.162$; $\pi_2 = 109.767$.
(c) $P = 17.457$; $q_1 = 5.458$; $q_2 = 10.899$; $\pi_1 = 63.490$; $\pi_2 = 129.873$.

Chapter 9

9.1: (a) $Q_c = 10$; $P_c = 5$.
(b) $Q_g = 12$; consumers pay $P_g = 4$; $CS = 11$; $PS = 11$; tax = 24; efficiency loss = 2.
(c) $Q_{sg} = 12$; consumers pay $P_g = 6$; $Q_{dg} = 8$; $CS = -9$; $PS = 11$; tax = 16; efficiency loss = 14; note that the bird seed market is ignored in these measures of welfare changes.
(d) $Q_{sg} = 12$; consumers pay $P_g = 5$; $Q_{dg} = 10$; $CS = 0$; $PS = 11$; tax = 18; efficiency loss = 7.
9.2: (a) $P = 3.85$; $D_u = 6.15$; $S_u = 15.38$; $D_w = 13.08$; $S_w = 3.85$; exports (imports) equal 9.23.
(b) $P = 3.6$; $D_u = 6.4$; $S_u = 16$; $D_w = 13.2$; $S_w = 3.6$; exports (imports) equal 9.6.
(c) $P_u = 4$; $D_u = 6$; $S_u = 16$; $P_w = 3.33$; $D_w = 13.33$; $S_w = 3.33$; exports (imports) equal 10.
9.3: Increase in total rent to land equals 0.11. Increase in labor surplus (benefit) equals 0.03.

Chapter 10

10.1: (a) $P = 6.968$; $Q_d = 3.032$; $Q_s = 1.161$; $M = 1.871$.
(b) $P = 7$; $Q_d = 3$; $Q_s = 1.167$; $M = 1.833$; $CS = -0.097$; $PS = 0.038$; efficiency loss = 0.06.
(c) $P = 7$; $Q_d = 3$; $Q_s = 1.167$; $M = 1.833$; $CS = -0.097$; $PS = 0.038$; tariff revenue = 0.611; efficiency gain = 0.551.
(d) $P_g = 6.963$; $Q_d = 3.037$; $Q_s = 1.167$; $M = 1.870$; $CS = 0.014$; $PS = 0.038$; tax = 0.043; efficiency gain = 0.009.
10.2: (a) $P_c = 40.8$; $Q_c = 59.2$; $q_{1c} = 38.8$; $q_{2c} = 20.4$.
(b) $\lambda = 0.62505$; $P_g = 62.997$; $Q_g = 37.003$; $q_{1g} = 24.252$; $q_{2g} = 12.751$; cartel residual = 1825.906.
10.3: (a) $q = 0.625P$; $\pi = 0.3125P^2$.
(b) $q = P^{0.5}$; $\pi = 2P^{1.5}/3$.
10.4: $P_i = 0.5P^*$; therefore, $P = 8q$ for all $P \leq 0.5P^*$ and $P = 2q$ for all $P \geq 0.5P^*$.
10.5: (a) P_i is specified implicitly by $\pi^*(P_i) = Ba$, where a equals the farmer's acreage. Let $P = s(q)$ be the farmer's competitive supply function. Then $q = 0$ for all $P \leq P_i$ and $P = s(q)$ for all $P \geq P_i$.
(b) The aggregate supply function is a stepped function; $Q = 0$ for $P \leq P_i$ and $P = S(Q)$ for $P \geq P_i$. At P_i the supply curve is flat.

Index

I

J

K

L

M

N

O

P